DNA–Protein Interactions

Methods in Molecular Biology

John M. Walker, SERIES EDITOR

37. **In Vitro Transcription and Translation Protocols,** edited by *Martin J. Tymms,* 1995
36. **Peptide Analysis Protocols,** edited by *Michael W. Pennington and Ben M. Dunn,* 1994
35. **Peptide Synthesis Protocols,** edited by *Ben M. Dunn and Michael W. Pennington,* 1994
34. **Immunocytochemical Methods and Protocols,** edited by *Lorette C. Javois,* 1994
33. ***In Situ* Hybridization Protocols,** edited by *K. H. Andy Choo, 1994*
32. **Basic Protein and Peptide Protocols,** edited by *John M. Walker, 1994*
31. **Protocols for Gene Analysis,** edited by *Adrian J. Harwood, 1994*
30. **DNA–Protein Interactions,** edited by *G. Geoff Kneale, 1994*
29. **Chromosome Analysis Protocols,** edited by *John R. Gosden, 1994*
28. **Protocols for Nucleic Acid Analysis by Nonradioactive Probes,** edited by *Peter G. Isaac, 1994*
27. **Biomembrane Protocols:** *II. Architecture and Function,* edited by *John M. Graham and Joan A. Higgins, 1994*
26. **Protocols for Oligonucleotide Conjugates,** edited by *Sudhir Agrawal, 1994*
25. **Computer Analysis of Sequence Data:** *Part II,* edited by *Annette M. Griffin and Hugh G. Griffin, 1994*
24. **Computer Analysis of Sequence Data:** *Part I,* edited by *Annette M. Griffin and Hugh G. Griffin, 1994*
23. **DNA Sequencing Protocols,** edited by *Hugh G. Griffin and Annette M. Griffin, 1993*
22. **Optical Spectroscopy, Microscopy, and Macroscopic Techniques,** edited by *Christopher Jones, Barbara Mulloy, and Adrian H. Thomas, 1994*
21. **Protocols in Molecular Parasitology,** edited by *John E. Hyde, 1993*
20. **Protocols for Oligonucleotides and Analogs,** edited by *Sudhir Agrawal, 1993*
19. **Biomembrane Protocols:** *I. Isolation and Analysis,* edited by *John M. Graham and Joan A. Higgins, 1993*
18. **Transgenesis Techniques,** edited by *David Murphy and David A. Carter, 1993*
17. **Spectroscopic Methods and Analyses,** edited by *Christopher Jones, Barbara Mulloy, and Adrian H. Thomas, 1993*
16. **Enzymes of Molecular Biology,** edited by *Michael M. Burrell, 1993*
15. **PCR Protocols,** edited by *Bruce A. White, 1993*
14. **Glycoprotein Analysis in Biomedicine,** edited by *Elizabeth F. Hounsell, 1993*
13. **Protocols in Molecular Neurobiology,** edited by *Alan Longstaff and Patricia Revest, 1992*
12. **Pulsed-Field Gel Electrophoresis,** edited by *Margit Burmeister and Levy Ulanovsky, 1992*
11. **Practical Protein Chromatography,** edited by *Andrew Kenney and Susan Fowell, 1992*
10. **Immunochemical Protocols,** edited by *Margaret M. Manson, 1992*

Earlier volumes are still available. Contact Humana for details.

Methods in Molecular Biology • 30

DNA–Protein Interactions
Principles and Protocols

Edited by

G. Geoff Kneale

*School of Biological Sciences; University of
Portsmouth, Portsmouth, UK*

Humana Press 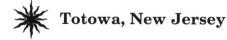 Totowa, New Jersey

© 1994 Humana Press Inc.
999 Riverview Drive, Suite 208
Totowa, New Jersey 07512

All rights reserved.

No part of this book may be reproduced, stored in a retrieval system, or transmitted in any form or by any means, electronic, mechanical, photocopying, microfilming, recording, or otherwise without written permission from the Publisher.

Photocopy Authorization Policy:
Authorization to photocopy items for internal or personal use, or the internal or personal use of specific clients, is granted by Humana Press, **provided** that the base fee of US $3.00 per copy, plus US $00.20 per page is paid directly to the Copyright Clearance Center at 27 Congress Street, Salem, MA 01970. For those organizations that have been granted a photocopy license from the CCC, a separate system of payment has been arranged and is acceptable to Humana Press. The fee code for users of the Transactional Reporting Service is: [0-89603-XXX-X/93 $3.00 + $00.20].

Printed in the United States of America. 10 9 8 7 6 5 4 3 2

Library of Congress Cataloging in Publication Data

Main entry under title:

Methods in molecular biology.

DNA–protein interactions: principles and protocols / edited by G. Geoff Kneale.
     p. cm.—(Methods in molecular biology; 30)
  Includes index.
  ISBN 0-89603-256-6
  1. DNA–protein interactions.   I. Kneale, G. Geoff.   II. Series:
Methods in molecular biology (Totowa, N.J.); 30.
QP624.75.P74D57   1994
574.87'3282—dc20   93-21448
     CIP
     rev.

Preface

The study of protein–nucleic acid interactions is currently one of the most rapidly growing areas of molecular biology. DNA binding proteins are at the very heart of the regulation and control of gene expression, replication, and recombination: Enzymes that recognize and either modify or cleave specific DNA sequences are equally important to the cell. Some of the techniques reported in this volume can be used to identify previously unknown DNA binding proteins from crude cell extracts. Virtually all are capable of giving direct information on the molecular basis of the interaction—the location of the DNA binding site; the strength and specificity of binding; the identities of individual groups on specific bases involved in binding; the specific amino acid residues of the protein that interact with the DNA; or the effects of protein binding on gross conformation and local structure of DNA.

The recognition of DNA sequences by proteins is a complex phenomenon, involving specific hydrogen bonding contacts to the DNA bases ("direct readout") and/or interactions with the sugar–phosphate backbone ("indirect readout"). The latter interactions can also be highly specific because of sequence-dependent conformational changes in the DNA. In addition, intercalation of planar aromatic amino acid side-chains between the DNA bases can occur, most notably with single-stranded DNA binding proteins. Furthermore, when bound, many DNA binding proteins induce drastic structural changes in the DNA as an integral part of their function. The advantages of using complementary techniques to investigate the complexities of protein–DNA interactions will be clear, since each has advantages and disadvantages for specific tasks.

The early chapters in this volume cover DNA footprinting and related techniques, including "protection" methods (which probe accessibility of a DNA sequence to a reagent) and "interference" techniques (which investigate the effects on DNA binding of modification of groups in the DNA recognition sequence). The results, even with the same reagent, are rarely identical, since the former picks up contacts that may not be essential for DNA binding. Ideally, both should be attempted. Although the "interference" techniques cannot be used in vivo, they are often more sensitive than "protection" techniques when applied to crude extracts. The variation between probes in sequence selectivity, site of reaction, and conformational dependence gives rise to the wide range of related techniques now in use.

Chapter 1 deals with DNase I footprinting and, like the ExoIII technique described in Chapter 2, is a relatively mild procedure that is most likely to identify the intact DNA binding site. However, because of the bulky size of the enzymatic probes, the use of chemical probes is recommended for higher resolution studies. Hydroxyl radical cleavage (Chapters 3 and 4) is relatively sequence-independent and attacks the DNA backbone. It can be used as either a protection or an interference method. Copper–phenanthroline footprinting (Chapter 5) is also relatively nonspecific regarding sequence since it cleaves the deoxyribose ring, probably in the minor groove. Dimethyl sulfate is another reagent that can be used for both protection and interference experiments (Chapter 6); it is highly specific for purines and can distinguish minor and major groove contacts, but favors N7 of guanine in the major groove.

Diethyl pyrocarbonate (Chapter 7) is a probe for protein-induced conformational changes in the DNA, being particularly useful in the analysis of Z-DNA formation; it preferentially attacks guanine bases. Osmium tetroxide, on the other hand, attacks thymines and has proved useful in the identification of cruciform-like structures (Chapter 8). Singlet oxygen provides a probe of structural deformation in DNA; it is not appreciably sequence selective in double helical DNA (Chapter 9). Both the ethylation interference technique (Chapter 10) and uranyl photofootprinting (Chapter 11) are available for the analysis of contacts with the phosphate groups of the DNA backbone.

The following six chapters deal with methods to investigate the protein component of a nucleoprotein complex. Chapters 12 and 13 describe chemical modification and proteolysis of nucleoprotein complexes, techniques that could be considered the analogs of chemical and enzymatic DNA footprinting techniques. The latter approach is useful for the investigation of DNA binding domains within a protein. The following two chapters cover methods for the overexpression of DNA binding proteins and their domains; both take advantage of fusion techniques. Chapter 14 describes a general strategy (making use of PCR) for large-scale expression of putative DNA binding domains that can be used in subsequent in vitro studies. Chapter 15 describes an alternative method using GST fusions, and its application to the overexpression of eukaryotic transcription factors. If the gene for a DNA binding protein has been cloned, then site-directed mutagenesis provides an invaluable tool for the dissection of structure–function relationships; two different approaches are reported (Chapters 16 and 17). The latter chapter includes the use of saturation mutagenesis for the analysis of functional requirements for a given amino acid residue at a particular site in the protein.

Crosslinking techniques for the investigation of protein–DNA interactions are presented in the following two chapters. Such methods are valuable in identifying specific DNA sequences that interact with a given protein, and can sometimes be used where other techniques fail. Indeed, subsequent biochemical analysis of the protein may also reveal specific amino acid residues in close proximity to the DNA binding site. A potential problem with all such methods is that the yield of crosslinked product is often low. In Chapter 18, this is overcome by use of UV laser irradiation. Chapter 19 describes the use of photoaffinity labeling, where the modified base 8-azidoadenine is introduced into the target DNA sequence.

Quantitative determination of DNA binding affinities can be established by a number of techniques. The filter binding assay remains a useful method for many purposes, and is described in Chapter 20. However, the gel-shift assay (Chapter 21) is of more general utility and can often allow the resolution of multiple species of DNA–protein complexes. Indeed, the gel-shift technique forms a part of many of the methods described in earlier chapters. A specific application of the gel-shift assay is for the analysis of protein-induced DNA bending, for which specialized vectors have been developed (Chapter 22). It also forms part of the binding site-selection technique covered in Chapter 23, one of a number of recently developed protocols for the determination of specific base requirements in a DNA binding sequence.

This protocol employs solid-phase chemical sequencing for the analysis of preferred binding sites.

Spectroscopic techniques are especially useful in the quantitative study of protein–DNA interactions when highly purified components are available. Fluorescence methods can give information on stoichiometry, binding constants, and frequently also on the interaction of aromatic amino acid residues with the DNA (Chapter 24). In certain cases, however, the intrinsic fluorescence of the protein cannot be used to monitor DNA binding. Chapter 25 describes an alternative method based on the displacement of the fluorescent probe ANS from the DNA binding site. Circular dichroism can monitor conformational changes in either the protein or the DNA component of a complex (Chapter 26). Again, it is often the combination of such techniques that provides particularly valuable insights into the nature of the interaction. For more direct structural information, especially for large nucleoprotein complexes, electron microscopy can provide a low resolution image (Chapter 27). Ultimately, X-ray crystallography is the preferred technique for high resolution structural data and Chapter 28 provides a useful source of information for getting started in this direction. Clearly, a detailed description of X-ray diffraction techniques is well beyond the scope of this volume. The same is true of NMR spectroscopy, which is also proving instrumental in our understanding of protein–nucleic acid interactions.

The remaining chapters in this volume describe functional assays for a variety of proteins that interact with DNA. Such studies are vital to complement the structural techniques, but are of necessity more specific. Chapters 29 and 30 deal with protocols for the assay of restriction enzyme activity using synthetic oligonucleotides and plasmids, respectively, as substrates. Chapter 31 describes a number of techniques for the assay of transcription factor activity, and Chapter 32 describes an assay for proteins involved in genetic recombination.

The aim in describing these techniques was to provide complete protocols wherever possible, so that they could be used by a relative newcomer to the field. Inevitably, this involves some duplication between chapters since there is some overlap of methodologies; nonetheless, this has been kept to a minimum. The introduction to each chapter provides an overview of the method, a brief theoretical back-

Preface

ground and, when relevant, a discussion of its advantages and drawbacks. In keeping with the format of other volumes in the series, there is a Materials section in each chapter listing all the reagents that will be required for the procedures. The individual procedures vary in length and complexity from chapter to chapter, but in each case they are broken down into a series of small steps. Finally, the Notes section provides hints or further explanation of the rationale underlying the technique, which should prove invaluable to the experimenter.

The reader should know that it was impossible to cover every variation on a given technique in one volume. Inevitably there will be omissions. However, the editor believes that the methods included will provide a sufficiently broad range that the experimenter will be able to tackle problems in every area of protein–nucleic acid interactions from a number of different perspectives, and so build up a more complete understanding of the molecular processes involved.

G. Geoff Kneale

Contents

Preface v
Contributors xv

Ch. 1. DNase I Footprinting,
  Benoît Leblanc and Tom Moss 1
Ch. 2. Footprinting with Exonuclease III,
  Willi Metzger and Hermann Heumann 11
Ch. 3. Hydroxyl Radical Footprinting,
  Peter Schickor and Hermann Heumann 21
Ch. 4. Hydroxyl Radical Interference,
  Peter Schickor and Hermann Heumann 33
Ch. 5. 1,10-Phenanthroline-Copper Ion Nuclease Footprinting of DNA–
  Protein Complexes *in Situ* Following Mobility-Shift
  Electrophoresis Assays,
  Athanasios G. Papavassiliou 43
Ch. 6. Identification of Protein–DNA Contacts with Dimethyl Sulfate:
  Methylation Protection and Methylation Interference,
  Peter E. Shaw and A. Francis Stewart 79
Ch. 7. Diethyl Pyrocarbonate as a Probe of Protein–DNA Interactions,
  Michael J. McLean 89
Ch. 8. Osmium Tetroxide Modification and the Study of DNA–Protein
  Interactions,
  James A. McClellan 97
Ch. 9. Diffusible Singlet Oxygen as a Probe of DNA Deformation,
  Malcolm Buckle and Andrew A. Travers 113
Ch. 10. Ethylation Interference,
  Iain Manfield and Peter G. Stockley 125
Ch. 11. Uranyl Photofootprinting of DNA–Protein Complexes,
  Peter E. Nielsen 141
Ch. 12. Nitration of Tyrosine Residues in Protein–Nucleic Acid Complexes,
  Simon E. Plyte 151
Ch. 13. Limited Proteolysis of Protein–Nucleic Acid Complexes,
  Simon E. Plyte and G. Geoff Kneale 161
Ch. 14. Cloning and Expression of DNA Binding Domains Using PCR,
  Daniel G. Fox and G. Geoff Kneale 169

Contents

CH. 15. Overexpression and Purification of Eukaryotic Transcription Factors as Glutathione-S-Transferase Fusions in *E. coli*,
Kevin G. Ford, Alan J. Whitmarsh, and David P. Hornby 185

CH. 16. Site-Directed Mutagenesis by the Cassette Method,
Andrew F. Worrall 199

CH. 17. Site-Directed and Site-Saturation Mutagenesis Using Oligonucleotide Primers,
Michael J. O'Donohue and G. Geoff Kneale 211

CH. 18. UV Laser-Induced Protein–DNA Crosslinking,
Stefan I. Dimitrov and Tom Moss 227

CH. 19. Ultraviolet Crosslinking of DNA–Protein Complexes via 8-Azidoadenine,
Rainer Meffert, Klaus Dose, Gabriele Rathgeber, and Hans-Jochen Schäfer 237

CH. 20. Filter-Binding Assays,
Peter G. Stockley 251

CH. 21. The Gel Shift Assay for the Analysis of DNA–Protein Interactions,
John D. Taylor, Alison J. Ackroyd, and Stephen E. Halford ... 263

CH. 22. Improved Plasmid Vectors for the Analysis of Protein-Induced DNA Bending,
Christian Zwieb and Sankar Adhya 281

CH. 23. Determination of Sequence Preferences of DNA Binding Proteins Using Pooled Solid-Phase Sequencing of Low Degeneracy Oligonucleotide Mixtures,
Joseph A. Gogos and Fotis C. Kafatos 295

CH. 24. Analysis of DNA–Protein Interactions by Intrinsic Fluorescence,
Mark L. Carpenter and G. Geoff Kneale 313

CH. 25. A Competition Assay for DNA Binding Using the Fluorescent Probe ANS,
Ian Taylor and G. Geoff Kneale 327

CH. 26. Circular Dichroism for the Analysis of Protein–DNA Interactions,
Mark L. Carpenter and G. Geoff Kneale 339

CH. 27. Electron Microscopy of Protein–Nucleic Acid Complexes: *Uniform Spreading and Determination of Helix Handedness*,
Carla W. Gray 347

CH. 28. Reconstitution of Protein–DNA Complexes for Crystallization,
Rachel M. Conlin and Raymond S. Brown 357

CH. 29. Assay of Restriction Endonucleases Using Oligonucleotides,
Bernard A. Connolly 371

CH. 30. Assays for Restriction Endonucleases Using Plasmid Substrates,
Stephen E. Halford, John D. Taylor, Christian L. M. Vermote, and I. Barry Vipond 385

Ch. 31. Assays for Transcription Factor Activity,
Stephen Busby, Annie Kolb, and Stephen Minchin 397

Ch. 32. An Assay for in Vitro Recombination Between Duplex DNA Molecules,
Berndt Müller and Stephen C. West 413

Index 425

Contributors

ALISON J. ACKROYD • *Department of Biochemistry, University of Texas at Dallas, Southwestern Medical Center, Dallas, TX*
SANKAR ADHYA • *Laboratory of Molecular Biology, National Cancer Institute, National Institutes of Health, Bethesda, MD*
RAYMOND S. BROWN • *Howard Hughes Medical Institute, Harvard University, Boston, MA*
MALCOLM BUCKLE • *Institute Pasteur, Paris, France*
STEPHEN BUSBY • *School of Biochemistry, University of Birmingham, Birmingham, UK*
MARK L. CARPENTER • *Sir William Dunn School of Pathology, University of Oxford, Oxford, UK*
RACHEL M. CONLIN • *Howard Hughes Medical Institute, Harvard University, Cambridge, MA*
BERNARD A. CONNOLLY • *Department of Biochemistry and Genetics, The University, Newcastle-upon-Tyne, UK*
STEFAN I. DIMITROV • *Centre de Recherche en Cancérologie de l'Université Laval, Québec, Canada*
KLAUS DOSE • *Institüt für Biochemie der Johannes Gutenberg-Universität, Mainz, Germany*
KEVIN G. FORD • *Krebs Institute, Department of Molecular Biology and Biotechnology, University of Sheffield, Sheffield, UK*
DANIEL G. FOX • *Biophysics Laboratories, School of Biological Sciences, University of Portsmouth, Portsmouth, UK*
JOSEPH A. GOGOS • *Department of Cellular and Developmental Biology, The Biological Laboratories, Harvard University, Cambridge, MA*
CARLA W. GRAY • *Program in Molecular and Cell Biology (FO31), The University of Texas at Dallas, Dallas, TX*

STEPHEN E. HALFORD • *Department of Biochemistry, Centre for Molecular Recognition, University of Bristol, Bristol, UK*
HERMANN HEUMANN • *Max Planck Institute of Biochemistry, Martinsried, Germany*
DAVID P. HORNBY • *Krebs Institute, Department of Molecular Biology and Biotechnology, University of Sheffield, Sheffield, UK*
FOTIS C. KAFATOS • *Institute of Molecular Biology and Biotechnology, Research Center of Crete, Crete, Greece; and Department of Cellular and Developmental Biology, The Biological Laboratories, Harvard University, Cambridge, MA*
G. GEOFF KNEALE • *Biophysics Laboratories, School of Biological Sciences, University of Portsmouth, Portsmouth, UK*
ANNIE KOLB • *Department de Biologie Moleculaire, Institut Pasteur, Paris, France*
BENOÎT LEBLANC • *Centre de Recherche en Cancérologie de l'Université Laval, Québec, Canada*
IAIN MANFIELD • *Department of Genetics, University of Leeds, Leeds, UK*
JAMES A. MCCLELLAN • *Biophysics Laboratories, School of Biological Sciences, University of Portsmouth, Portsmouth, UK*
MICHAEL J. MCLEAN • *Cambridge Research Biochemicals, Cheshire, UK*
RAINER MEFFERT • *Institüt für Biochemie der Johannes Gutenberg-Universität, Mainz, Germany*
WILLI METZGER • *Max Planck Institute of Biochemistry, Martinsried, Germany*
STEPHEN MINCHIN • *School of Biochemistry, University of Birmingham, Birmingham, UK*
TOM MOSS • *Centre de Recherche en Cancérologie de l'Université Laval, Québec, Canada*
BERNDT MÜLLER • *Imperial Cancer Research Fund, Clare Hall Laboratories, South Mimms, UK*
PETER E. NIELSEN • *Research Center for Medical Biotechnology, Department of Biochemistry, The Panum Institute, Copenhagen, Denmark*
MICHAEL J. O'DONOHUE • *Laboratory of Organic Chemistry, University of Paris V, Paris, France*

ATHANASIOS G. PAPAVASSILIOU • *European Molecular Biology Laboratory, Heidelberg, Germany*
SIMON E. PLYTE • *Ludwig Institute for Cancer Research, London, UK*
GABRIELE RATHGEBER • *Institüt für Biochemie der Johannes Gutenberg-Universität, Mainz, Germany*
HANS-JOCHEN SCHÄFER • *Institüt für Biochemie der Johannes Gutenberg-Universität, Mainz, Germany*
PETER SCHICKOR • *Max Planck Institute of Biochemistry, Martinsried, Germany*
PETER E. SHAW • *Max Planck Institüt für Immunbiologie, Freiburg - Zähringen, Germany*
A. FRANCIS STEWART • *European Molecular Biology Laboratories, Heidelberg, Germany*
PETER G. STOCKLEY • *Department of Genetics, University of Leeds, Leeds, UK*
IAN TAYLOR • *Biophysics Laboratories, School of Biological Sciences, University of Portsmouth, Portsmouth, UK*
JOHN D. TAYLOR • *Department of Biochemistry, Duke University Medical Center, Durham, NC*
ANDREW A. TRAVERS • *MRC Laboratory of Molecular Biology, Cambridge, UK*
CHRISTIAN L. M. VERMOTE • *Department of Biochemistry, Centre for Molecular Recognition, University of Bristol, Bristol, UK*
I. BARRY VIPOND • *Department of Biochemistry, Centre for Molecular Recognition, University of Bristol, Bristol, UK*
STEPHEN C. WEST • *Imperial Cancer Research Fund, Clare Hall Laboratories, South Mimms, UK*
ALAN J. WHITMARSH • *Krebs Institute, Department of Molecular Biology and Biotechnology, University of Sheffield, Sheffield, UK*
ANDREW F. WORRALL • *Department of Biochemistry, University of Southampton, Southampton, UK*
CHRISTIAN ZWIEB • *Department of Molecular Biology, The University of Texas Health Center at Tyler, Tyler, TX*

CHAPTER 1

DNase I Footprinting

Benoît Leblanc and Tom Moss

1. Introduction

DNase I footprinting was developed by Galas and Schmitz in 1978 as a method to study the sequence-specific binding of proteins to DNA *(1)*. In this technique a suitable uniquely end-labeled DNA fragment is allowed to interact with a given DNA-binding protein and then the complex is partially digested with DNase I. The bound protein protects the region of the DNA with which it interacts from attack by the DNase. Subsequent molecular weight analysis of the degraded DNA by electrophoresis and autoradiography identifies the region of protection as a gap in the otherwise continuous background of digestion products (for examples, *see* Fig. 1). The technique can be used to determine the site of interaction of most sequence-specific DNA-binding proteins but has been most extensively applied to the study of transcription factors. Since the DNase I molecule is relatively large compared to other footprinting agents (*see* Chapters 3 and 7 in this volume), its attack on the DNA is more readily prevented by steric hindrance. Thus DNase I footprinting is the most likely of all the footprinting techniques to detect a specific DNA–protein interaction. This is clearly demonstrated by our studies on the transcription factor xUBF (*see* Fig. 1B). The xUBF interaction with the *Xenopus* ribosomal DNA enhancer can be easily detected by DNase I footprinting but has not yet been detected by other footprinting techniques.

DNase I footprinting can not only be used to study the DNA interactions of purified proteins but also as an assay to identify proteins of

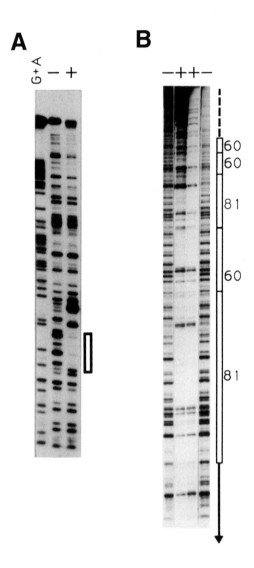

Fig. 1. Examples of DNase I footprints. **A.** Footprint (open box) of a chicken erythrocyte DNA binding factor on the promoter of the H5 gene *(2)* (figure kindly donated by A. Ruiz-Carrillo). **B.** Interaction of the RNA polymerase I transcription factor xUBF with the tandemly repeated 60 and 81 bp. *Xenopus* ribosomal gene enhancers. Both **A** and **B** used 5' end-labeled fragments. (–) and (+) refer to naked and complexed DNA fragments and (G + A) to the chemical sequence ladder.

interest within a crude cellular or nuclear extract *(2)*. Thus it can serve much the same function as a gel-shift analysis in following a specific DNA-binding activity through a series of purification steps. Since DNase I footprinting can often be used for proteins that do not "gel-shift," it has more general applicability. However, because of the need for a protein excess and the visualization of the footprint by a partial DNA digestion ladder, the technique requires considerably more material than would a gel-shift.

DNase I (EC 3.1.4.5) is a protein of roughly 40 Å diameter. It binds in the minor groove of the DNA and cuts the phosphodiester backbone of both strands independently *(3)*. Its bulk helps to prevent it from cutting the DNA under and around a bound protein. However, a bound protein will usually have other effects on the normal cleavage by DNase I, resulting in some sites becoming hypersensitive to DNase I (*see* Figs. 1 and 2). It is also not uncommon to observe a change in the pattern of DNase cleavage without any obvious extended protection (Fig. 2).

Unfortunately, DNase I does not cleave the DNA indiscriminately, some sequences being very rapidly attacked whereas others remain unscathed even after extensive digestion *(4)*. This results in a rather uneven "ladder" of digestion products after electrophoresis, something that limits the resolution of the technique (*see* naked DNA tracks in Figs. 1 and 2). However, when the protein-protected and naked DNA ladders are run alongside each other, the footprints are normally quite apparent. To localize the position of the footprints, G + A and/or C + T chemical sequencing ladders of the same end-labeled DNA probe *(5)* should accompany the naked and protected tracks (*see* Note 9). Since a single end-labeled fragment allows one to visualize interactions on one strand only of the DNA, it is usual to repeat the experiment with the same fragment labeled on the other strand. DNA fragments can be conveniently 5' labeled with T_4 kinase and 3' labeled using Klenow, T_4 polymerase (fill out), or terminal transferase *(6)*. A combination of the 5' and 3' end-labeling allows both DNA strands to be analyzed side by side from the same end of the DNA duplex.

DNase I footprinting requires an excess of DNA-binding protein over the DNA fragment used. The higher the percent occupancy of a site on the DNA, the clearer a footprint will be observed. It is there-

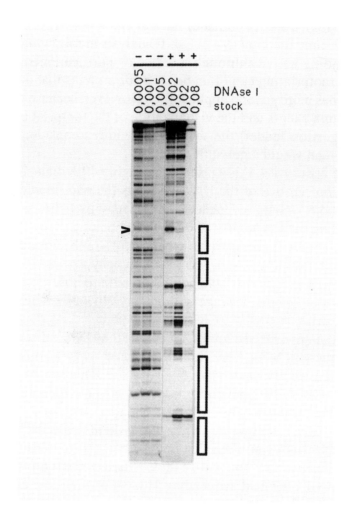

Fig. 2. Course of digestion with increasing amounts of DNase I. Here xUBF was footprinted on the *Xenopus* ribosomal promoter using a 5' end-labeled fragment. The numbers above the tracks refer to the DNase I dilution, in U/µL employed, and (−) and (+) refer to the naked and complexed DNAs respectively. The predominant footprints are indicated by open boxes.

fore important not to titrate the available proteins with too much DNA. This limitation can in part be overcome when a protein also generates a gel-shift. It is then feasible to fractionate the partially DNase digested protein–DNA complex by nondenaturing gel electrophoresis and to excise the shifted band (which is then a homogeneous protein-DNA complex) before analyzing the DNA by denaturing gel electrophoresis as in the standard footprint analysis (*see* Chapters 4, 6, and 21 in this volume).

Footprinting crude or impure protein fractions usually requires that an excess of a nonspecific competitor DNA be added. The competitor binds nonspecific DNA-binding proteins as effectively as the specific labeled target DNA fragment and hence, when present in sufficient excess, leaves the main part of the labeled DNA available for the sequence-specific protein. Homogeneous and highly enriched protein fractions usually do not require the presence of a nonspecific competitor during footprinting. When planning a footprinting experiment, it is a prerequisite to start by determining the optimal concentration of DNase I to be used. This will be a linear function of the amount of nonspecific DNA competitor but more importantly and less reproducibly, this will be a function of the amount and purity of the protein fraction added. As a general rule, more DNase is required if more protein is present in the binding reaction, whether or not this protein binds specifically. Thus, very different DNase concentrations may be required to produce the required degree of digestion on naked and protein-bound DNA. A careful titration of the DNase concentration is therefore essential to optimize the detection of a footprint and can even make the difference between the detection or lack of detection of a given interaction.

The following protocol was developed to study the footprinting of the *Xenopus* ribosomal transcription factor xUBF, which is a rather weak DNA-binding protein, with a rather broad sequence specificity. The protocol is not original, being derived from several articles *(1,7)*. It does, however, represent a very practical approach that can be broadly applied. We recommend that the reader also refers to the available literature for more information on the quantitative analysis of protein-DNA interactions by footprinting *(8)*.

2. Materials

1. 2X Binding buffer: 20% glycerol, 0.2 mM EDTA, 1 mM DTT, 20 mM HEPES, pH 7.9, and 4% polyvinyl alcohol (*see* Note 1).
2. Poly d(AT): 1 mg/mL in TE (10 mM Tris-HCl, pH 8.0, 1 mM EDTA). Keep at –20°C (*see* Note 2).

3. End-labeled DNA fragment of high-specific activity (*see* Note 3).
4. Cofactor solution: 10 mM MgCl$_2$, 5 mM CaCl$_2$.
5. DNase I stock solution: A standardized vial of DNase I (D–4263, Sigma, St. Louis, MO) is dissolved in 50% glycerol, 135 mM NaCl, 15 mM sodium acetate, pH 6.5, at 10 Kunitz U/μL. This stock solution can be kept at –20°C for many months (*see* Note 4).
6. 1M KCl.
7. Reaction stop buffer: 1% SDS, 200 mM NaCl, 20 mM EDTA, pH 8.0, 40 μg/mL tRNA (*see* Note 5).
8. 10X TBE buffer: 900 mM Tris-borate, pH 8.3, 20 mM EDTA.
9. Loading buffer: 7M urea, 0.1X TBE, 0.05% of xylene cyanol, and bromophenol blue.
10. Sequencing gel: 6% acrylamide, 7M urea, 1X TBE.
11. Phenol-chloroform (1:1) saturated with 0.3M TNE (10 mM Tris-HCl, pH 8.3, 1 mM EDTA, 0.3M NaCl).
12. Ethanol 99% and ethanol 80%. Keep at –20°C.
13. 1M pyridine formate, pH 2.0. Keep at 4°C.
14. 10M piperidine.

3. Methods

The footprinting reaction is done in three stages: binding of the protein to the DNA, partial digestion of the protein–DNA complex with DNase I, and separation of the digestion fragments on a DNA sequencing gel.

1. The binding reaction is performed in a total volume of 50 μL containing 25 μL of 2X binding buffer, 0.5 μL of 1 mg/mL poly d(AT), 2–3 ng of end-labeled DNA fragment (~15,000 cpm) (*see* Note 6), the protein fraction and 1M KCl to bring the final KCl concentration to 60 mM. The maximum volume of the protein fraction that can be used will depend on the salt concentration of this solution. The reaction is performed in a 1.5-mL Eppendorf tube.
2. Incubate on ice for 20 min.
3. During the binding reaction, dilute the DNase I stock solution in distilled water at 0°C. We suggest working concentrations of about 0.0005–0.1 Kunitz U/μL, depending on the level of protein present (*see* Note 7 and step 5). A good range is the following: 0.0005; 0.001; 0.005; 0.002; 0.02; 0.08 Kunitz U/μL.
4. After the incubation, transfer the reaction tubes in batches of eight to a rack at room temperature and add 50 μL of the cofactor solution to each.
5. Add 5 μL of the appropriate DNase I dilution to a tube every 15 s (from

the 0.0005–0.005 Kunitz U/µL stocks for naked DNA; from the 0.002–0.08 ones for DNA + proteins).
6. After 2 min digestion, each reaction is stopped by the addition of 100 µL of the stop solution (see Note 8).
7. After all the reactions have been processed, extract each reaction once with phenol-chloroform as follows: add 1 vol phenol-chloroform (1:1) saturated with $0.3M$ TNE, vortex briefly, and centrifuge in a desktop microcentrifuge for about 10 min. Recover the top phase and transfer to a new microcentrifuge tube.
8. Add 2 vol (400 µL) of ethanol 99% (–20°C) and allow nucleic acids to precipitate at –80°C for 20 min.
9. Microcentrifuge for 15 min, ~10,000g, and remove the supernatant with a Pasteur pipet. Check the presence of a radioactive pellet with a Geiger counter before discarding the ethanol.
10. Add 200 µL of 80% ethanol (–20°C) to the pellet and microcentrifuge again for 5 min. After removing the supernatant, dry the pellets in a vacuum dessicator.
11. Resuspend each pellet in 4.5 µL loading buffer, vortex, and centrifuge briefly.
12. A G + A ladder and a molecular weight marker should be run in parallel with the samples on a sequencing gel (see Note 9). The G + A ladder can be prepared as follows (5): ~200,000 cpm of end-labeled DNA are diluted into 30 µL H$_2$O (no EDTA). 2 µL of $1M$ pyridine formate, pH 2.0, are added and the solution incubated at 37°C for 15 min. One hundred fifty microliters of $1M$ piperidine are added directly and the solution incubated at 90°C for 30 min in a well sealed tube (we use a 500-µL microcentrifuge tube in a thermal cycler). Add 20 µL of $3M$ sodium acetate and 500 µL of ethanol and precipitate at –80°C for 10–20 min. Microcentrifuge (10,000g, 10 min) and repeat the precipitation. Finally, redissolve the pellet in 200 µL of H$_2$O and lyophilize. Resuspend in loading buffer and apply about 5,000 cpm/track.
13. Prerun a standard 6% acrylamide, sequencing gel (43 × 38 cm, 0.4 mm thick, 85 W) for 30 min before loading each of the aliquots from the DNase I digestion, plus the markers. Running buffer is 1X TBE. Wash the wells thoroughly with a syringe, denature the DNA for 2 min at 90°C, and load with thin-ended micropipet tips. Run the gel hot to keep the DNA denatured (see Note 10). After the run, cover the gel in plastic wrap and expose it overnight at –70°C with an intensifying screen. We use either a Cronex Lightning Plus (Dupont, Wilmington, DE) or Kyokko Special (Fuji, Japan) screens, the latter being about 30% less sensitive but also less expensive. Several different exposures will probably be required to obtain suitable band densities.

4. Notes

1. This binding buffer has been shown to work well for the transcription factor NF-1 *(6)*, and in our laboratory for both the hUBF and xUBF factors and thus should work for many factors. Glycerol and PVA (an agent used to reduce the available water volume and hence concentrate the binding activity) are not mandatory. The original footprinting conditions of Galas and Schmitz *(1)* for the binding of the *lac* repressor on the *lac* operator were 10 mM cacodylate buffer, pH 8.0, 10 mM MgCl$_2$, 5 mM CaCl$_2$, and 0.1 mM DTT. Particular conditions of pH, cofactors, and ionic strength may need to be determined for an optimal binding of different factors to DNA.
2. Since poly d(IC), another nonspecific general competitor, has been shown to compete quite efficiently with G-C rich DNA sequences, poly d(AT) is prefered here. The choice of an appropriate nonspecific competitor (whether it is synthetic, as in this case, or natural, e.g., pBR322 or calf thymus DNA) may have to be determined empirically for the protein studied. When working with a pure or highly enriched protein, no competitor is usually needed. The DNase I concentration must then be reduced accordingly (to about naked DNA values).
3. Single-stranded breaks in the end-labeled DNA fragment must be avoided because they give false signals indistinguishable from genuine DNase I cleavage and hence can mask an otherwise good footprint. It is therefore advisable to check the fragment on a denaturing gel before use. Always use a freshly labeled fragment (3–4 d at the most) because radiochemical nicking will degrade it.
4. These standardized vials allow for very reproducible results. Glycerol will keep the enzyme from freezing, as repeated freeze–thaw cycles will greatly reduce its activity.
5. Do not be tempted to use too much RNA since it causes a very annoying fuzziness of the gel bands that prevents resolution of the individual bands.
6. The use of 5'end-labeling with kinase in the presence of crude protein extracts can sometimes lead to a severe loss of signal because of the presence of phosphatases. In these cases 3'end-labeling by "fill out" with Klenow or T$_4$ polymerase is to be prefered.
7. For naked DNA and very low amounts of protein, working stocks diluted to 0.0005–0.005 Kunitz U/µL give a good range of digestion.
8. It is convenient to work with groups of eight samples during the DNase I digestion. Cofactor solution is added to eight samples at a time and then the DNase I digestions begun at 15 s intervals: 15 s after adding DNase to the eighth sample, stop solution is added to sample 1 and then to the other samples at 15 s intervals.

9. In comparing a chemical sequencing ladder with the products of DNase I digestion, one must bear in mind that each band in the sequencing ladder corresponds to a fragment ending in the base *preceding* the one read because chemical modification and cleavage destroys the target base. For example, if a DNase I gel band corresponds in mobility to the sequence ladder band read as G in the sequence ACGT, then the DNase I cleavage occured between the bases C and G. DNase I cleaves the phosphodiester bond, leaving a 3'-OH, whereas the G + A and C + T sequencing reactions leave a 3'-PO_4, causing a mobility shift between the two types of cleavage ladders. This is a further potential source of error. However, in our experience the shift is less than half a base and hence cannot lead to an error in the deduced cleavage site.

10. Sequencing gels are not denaturing unless run hot ($7M$ urea produces only a small reduction in the T_m of the DNA). A double-stranded form of the DNA fragment is therefore often seen on the autoradiogram, especially at low levels of DNase I digestion (*see* Fig. 2) and can sometimes be misinterpreted as a hypersensitive cleavage. By running a small quantity of undigested DNA fragment in parallel with the footprint this error can be avoided.

Acknowledgments

The authors wish to thank A. Ruiz-Carrillo for providing the autoradiogram in Fig. 1A. The work was supported by the Medical Research Council of Canada (MRC). T. Moss is presently an F.R.S.Q. "Chercheur-boursier" and B. Leblanc was until recently supported by a grant from the F.C.A.R. of Québec.

References

1. Schmitz, A. and Galas, D. J. (1978) DNase I footprinting: a simple method for the detection of protein-DNA binding specificity. *Nucleic Acids Res.* **5,** 3157–3170.
2. Rousseau, S., Renaud, J., and Ruiz-Carrillo, A. (1989) Basal expression of the histone H5 gene is controlled by positive and negative *cis*-acting sequences. *Nucleic Acids Res.* **17,** 7495–7511.
3. Suck, D., Lahm, A., and Oefner, C. (1988) Structure refined to 2 Å of a nicked DNA octanucleotide complex with DNase I. *Nature* **332,** 464–468.
4. Drew, H. R. (1984) Structural specificities of five commonly used DNA nucleases. *J. Mol. Biol.* **176,** 535–557.
5. Maxam, A. M. and Gilbert, W. (1980) Sequencing end-labeled DNA with base-specific chemical cleavages, in *Methods in Enzymology,* vol. 65 (Grossman, L. and Moldave, K., eds), Academic, New York, pp. 499–560.
6. *Current Protocols in Molecular Biology*, Chapter 3 (1991) (Ausubel, F. M.,

Brent, R., Kingston, R. E., Moore, D. E., Smith, S. A., and Struhl, K., eds.), Greene and Wiley-Interscience, New York.
7. Walker, P. and Reeder, R. H. (1988) The *Xenopus laevis* ribosomal gene promoter contains a binding site for nuclear factor-1. *Nucleic Acids Res.* **16,** 10,657–10,668.
8. Brenowitz, M., Senear, D. F., and Kingston, R. E. (1991) DNase footprint analysis of protein-DNA binding, in *Current Protocols in Molecular Biology* (Ausubel, F. M., Brent, R., Kingston, R. E., Moore, D. E., Smith, S. A., and Struhl, K., eds.), Greene and Wiley-Interscience, New York, pp. 12.4.1–12.4.11.

CHAPTER 2

Footprinting with Exonuclease III

Willi Metzger and Hermann Heumann

1. Introduction

Within the last few years footprinting techniques have become increasingly important in the study of protein-nucleic acid interactions. This is partly the result of a fast-growing number of known nucleic acid binding proteins but also because of an increase in the available probes that can be chosen in order to tackle a specific problem. There are two major groups of probes—the chemical probes and the enzymatic probes. These enzymatic probes, such as DNase I or exonuclease III, have the advantage of acting specifically at the DNA. Chemical probes are often less specific and react also with the protein. This can disturb the correct interaction of protein and DNA. For the study of very fragile protein-DNA complexes, therefore, enzymatic probes are often preferable.

The exploitation of a specific enzymatic function can also be a reason for choosing an enzyme as probe. The exonuclease activity in combination with the processivity of exonuclease III can be the reason to choose this enzyme as a probe when information about the position of a sequence-specific bound protein is required. A prerequisite for the successful use of exonuclease III as a footprinting probe is that the half-life of the protein-DNA complex is longer than the application time of exonuclease III.

1.1. Description of the Enzymatic Function of Exonuclease III

Exo III is a monomeric enzyme with a mol wt of 28,000 kDa. It contains several distinct activities: a 3'–5' exonuclease activity, a DNA 3' phosphatase activity, an AP endonuclease activity, and an RNase H activity *(1)*.

1.2. Principle of the Procedure

Footprinting with Exo III makes use of the 3'–5' activity *(2)*. After a protein has been bound specifically to a DNA fragment containing the signal sequence of the protein, Exo III removes mononucleotides of both DNA strands in a processive way, starting from the 3' ends. Free DNA is fully digested. This is advantageous compared to other probes, since there are no background problems caused by the presence of free DNA in the assay. The specifically bound protein blocks the action of Exo III and leaves a double-stranded DNA only in that region where the protein is bound (Fig. 1). The length of the remaining DNA strands is determined by electrophoresis on a sequencing gel using proper DNA length standards. The DNA strands are detected by radioactive 5'-end-labeling of the DNA. A clear allocation of the two resulting bands to the lower or upper strands is sometimes difficult. This problem can be solved by removal of one label (e.g., by a restriction cut before application of Exo III). Figure 1A shows schematically the procedure and Fig. 1B the resulting bands in a denaturing acrylamide gel.

1.3. Interpretation

From the length of the two DNA strands remaining after Exo III treatment the size of the protected area can be determined. If the DNA fragment used has a length of k base pairs and the remaining strands have a length of m and n base pairs, respectively, the size of the protected region (x) is

$$x = m + n - k.$$

Correct interpretation of the footprinting data requires a critical assessment of the action of the probing enzyme on the protein-DNA complex.

The interpretation is straightforward if the protein is strongly bound to DNA. The protected region is an upper limit of the DNA region

Footprinting with Exonuclease

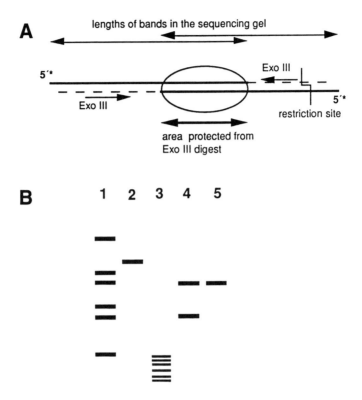

Fig. 1. **A.** Schematic representation of the Exo III footprinting procedure. **B.** Exo III footprint of the complex shown in **A.** Lane 1: Size markers. Lane 2: Labeled DNA-fragment. Lane 3: Labeled DNA-fragment, after Exo III treatment. Lane 4: Pattern of DNA-protein complex, obtained after Exo III treatment with both 5'-ends labeled. Lane 5: Pattern of DNA-protein complex, obtained after Exo III treatment. One label was removed beforehand by cutting with a restriction enzyme as indicated in **A.**

interacting with the protein, since the protein acts as a steric hindrance for Exo III. An assessment of the protection is more complicated if the strength of interaction between protein and DNA varies within the interacting domain. Owing to the processivity of Exo III, this enzyme can nibble into the DNA region, which has a low affinity to the sequence binding protein, resulting in a smaller size of the protected region. In order to decide if such a process occurs, a time dependent measurement of the Exo III digestion is essential.

1.4. Examples for the Application of Exo III as Footprinting Probe

Exo III was used to follow the movement of *E. coli* RNA polymerase during RNA synthesis *(3,4)*. For that purpose the RNA synthesis was stopped in specific frames and the arrested complex was subjected to Exo III digestion. The result was a set of bands marking the boundaries of RNA polymerase on the DNA at each step of RNA synthesis. Exo III offers the possibility of detecting specific protein DNA interactions even in crude extracts. Only those proteins whose half-life is longer than the application time of the probing Exo III act as a block for Exo III *(5–7)*.

2. Materials

2.1. Exonuclease III

Exonuclease III is available from BRL (Gaithersburg, MD) or Boehringer, Mannheim (Indianapolis, IN). It can be stored over months at –20°C.

2.2. Reagents and Materials for the Sequencing Gel

1. Acrylamide solution: 40% acrylamide, 0.66% *bis*-acrylamide in H_2O.
2. 10X TBE: $1M$ Tris-HCl, pH 8.6; 840 mM boric acid; 10 mM EDTA.
3. 10% Ammonium persulfate (freshly prepare before use).
4. TEMED.
5. Preparation of an 8% acrylamide gel: Weigh 21 g of urea, add 5 mL of 10X TBE solution and 10 mL of acrylamide solution. Dissolve under mild heating. Add double-distilled water to a final vol of 50 mL. Filter and degas the solution (filter pore size 0.2 µm). Add 500 µL of 10% ammonium persulfate solution and 30 µL of TEMED. Pour gel immediately.
6. Loading buffer for the sequencing gel: 100 mL deionized formamide, 30 mg xylenecyanol FF, 30 mg bromophenol blue, 750 mg EDTA.
7. Electrophoresis buffer: 1X TBE.

2.3. The Nondenaturing Gel for the Band-Shift Assay

1. Acrylamide solution: 30% acrylamide, 0.8% *bis*-acrylamide in H_2O.
2. $1M$ Tris-HCl, pH 7.9.
3. 10% Ammonium persulfate (freshly prepare before use).
4. 5% TEMED (diluted in water).

5. Preparation of nondenaturing gel: Mix 240 μL 1M Tris-HCl, pH 7.9, 2.75 mL acrylamide solution, and 25.7 mL of H_2O and degas. Add 300 μL of 10% ammonium persulfate and 70 μL of 5% TEMED. Pour gel.
6. Loading buffer for the nondenaturing gel (10X solution): 40% sucrose, 0.1% bromophenol blue.
7. 5% Dichloro-dimethylsilane solution (in chloroform).
8. 0.3% γ-Methacryl-oxypropyl-trimethoxy-silane, 0.3% acetic acid dissolved in ethanol.
9. Electrophoresis buffer: 8 mM Tris-HCl, pH 7.9. Store the solutions protected against light at 4°C. Dilute buffers with bidistilled water.

2.4 Other Items

1. Sequencing gel apparatus (Pharmacia, Piscataway, NJ).
2. Filters for drop dialysis VS, 0.025 μm (Millipore, Bedford, MA).
3. Peristaltic pump.
4. SpeedVac concentrator.

3. Methods

3.1. Establishing the Conditions for Obtaining Optimum Yield of Specific Protein-DNA Complexes

A very elegant method for establishing the optimum conditions for the formation of a specific protein-DNA complex is acrylamide gel electrophoresis under nondenaturing conditions. This "band shift assay" allows one to differentiate between bound and unbound DNA *(8,9)*. It permits the determination of the stoichiometry of the components and the optimum salt conditions for specific binding (*see* Chapter 21). This method can be applied even for high molecular weight protein DNA complexes if the concentration of the acrylamide is low enough to enable the complex to enter the gel matrix. The gel composition described in Section 2.3. is optimal for the study of high-molecular weight complexes (*see* Note 6).

These low concentration gels are difficult to handle. Therefore, the glass plates must be subjected to a special treatment by which the gel is bound to one of the plates:

1. Wash the glass plates (20 × 20 cm) with ethanol.
2. Treat one plate with γ-methacryl-oxypropyl-trimethoxy-silane solution. Wash this plate carefully four times with ethanol to avoid sticking of the other plate to the gel.

3. Treat the second plate with dichloro-dimethylsilane solution.
4. Form the protein-DNA complex in a volume of about 15 µL under the desired conditions.
5. Dialyze the complex, if necessary, against the electrophoresis buffer by drop dialysis in the following way:
   a. Pour the dialysis buffer into a Petri dish.
   b. Put the filter (Millipore, see Section 2.4.) with the glossy side upward onto the buffer so that it can float freely.
   c. Put the sample as a drop on the filter, and remove the drop after 1 h of dialysis.
6. Add 1/10 vol of the 10X loading buffer to the dialyzed complex and apply the sample onto the nondenaturing acrylamide gel.
7. Run the gel at 20 V/cm for about 2 h. Pump the buffer from the anode to the cathode chambers and back again to avoid a pH-decrease in the anode chamber (see Note 6).

3.2. Establishing the Conditions for the Digestion of the DNA

1. Label the DNA at the 5'-ends, using T_4 polynucleotide kinase and γ-[^{32}P]-ATP. Take an aliquot of the 5'-end-labeled DNA and remove one 5'-label by cutting with an appropriate restriction enzyme. Use both types of DNA in the subsequent steps.
2. Take care that the total amount of DNA in an assay of 20 µL is not below 100 ng. The total amount of radioactivity in one assay should be around 20,000 cpm. Use the salt conditions that are optimal for the binding conditions of the probed DNA binding protein. Add 6 mM Mg^{2+}, if not already present in the incubation assay.
3. Add Exo III. In order to establish the optimum conditions, perform a series of experiments using different Exo III concentrations varying between 1 and 200 U, and different incubation periods varying between 1 and 45 min. Exo III seems to be rather stable over a wide range of ionic strengths; at least no big changes could be observed in the range between 0 and 100 mM NaCl (or KCl) in the incubation mixture (see Note 1).
4. Add EDTA to a final concentration of 20 mM in order to stop the reaction at the appropriate time.
5. Add sodium-acetate to a final concentration of 0.3M followed by 2.5 vol of ice-cold 100% ethanol to precipitate the digested DNA. Keep the solution at –70°C for 20 min.
6. Spin down the solution in a microcentrifuge for 15 min. Wash the pellet with ice-cold 75% ethanol, dry under vacuum, and dissolve in the loading (formamide) buffer.

7. Boil the sample for 2 min and apply onto a 6–10% sequencing gel. (For the analysis of fragments in the range 50–150 bases a gel consisting of 8% acrylamide is adequate, as described in Section 2.2.). Use as a length standard a Maxam-Gilbert sequencing reaction of the 5'-end-labeled DNA-fragment.
8. Run the gel at 50 W at a temperature of 60°C for 2 h.
9. Expose the gel after electrophoresis to an X-ray film using an intensifying screen at –70°C overnight.
10. Make sure that the free DNA is fully digested (usually much shorter DNA-pieces than the predicted "half-cut" fragment size [2] are obtained).

3.3. Exo III Digestion of the Protein-DNA Complex

1. Form the complex using the conditions established under Section 3.1.
2. Subject the complex to Exo III digestion using the conditions established under Section 3.2. (see Notes 2–6).
3. Add 20 mM EDTA 1% SDS (final concentration) to stop the reaction. SDS is necessary in order to destroy the protein-DNA complex.
4. Proceed as described in Section 3.2. (steps 5–9) for recovery and gel electrophoretic analysis of the DNA. For recovery of the DNA a phenol extraction before the ethanol precipitation might be advisable if the complex is collected from a crude extract.

3.4. Modifications of the Procedure

Depending on the kind of protein-DNA complexes you are investigating, several modifications of the procedure described in the previous section may be useful or necessary. If the protein-DNA complexes are not homogeneous (e.g., part of the DNA-fragments are complexed with more than one protein molecule), the desired complex can be purified by a nondenaturing acrylamide gel as described in Section 3.1., provided the different species of complexes show different gel mobilities and have a half-life long enough to survive the electrophoresis procedure. Such a purification step requires the use of ten times the amount of radioactively labeled DNA.

The procedure is as follows:

1. Form the complex.
2. Subject the complex to digestion with Exo III according to Section 3.3., step 2.
3. Dialyze the complex by drop dialysis described in Section 3.1., step 5 against a low salt buffer (e.g., 10 mM Tris-HCl, pH 7.9) in order to avoid salt effects during electrophoresis.

4. Apply the complex to the nondenaturing gel as described in Section 3.1., steps 6–7. The half-life of the complex is in most cases not changed by the Exo III digestion of the DNA.
5. Expose the gel to an X-ray film with an enhancer screen at –70°C. An hour must be enough to recognize the complex bands. If not, the recovery is too small for a subsequent sequencing gel.
6. Mark exactly the position of the film in order to find the positions of the bands of interest.
7. Excise the complex-bands of interest with a spatula. The band representing the free DNA will be visible as a smear after Exo III digestion.
8. To elute the complex DNA put the excised gel slice in 600 µL of bidistilled water. Heat the complex to 90°C for 3 min and shake overnight at room temperature. The effectiveness of the elution can be easily monitored by comparing the radioactivity of the eluate with the radioactivity of the gel slice.
9. Vacuum-dry the eluate in a SpeedVac concentrator.
10. Dissolve the pellet in 10 µL formamide-buffer and spin down the gel residue.
11. Transfer the supernatant to a new Eppendorf cup and apply the sample to a sequencing gel as described in Section 3.2., steps 7–9.

4. Notes

1. It has been observed that many batches of commercially available Exo III contain an activity that removes the 5'-label. A 5'-phosphatase or a 5'–3' exonuclease activity could account for this phenomenon. Filling in the 5'-protruding ends with α-thio-dNTPs as described by some authors *(10,11)* may eliminate the problem. Addition of *E. coli* t-RNA can reduce the effect, but not completely avoid it.
2. Investigation of the complexes of specific binding of proteins and DNA in crude extracts by Exo III requires additional precautions in order to avoid problems caused by nuclease activity during Exo III exposure. One should add to the incubation assay *(5)* sodium-phosphate, t-RNA, deoxyoligonucleotides and fragmented phage DNA (e.g., 2 mM sodium-phosphate, 1 µg of ΦX 174 DNA cut with *Hae*III, 10 µg of yeast t-RNA, 1 µg mixed p(dN)$_5$). The author *(5)* claims this suppresses nuclease activities contained in the crude extracts, but it may well be that the fact mentioned under Note 1 plays an additional role.
3. Testing different concentrations of Exo III and different incubation periods can provide additional information about the nature of the protein-DNA complex under study. If Exo III is able to "nibble" further into a protected area with increasing exposure time, this indicates differences in the strength of protein-DNA interaction *(10)*.

4. Different binding sites for one or more proteins may be detected as distinct stop points for Exo III as shown for refs. *12–14.* This applies when working with crude extracts as for the use of purified factors. It is necessary, however, that the ratio of DNA:binding proteins is >1.
5. Heparin, which is often used as a DNA competitor for *E. coli* RNA-polymerase and other DNA-binding proteins, also interacts with Exo III and reduces its activity markedly.
6. The gel concentration has to be adjusted according to the molecular weight of the protein-DNA complex. Here we describe the conditions established for the study of the *E. coli* RNA polymerase (MW 455 000) and a DNA fragment of 130 bp carrying a promoter *(9).* For some applications another widely used nondenaturing gel may be appropriate: 1X TBE-buffer, 4% acrylamide, 0.1% *bis*-acrylamide. Circulation of the buffer is not necessary here.

Further Reading

Kow, Y. W. (1989) Mechanism of action of *Escherichia coli* Exonuclease III. *Biochemistry* **28,** 3280–3287.

References

1. Rogers, S. G. and Weiss, B. (1980) Exonuclease III of *Escherichia coli* K-12, an AP endonuclease. *Meth. Enzymol.* **65,** 201–211.
2. Shalloway, D., Kleinberger, T., and Livingston, D. M. (1980) Mapping of SV 40 DNA replication origin region binding sites for the SV 40 DNA replication antigen by protection against Exonuclease III digestion, *Cell* **20,** 411–422.
3. Metzger, W., Schickor, P., and Heumann, H. (1989) A cinematographic view of *Escherichia coli* RNA polymerase translocation. *EMBO J.* **8,** 2745–2754.
4. Pavco, P. A. and Steege, D. A. (1990) Elongation by *Escherichia coli* RNA polymerase is blocked *in vitro* by a site specific DNA binding protein. *J. Biol. Chem.* **265,** 9960–9969.
5. Wu, C. (1985) An exonuclease protection assay reveals heat-shock element and TATA box binding proteins in crude nuclear extracts. *Nature,* **317,** 84–87.
6. Loh, T. P., Sievert, L. L., and Scott, R. W. (1990) Evidence for a stem cell-specific repressor of Moloney murine leukemia virus expression in embryonic carcinoma cells. *Mol. Cell. Biol.* **10,** 4045–4057.
7. Carnevali, F., La Porta, C., Ilardi, V., and Beccari, E. (1989) Nuclear factors specifically bind to upstream sequences of a *Xenopus laevis* ribosomal protein gene promoter. *Nucleic Acids Res.* **17,** 8171–8184.
8. Fried, M. and Crothers, D. M. (1981) Equilibria and kinetics of lac repressor-operator interactions by polyacrylamide gel electrophoresis. *Nucleic Acids Res.* **9,** 6505–6525.
9. Heumann, H., Metzger, W., and Niehörster, M. (1986) Visualization of intermediary transcription states in the complex between *Escherichia coli* DNA-

dependent RNA polymerases and a promoter-carrying DNA fragment using the gel retardation method. *Eur. J. Biochem.* **158,** 575–579.
10. Straney, D. C. and Crothers, D. M. (1987) A stressed intermediate in the formation of stably initiated RNA chains at the *Escherichia coli* lac UV 5 promoter. *J. Mol. Biol.* **193,** 267–278.
11. Straney, D. C. and Crothers, D. M. (1987) Comparison of the open complexes formed by RNA polymerase at the *Escherichia coli* lac UV 5 promoter. *J. Mol. Biol.* **193,** 279–292.
12. Gaur, N. K., Oppenheim, J., and Smith, I. (1991) The *Bacillus subtilis sin* gene, a regulator of alternate developmental processes, codes for a DNA-binding protein. *J. Bact.* **173,** 678–686.
13. Owen, R. D., Bortner, D. M., and Ostrowski, M. C. (1990) *ras* oncogene activation of a VL30 transcriptional element is linked to transformation. *Mol. Cell. Biol.* **10,** 1–9.
14. Wilkison, W. O., Min, H. Y., Claffey, K. P., Satterberg, B. L., and Spiegelman, B. M. (1990) Control of the adipisin gene in adipocyte differentiation. *J. Biol. Chem.* **265,** 477–482.

CHAPTER 3

Hydroxyl Radical Footprinting

Peter Schickor and Hermann Heumann

1. Introduction

The basic principle of the DNA footprinting technique is the measuring of the accessibility of the DNA by a probe. The probe can be an enzyme or a chemical reagent that is able to cut the DNA backbone. The target is a DNA fragment with a signal-sequence for a sequence-specific binding protein. The sites on the DNA that interact with the protein are excluded from cutting by the probe. These sites appear after electrophoresis as blanks in the otherwise regular cutting pattern, which represent the characteristic footprint of the protein.

It is obvious that the footprinting pattern is determined by the type of probe being used. Hydroxyl radicals as probes are very convenient to handle. They are distinguished by a number of advantages compared to other probes:

1. Hydroxyl radicals cut the DNA with almost no sequence dependence.
2. The resolution of the footprint is very high (1 bp), because the probe is very small.
3. The cutting reaction is compatible over a wide range of buffer compositions, salt, pH, and temperature. Only glycerol, a radical scavenger, interferes with the cutting when present at concentrations higher than 0.5%.
4. All chemicals needed are easily available and uncomplicated in handling.

1.1. Generation and Action of Hydroxyl Radicals

Hydroxyl radicals are generated according to the Fenton reaction by reduction of iron (II) with hydrogen peroxide as follows:

$$\text{Fe(EDTA)}^{2-} + H_2O_2 \longrightarrow \text{Fe(EDTA)}^- + OH^- + OH\bullet$$

ascorbate

The resulting iron (III) is reduced by ascorbate back to iron (II), which can start a new cycle. The use of a negatively charged iron (EDTA)-complex prevents the iron from interacting electrostatically with DNA, so the only reactant interacting with DNA is the hydroxyl radical *(1)*.

The exact manner in which hydroxyl radicals act on DNA is still not known. The radicals are thought to abstract an H-atom from the sugar moiety of the DNA backbone, and secondary reactions of the resulting sugar radical cause the backbone to break, leaving a gap in one strand of the double helix with phosphate groups on either side *(2)*.

1.2. Principle of the Procedure

After formation of the complex of a sequence-specific binding protein and a DNA fragment carrying the binding sequence (*see* Fig. 1), the complex is subjected to hydroxyl radical treatment. Hydroxyl radicals introduce single base deletions randomly distributed in the DNA. The concentration of the hydroxyl radicals is adjusted so that the yield of deletions is less than one per DNA, that is, approx 10% of the DNA fragments are affected. Cutting of the DNA is prevented at those sites on the DNA where the protein is bound. This partially cut DNA is applied to a sequencing gel. The DNA is detected by radioactive labeling. If the label is specifically fixed to one end of one DNA strand, a DNA ladder is produced that is similar to that obtained by sequence analysis. Blanks within this regular ladder indicate the sites where the protein is bound. This becomes more obvious if a reference DNA is included that has been subjected to the same procedure but without previous protein binding. If complex formation is incomplete (i.e., the assay contains free DNA), the obtained footprinting pattern becomes blurred. This can be avoided by separating the hydroxyl radical treated protein DNA complex from free DNA by a nondenaturing gel electrophoresis (Fig. 2) before application to a sequencing gel. Figure 3 shows schematically the obtained footprinting pattern of a protein DNA complex depicted in Fig. 1.

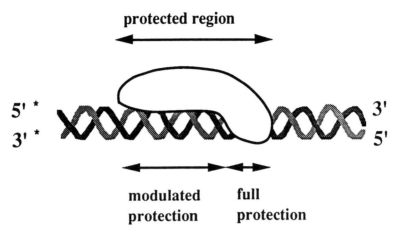

Fig. 1. A putative DNA-binding protein interacts with the DNA over three helical turns. The major portion of the protein interacts with only one side of the DNA over two helical turns. A minor portion of the protein wraps fully around the DNA. (*) Indicates the position of the radioactive label.

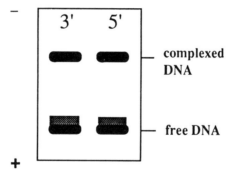

Fig. 2. Protein–DNA complexes are separated from free DNA by nondenaturing gel electrophoresis. The two lanes represent the same complex with a single label at one end of the DNA, at the 3' end, and at the 5' end, respectively. Hydroxyl radicals create gaps in the DNA, which cause a larger retardation of the modified fragments within the gel. This effect is visible only in the free DNA.

1.3. Interpretation of the Footprinting Pattern

Figure 3 shows that a full experimental set contains six DNA ladders, length standards, free DNA as reference, and the complex. Blanks in the DNA ladder indicate exclusion of radical attack of the DNA because of the presence of the bound protein. These blanks can be

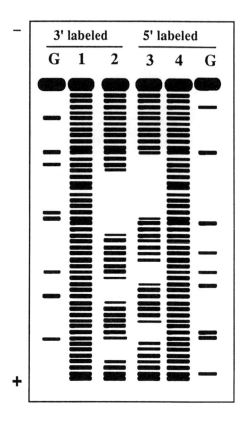

Fig. 3. The bands of the nondenaturing gel in Fig. 2 containing the complex and the free DNA are eluted and applied (under denaturing conditions) on a sequencing gel. Lanes 1 and 4 show the sequence ladder of free DNA labeled at the 3' and the 5' end, respectively. Lanes 2 and 3 show the complex containing the bands. The complex depicted schematically in Fig. 1 would result in the footprint displayed in lanes 2 and 3. Lanes G contain the length standards obtained by a G-specific Maxam-Gilbert sequencing reaction.

assigned to a specific sequence by means of the previous assignment procedure.

The following information can be extracted from the hydroxyl radical footprinting pattern:

1. The total size of the DNA sequence interacting with the protein can be read from the the position of the blanks.
2. A variation of the intensity of the bands within the interacting sequence

reflects differences in the modes of interaction. By a comparison of the footprints on both strands the modes of interaction can often be interpreted:
a. If both strands show a blank at the same region it indicates that the protein wraps around the DNA.
b. A blank at only one strand indicates single-strand formation with one strand protected by interaction with the protein.
c. Modulation of the intensity of the bands with a regular phasing according to the helix repeat (e.g., 10.3 bp for B-DNA) indicates binding of the protein to one side of the DNA. This interpretation is supported if the complementary strand shows the same pattern but with an offset of two or three bases in the 3' label strand. This offset is a consequence of the helicity of the DNA, as shown schematically in Fig. 1.

1.4. Examples for the Application of Hydroxyl Radicals as Footprinting Probes

1.4.1. Protein DNA Complexes

Hydroxyl radicals were used to follow the formation of the transcriptionally active "open" complex between the DNA-dependent RNA polymerase of *E. coli* and a specific promoter. Schickor et al. *(3)* could clearly show the transition from a "closed" complex to an "intermediate" to the final "open" complex. Each complex is characterized by a specific "footprint."

The movement of *E. coli* RNA polymerase during mRNA synthesis was followed by probing a series of specifically arrested transcribing complexes with hydroxyl radicals *(4,5)*.

The interaction of DNA with histones in a nucleosome was investigated by Hayes et al. *(6)*. Using hydroxyl radicals, they showed that the helical periodicity of the DNA changes on nucleosome formation. Furthermore, they could detect two distinct regions of DNA with different helical periodicities within the nucleosome.

Specific contacts between λ-repressor and cro-protein with the cognate operator DNA were revealed by probing these complexes with hydroxyl radicals. Both proteins display the same mode of binding, interacting with only one side of the operator sequence *(7)*.

1.4.2. Antibiotic DNA Complexes

Mithramycin, a small antitumor antibiotic drug, was shown to bind to the minor groove of GC-rich DNA sequences, thereby protecting only 3 bases from hydroxyl radical attack *(8)*.

1.4.3. DNA Structures

The accessibility of bent DNA was studied using hydroxyl radicals. The bend was derived by A-tracts in phase with the helical repeat *(9)*.

Hydroxyl radicals can be used to measure the number of base pairs per helical turn along any DNA molecule. The DNA is adsorbed onto crystalline calcium phosphate before being subjected to radical treatment. From the variation of the intensity of the bands, the helical periodicity of the DNA can be directly obtained *(10)*.

1.4.4. RNA Protein Complexes

Splicing-specific ribonucleoprotein complexes were analyzed by hydroxyl radical treatment. These studies revealed that several regions of the 3' splice site of mRNA precursors are not accessible for hydroxyl radicals, for example, the 3' intron/exon junction, the polypyrimidine tract, and the site of branch formation *(11)*.

1.4.5. RNA Structures

By using hydroxyl radicals, Celander and Cech *(12)* demonstrated that at least three magnesium ions are necessary for formation of a catalytically active RNA molecule.

2. Materials

2.1. Chemicals Used for the Cutting Reaction

Prepare the following solutions separately:

1. 20 mM sodium ascorbate.
2. 0.6% hydrogen peroxide.
3. Iron (II)-EDTA-mix: Mix equal volumes of 2 mM ammonium iron (II) sulfate hexahydrate $(NH_4)_2Fe(SO_4)_2 \cdot 6 H_2O$ and 4 mM EDTA (*see* Note 1).
4. Stop mix: 0.1M Thiourea, 20 mM EDTA, pH 8.0 (*see* Note 2).

2.2. The Sequencing Gel

1. Urea (ultra pure).
2. 10X TBE: 1M Tris-HCl, pH 8.6, 0.84M boric acid, 10 mM EDTA.
3. Acrylamide solution: 40% acrylamide, 0.66% *bis*-acrylamide.
4. 10% ammonium persulfate (prepare immediately before use).
5. TEMED.
6. Sequencing gel (8%): 21 g urea, 5 mL of 10X TBE, and 10 mL of the acrylamide solution are mixed under mild heating. When the urea is dissolved the solution is made up to 50 mL with bidistilled H_2O. The solution is filtered (filter pore size: 0.2 µm) and degassed for 5 min.

500 µL of 10% ammonium persulfate and 30 µL of TEMED are added immediately before pouring the solution between the glass plates.
7. Loading buffer for the sequencing gel (stock solution): 100 mL formamide (deionized), 30 mg xylenecyanol FF, 30 mg bromophenol blue, 750 mg EDTA.
8. Electrophoresis buffer: 1X TBE.

2.3. The Nondenaturing Gel for the Band-Shift Assay

1. $1M$ Tris-HCl, pH 7.9.
2. Acrylamide solution: 30% acrylamide, 0.8% *bis*-acrylamide.
3. 10% Ammonium persulfate (prepare immediately before use).
4. 5% TEMED (aqueous solution).
5. Nondenaturing gel: 240 µL of $1M$ Tris-HCl, pH 7.9, 2.75 mL of the acrylamide solution, and 25.7 mL of bidistilled water are mixed and degassed for 5 min. Three hundred microliters of 10% ammonium persulfate and 70 µL of 5% TEMED are added before pouring the solution between the glass plates.
6. Loading buffer for the nondenaturing gel (stock solution): 40% sucrose, 0.1% bromophenol blue (*see* Note 3).
7. Electrophoresis buffer: 8 mM Tris-HCl, pH 7.9.

2.4. Other Items

1. Sequencing gel apparatus.
2. Apparatus for nondenaturing gel electrophoresis.
3. Filters for drop dialysis, VS, 0.025 µm (Millipore, Bedford, MA).
4. Peristaltic pump.
5. SpeedVac concentrator.

3. Method

3.1. Establishing the Conditions for Obtaining Optimum Yield of Specific Protein–DNA Complexes

The method of establishing the conditions for complex formation using the band shift assay is described in Chapter 21.

3.2. Establishing the Conditions for Cutting the DNA by Hydroxyl Radicals

1. Endlabel the DNA fragment either at the 5'-positions, for example, by T4 polynucleotide kinase with γ-[^{32}P]ATP or at the 3'-positions using the proper α-[^{32}P]dNTP, and Klenow fragment under standard conditions

(13). Remove one label by restriction of the DNA fragment with the appropriate endonuclease. Purify your uniquely endlabeled DNA fragment by a nondenaturing gel electrophoresis (*see* Section 2.3.). The total amount of radioactivity in one assay should be around 10,000–15,000 cpm. The total amount of DNA in an assay of 15 µL should not be below 100 ng. The optimum length of the DNA fragment is between 100 and 200 bp (*see* Note 4).

2. For the cutting reaction, use the salt conditions that are optimum for formation of the protein-DNA complex. Add to the incubation assay 3.5 µL each of the previously prepared solutions of ascorbate, hydrogen peroxide, and the iron (II)-EDTA mix (*see* Section 2.1.) by putting single drops on the inner wall of the tube and by rapidly mixing the three drops before combining them with the sample using a micropipet.
3. Incubate 3–4 min at room temperature.
4. Add 7 µL stop-mix (*see* Section 2.2.).
5. Add sodium acetate to a final concentration of $0.3M$, followed by 2.5 vol of ice-cold 100% ethanol, to precipitate the DNA. Keep the solution at –70°C for 30 min.
6. Spin down the solution in a microcentrifuge for 30 min. Wash the pellet with ice-cold 80% ethanol, dry the pellet under vacuum, and dissolve the pellet in formamide buffer.
7. Adjust the amount of radioactivity and volume in each sample to about 5000–6000 cpm in 4–5 µL.
8. Heat the sample for not longer than 2 min at 90°C and put it on ice (*see* Note 5). Apply the sample onto a 6–10% sequencing gel, (for the analysis of fragments in the range of 50–150 bases a gel consisting of 8% acrylamide is adequate, as described in Section 2.2.). Use as length standards a Maxam-Gilbert sequencing reaction of the 5'- and 3'- labeled DNA-fragment.
9. Run the gel at 50 W at a temperature of 60°C for 1.75 h.
10. Expose the gel after electrophoresis to an X-ray film using an intensifying screen at –70°C overnight.

3.3. Footprinting of Protein–DNA Complexes

1. Prepare two 15-µL samples of the complexes, one with the DNA labeled at the 3'-end, and another with the label at the 5'-end of the DNA, as described in Section 3.2., step 1, for free DNA. Use the conditions established for optimal complex formation from Section 3.1. The total amount of radioactivity in one assay should be around 60,000–80,000 cpm.
2. Pour 30–40 mL of the dialysis buffer containing 8 mM Tris-HCl, pH 7.9, into a Petri dish (*see* Note 6). Place a Millipore filter (*see* Section

2.5.) on the surface of the buffer, shiny side (hydrophobic side) up. Put the samples containing the complexes onto the filter for 1 h in order to remove glycerol and salt (*see* Note 7). Remove the samples from the filter by a micropipet and transfer them to a fresh 1.5-mL Eppendorf tube.

3. Subject the two samples to hydroxyl radical treatment described in Section 3.2., steps 2–4.
4. Separate the complex and the free DNA by nondenaturing acrylamide gel electrophoresis. The conditions for studying a high molecular weight complex are described below (*see* Note 8). Use the conditions for preparing the nondenaturing gel as described in Section 2.3. These low concentration gels are difficult to handle, therefore, the glass plates must be subjected to a special treatment by which the gel is bound to one of the plates. Wash the glass plates (16.5 × 13 cm) with ethanol. Treat one plate with γ-methacryl-oxypropyl-trimethoxy-silane. Wash this plate carefully four times with ethanol to avoid sticking of the other plate to the gel. Treat the second plate with dichlorodimethylsilane. Use spacers 1–1.5 mM thick. Use combs that allow you to apply amounts of 50 µL sample.
5. Add loading buffer (1/10 of the sample volume, *see* Section 2.3.) to the sample. Apply the sample onto the nondenaturing acrylamide gel. Run the gel at 20 V/cm for about 2 h. Pump the buffer from the anode to the cathode chamber and back again to avoid a pH decrease in the anode chamber (*see* Note 9).
6. Remove one of the plates by lifting it carefully with a spatula, and cover the gel (which sticks to the other glass plate) with plastic wrap. Place an X-ray film on the gel. Put the gel and the film into a cassette, and expose the film 1–2 h at 4°C. Mark the bands containing the complex and, if visible, the free DNA. This is possible by placing the film after developing on the gel again in exactly the same way as before and mark the positions on the glass plate.
7. Remove the plastic wrap from the gel, cut out the marked bands with a scalpel or spatula, and cut them into small pieces. Put the slices into a new reaction cup with 700 µL of elution buffer (0.3M sodium acetate, 1 mM EDTA) and shake overnight at room temperature.
8. Spin the tube for few seconds in a microcentrifuge and transfer the liquid to a new tube. Avoid transferring gel pieces to the new tube. Add 1 mL of ice-cold 100% ethanol, shake the tube for a few seconds, and put into –70°C for at least 30 min.
9. Spin the sample in a microcentrifuge for 30 min and remove the supernatant. Wash once with 1 mL of ice-cold 80% ethanol.
10. Dry the sample under vacuum. Dissolve the large pellet in 70 µL of

bidistilled water and centrifuge again for 30 min (*see* Note 10). Transfer the liquid to another tube with a glass capilliary to ensure that no remaining gel pieces are transferred.

11. Dry the sample in a SpeedVac concentrator for 1.5 h. Dissolve the remainder in a small volume of formamide buffer (usually 5–10 µL, depending on the amount of radioactivity).
12. Analyze the DNA by gel electrophoresis as described for free DNA in Section 3.2., steps 7–10. Load the following samples on the gel: 3' endlabeled DNA—the complex, the free DNA (either recovered from the gel and/or prepared separately as in Section 3.2.), and the G-specific Maxam-Gilbert reaction as length standard. For 5' endlabeled DNA use the corresponding samples as for the 3' endlabeled DNA.

4. Notes

1. The iron (II)- and the H_2O_2-solutions should be freshly made before use.
2. The solutions of ascorbate (20 m*M*), EDTA (4 m*M*), H_2O_2 (as a 30% stock solution), and the stop-mix are stable for months stored in a dark bottle at 4°C.
3. The solutions should be protected from light and kept at 4°C. Dilute buffers with bidistilled water.
4. It is strongly recommended to check the quality of the labeled DNA before use on a sequencing gel. Nicks in the double strand, which could be derived from DNase activities during the preparation procedure, will appear as additional bands in the sequencing gel. This admixture of bands would spoil the whole footprint, even when present in only small amounts. Furthermore, it is recommended not to store the pure, labeled DNA longer than 2 wk, because the radiation of the label also creates nicks in the DNA.
5. Longer heating or boiling creates additional cuts in the DNA.
6. The buffer conditions can be varied, for example, pH, but the ionic strength should not be too high (max 50 m*M* NaCl) in order to obtain sharp bands during the following electrophoresis. Many protein–DNA complexes are very stable at low ionic strength (e.g., complexes between RNA polymerase and promoters *[14]*), therefore, in most cases the stability of the pH in the following electrophoresis limits the ionic strength.
7. The purpose of the dialysis is twofold: Removal of glycerol interferes with the cutting reaction and removal of salt, and lowers the quality of the electrophoresis pattern. As a rough approximation, one can remove up to 80–90% of the glycerol and salt present in the sample within 1 h of drop dialysis.

8. The gel concentration has to be adjusted according to the molecular weight of the protein–DNA complex. Here we describe the conditions established for the study of the *E. coli* RNA polymerase (mol wt 490,000) and a DNA fragment of 130 bp carrying a promoter *(14)*.
9. The gel is ready when the dye marker is about 3–5 cm above the end of the gel.
10. Often there is a very big pellet visible, which is caused by some remaining gel pieces.

Further Reading

Tullius, Th. D. (1989) Physical studies of protein-DNA complexes by footprinting. *Ann. Rev. Biophys. Biophys. Chem.* **18,** 213–237.

Tullius, Th. D. (1989) Structural studies of DNA through cleavage by the hydroxyl radical. *Nucleic Acids and Molecular Biology* **3,** 1–12.

References

1. Tullius, Th. D., Dombroski, B. A., Churchill, M. E. A., and Kam, L. (1987) Hydroxyl radical footprinting: a high resolution method for mapping protein-DNA contacts. *Meth. Enzymol.* **155,** 537–558.
2. Shafer, G. E., Price, M. A., and Tullius, Th. D. (1989) Use of the hydroxyl radical and gel electrophoresis to study DNA structure. *Electrophoresis* **10,** 397–404.
3. Schickor, P., Metzger, W., Werel, W., Lederer, H., and Heumann, H. (1990) Topography of intermediates in transcription initiation of *E. coli*. *EMBO J.* **9,** 2215–2220.
4. Metzger, W., Schickor, P., and Heumann, H. (1989) A cinematographic view of *Escherichia coli* RNA polymerase translocation. *EMBO J.* **8,** 2745–2754.
5. Schickor, P. (1992) Ph.D. Thesis, University of Munich, Germany.
6. Hayes, J. J. H., Tullius, Th. D., and Wolffe, A. P. (1990) The structure of DNA in a nucleosome. *Proc. Natl. Acad. Sci. USA* **87,** 7405–7409.
7. Tullius, Th. D. and Dombroski, B. A. (1986) Hydroxyl radical "footprinting": High-resolution information about DNA-protein contacts and application to lambda repressor and cro protein. *Proc. Natl. Acad. Sci. USA* **83,** 5469–5473.
8. Cons, B. M. G. and Fox, K. R. (1989) High resolution hydroxyl radical footprinting of the binding of mithramycin and related antibiotics to DNA. *Nucleic Acids Res.* **17,** 5447–5459.
9. Burkhoff, A. M. and Tullius, Th. D. (1988) Structural details of an adenine tract that does not cause DNA to bend. *Nature* **331,** 455–457.
10. Tullius, Th. D. and Dombroski, B. A. (1985) Iron (II) EDTA used to measure the helical twist along any DNA molecule. *Science* **230,** 679–681.
11. Wang, X. and Padgett, R. A. (1989) Hydroxyl radical "footprinting" of RNA: Application to pre-mRNA splicing complexes. *Proc. Natl. Acad. Sci. USA* **86,** 7795–7799.

12. Celander, D. W. and Cech, Th. R. (1991) Visualizing the higher order folding of a catalytic RNA molecule. *Science* **251,** 401–407.
13. Maniatis, T., Fritsch, E. F., and Sambrock, J. (1982) in *Molecular Cloning. A Laboratory Manual,* Cold Spring Harbor Laboratory, Cold Spring Harbor, NY.
14. Heumann, H., Metzger, W., and Niehörster, M. (1986) Visualization of intermediary transcription states in the complex between *Escherichia coli* DNA-dependent RNA polymerase and a promoter-carrying DNA fragment using the gel retardation method. *Eur. J. Biochem.* **158,** 575–579.

CHAPTER 4

Hydroxyl Radical Interference

Peter Schickor and Hermann Heumann

1. Introduction

Interference studies are just the inverse approach of "footprinting" experiments. In both types of experiments the effect of a chemical modification of a single base on the binding of a sequence-specific protein is determined, but the experiments differ in the sequence of the binding event and the modification procedure. The "interference" approach is characterized first by chemical modification of the DNA and then by subsequent protein binding. These studies provide information about the change of the binding strength following single base modification. This change can be positive or negative. A quantification of the effect is possible by the gel shift assay *(1)*.

Here we describe the use of hydroxyl radicals as the modifying reagent. This probe has a number of advantages compared to the most commonly used probes, such as dimethylsulfate (DMS) or ethylnitrosourea. Hydroxyl radicals cleave the backbone of DNA with almost no sequence dependence, whereas most other reagents react in a highly sequence dependent manner with the bases of the DNA. Furthermore, hydroxyl radicals modify the DNA by elimination of a nucleoside (*see* Chapter 3). This allows the study of two kinds of effects:

1. The effect on protein binding caused by missing contacts. Other reagents prevent binding by introducing bulky groups into a base of the DNA. Whether this reflects the importance of a base for protein–DNA interaction or whether the bulky group is just a steric hindrance is difficult to determine.

2. The effect on the DNA structure because of the deletion. The missing nucleoside is a center of enhanced flexibility in the DNA. Therefore, the effect of DNA flexibility on the protein–DNA interaction can be studied. The modification of the DNA by hydroxyl radicals is a very fast and highly reproducible experiment in contrast to other methods. The reagents needed are easily available.

1.1. Generation and Action of Hydroxyl Radicals

Hydroxyl radicals introduce randomly distributed base deletions in the DNA. The generation and action of hydroxyl radicals is described in detail in Chapter 3.

1.2. Principle of the Procedure

A DNA fragment labeled either at the 3' or the 5' end is subjected to hydroxyl radical treatment. The concentration of the hydroxyl radicals is adjusted so that the number of deletions is less than one per DNA, that is, approx 10% of the DNA fragments are affected. This population of randomly modified DNA molecules is incubated with the protein under study (*see* Fig. 1). Those DNA molecules that are still able to bind the protein can be separated from those DNA molecules that are no longer able to bind the protein by nondenaturing gel electrophoresis. The bands containing free DNA and the complexed DNA (Fig. 2) are eluted and applied on a sequencing gel in order to determine the position of the modification (Fig. 3). The relative effect on the binding strength caused by a single base deletion can be quantified by determining the intensity change of the different bands by scanning.

1.3. Interpretation of the Interference Pattern

The interpretation of an electrophoresis pattern obtained by hydroxyl radical interference studies (Fig. 3) is not as straightforward as hydroxyl radical footprinting studies. The reason is that a single base deletion generated by hydroxyl radicals can cause two effects on protein binding:

1. Lack of specific contacts between the protein and the DNA. This leads to a decrease of the binding strength.
2. Enhanced flexibility of the DNA at the position of the deletion. This can lead to an increase *or* a decrease of the binding strength.

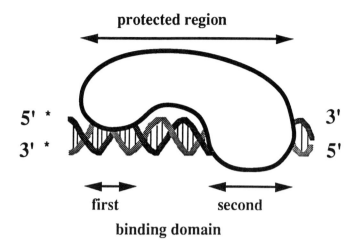

Fig. 1. A schematic representation of a protein–DNA complex with two distinct interaction sites. At the first binding site the protein interacts with only one side of the DNA. At the second binding site the protein wraps fully around the DNA. (*) Indicates the position of the radioactive label.

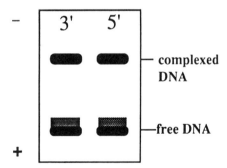

Fig. 2. Nondenaturing gel electrophoresis of the complex with hydroxyl radical pretreated DNA as target for the protein binding. The two lines show the same complex labeled at the 3' and the 5' ends, as indicated in Fig. 1. Note: The single base deletion leads to an enhanced flexibility of the DNA indicated by the "smear" of the band representing the free DNA.

A quantitative interpretation of the interference pattern is not always possible, since the intensity of the bands of the interference pattern reflects the sum of the different effects contributing to the binding strength. Additional information concerning the protection of the DNA

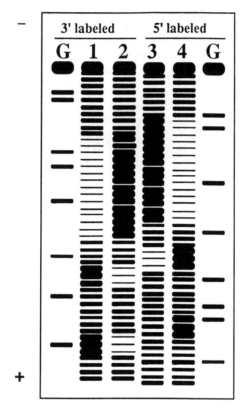

Fig. 3. A putative interference pattern of the protein DNA complex (Fig. 1) obtained after separation of "free" DNA and "complexed" DNA by a nondenaturing gel electrophoresis (Fig. 2). Lanes 1 and 4 show the pattern of the "free" DNA labeled at the 3' and the 5' ends, respectively. Lanes 2 and 3 show the pattern of the "complexed" DNA. Lanes G contain the length standards obtained by a G-specific Maxam-Gilbert sequencing reaction.

by the protein (e.g., by using hydroxyl radical footprinting) is necessary in order to differentiate between the two effects. The examples in the following paragraph might be used as a guideline for interpretation.

1.4. Examples for the Application of Hydroxyl Radicals as Interference Probes

1.4.1. Transcription Factors

The interference of binding of some eubacterial and eukaryotic transcription factors was investigated using hydroxyl radical pretreated DNA binding sequences. In all cases the single base deletion led to a decrease of the binding strength indicating a loss of contacts. Examples are: progesterone *(2)*, λ-repressor, cro-protein *(3)*, NF-κB *(4)*, and GCN4 transcription factor *(5)*.

1.4.2. RNA Polymerase-Promoter Interaction

A strong eubacterial promotor was subjected to hydroxyl radical treatment for investigating the influence of single base deletions on binding of *E. coli* RNA polymerase *(6)*. This study revealed three patterns of interaction that could be attributed to different sites of the promoter:

1. Direct base contact with the template strand in the "-35 region:" This was concluded from the reduced affinity of the polymerase to a promoter having a "gap" in this region together with the result from hydroxyl radical footprinting studies *(7)* that revealed close contact between the bases of this sequence and the protein. This is an example in which a base deletion leads to missing contacts between protein and DNA.
2. A DNA structure dependent interaction in the "-10 region:" This was inferred from the increased binding affinity of the polymerase to a promoter having deletions in this region, and from supplementary footprinting studies that revealed that this region is in close contact with the protein *(7)*. This is an example where a deletion causes a structural change of the DNA because of an enhancement of DNA flexibility favorable for protein binding.
3. An interaction that is based on a defined spatial relationship between the two domains, namely the "-35 region" and the "-10 region:" This conclusion was drawn from the finding that:
   a. A "gap" in the promoter region between the two domains reduces the binding affinity of RNA polymerase.
   b. These bases have no contact with the protein (as shown by hydroxyl radical footprinting studies *[7]*).
   c. The effect is observed in both strands. This is an example in which a deletion leads to a loss of the defined spatial relationship between two functionally important sites because of an enhancement of the DNA flexibility.

2. Materials

2.1. Chemicals Used for the Cutting Reaction

Prepare the following solutions separately:

1. 20 mM Sodium ascorbate.
2. 0.6% Hydrogen peroxide.

3. Iron (II)-EDTA-mix: Mix equal volumes of 2 mM ammonium iron (II) sulfate hexahydrate $(NH_4)_2Fe(SO_4)_2 \cdot 6H_2O$ and 4 mM EDTA (see Note 1).
4. Stop-mix: 0.1M Thiourea, 20 mM EDTA, pH 8.0 (see Note 2).

2.2. The Sequencing Gel

1. Urea (ultra pure).
2. 10X TBE: 1M Tris-HCl, pH 8.6, 0.84M boric acid, 10 mM EDTA.
3. Acrylamide solution: 40% acrylamide, 0.66% bis-acrylamide.
4. 10% Ammonium persulfate (prepare immediately before use).
5. TEMED.
6. Sequencing gel (8%): 21 g urea, 5 mL of 10X TBE, and 10 mL of the acrylamide solution are mixed under mild heating. When the urea is dissolved the solution is made up to 50 mL with bidistilled H_2O. The solution is filtered (filter pore size: 0.2 µm) and degassed for 5 min. 500 µL of 10% ammonium persulfate and 30 µL of TEMED are added immediately before pouring the solution between the glass plates.
7. Loading buffer for the sequencing gel (stock solution): 100 mL formamide (deionized), 30 mg xylenccyanol FF, 30 mg bromophenol blue, 750 mg EDTA.
8. Electrophoresis buffer: 1X TBE.

2.3. The Nondenaturing Gel for the Band-Shift Assay

1. 1M Tris-HCl, pH 7.9.
2. Acrylamide solution: 30% acrylamide, 0.8% bis-acrylamide.
3. 10% Ammonium persulfate (prepare immediately before use).
4. 5% TEMED (aqueous solution).
5. Nondenaturing gel: 240 µL of 1M Tris-HCl, pH 7.9, 2.75 mL of the acrylamide solution, and 25.7 mL of bidistilled water are mixed and degassed for 5 min. Three hundred microliters of 10% ammonium persulfate and 70 mL of 5% TEMED are added before pouring the solution between the glass plates.
6. Loading buffer for the nondenaturing gel (stock solution): 40% sucrose, 0.1% bromophenol blue (see Note 3).
7. Electrophoresis buffer: 8 mM Tris-HCl, pH 7.9.

2.4. Other Items

1. Sequencing gel apparatus.
2. Apparatus for nondenaturing gel electrophoresis.
3. Filters for drop dialysis, VS, 0.025 µm (Millipore, Bedford, MA).

4. Peristaltic pump.
5. SpeedVac concentrator.
6. 1X TE: 10 mM Tris-HCl, pH 7.9, 1 mM EDTA.

3. Methods

3.1. Establishing the Conditions for Obtaining Specific Protein–DNA Complexes

The method of establishing the conditions for complex formation using the band shift assay is described in Chapter 21. The enzyme to DNA ratio should be adjusted so that the ratio of complexed to free DNA is about 1:1 (*see* Note 4).

3.2. Establishing the Conditions for Cutting the DNA by Hydroxyl Radicals

1. Endlabel the DNA fragment either at the 5'-positions, for example, by T4 polynucleotide kinase with γ-[^{32}P]ATP or at the 3'-positions using the proper α-[^{32}P]dNTP and Klenow fragment under standard conditions *(8)*. Remove one label by restriction of the DNA fragment with the appropriate endonuclease. Purify your uniquely endlabeled DNA fragment by a nondenaturing gel electrophoresis (*see* Section 2.3.).
2. Dissolve both DNA preparations in 15 µL of 1X TE buffer, respectively. Add to both samples 3.5 µL each of the previously prepared solutions of ascorbate, hydrogen peroxide, and the iron (II)-EDTA mix (*see* Section 2.1.) by putting single drops on the inner wall of the tube and by rapidly mixing the three drops before combining them with the sample using a micropipet.
3. Incubate for 3–4 min at room temperature.
4. Add 7 µL stop-mix (*see* Section 2.1.).
5. Add sodium acetate to a final concentration of 0.3M, followed by 2.5 vol of ice-cold 100% ethanol, to precipitate the DNA. Keep the solution at –70°C for 30 min.
6. Spin down the solution in a microcentrifuge for 30 min. Wash the pellet with ice-cold 80% ethanol, dry the pellet under vacuum, and dissolve the pellet in 20–50 µL of 1X TE buffer.

3.3. Interference Studies on Protein–DNA Complexes

1. Prepare two 15 µL samples of the complexes, one with the DNA previously cut and labeled at the 3' and the 5' end, as described in Section 3.2. Use the conditions established for optimal complex formation from

Section 3.1. The total amount of radioactivity in one assay should be around 60,000–80,000 cpm.
2. Pour 30–40 mL of the dialysis buffer containing 8 mM Tris-HCl, pH 7.9, into a Petri dish (*see* Note 6). Place a Millipore filter (*see* Section 2.4.) on the surface of the buffer, shiny side (hydrophobic side) up. Put the samples containing the complexes onto the filter for 1 h in order to remove salt (*see* Note 7). Remove the samples from the filter by a micropipet and transfer them to a fresh 1.5-mL Eppendorf tube.
3. Separate the complex and the free DNA on a nondenaturing acrylamide gel electrophoresis. The method for separation of complexes from free DNA is described in Chapter 3 (Hydroxyl radical footprinting, Section 3.3., steps 4 and 5).
4. The whole procedure for the recovery of DNA from the gel is described in Chapter 13 (Section 3.3., steps 6–11).
5. Adjust the amount of radioactivity and volume in each sample to about 5000–6000 cpm in 4–5 µL. Heat the samples for not longer than 2 min at 90°C and put them on ice (*see* Note 8).
6. Analyze the DNA by gel electrophoresis. Apply the samples onto a 6–10% sequencing gel (for the analysis of fragments in the range of 50–150 bases a gel consisting of 8% acrylamide is adequate, as described in Section 2.2.). Load the following samples on the gel: 3' endlabeled DNA—The G-specific Maxam-Gilbert reaction as length standard, the free DNA recovered from the gel, and the complex. For 5' endlabeled DNA use the corresponding samples as for the 3' endlabeled DNA.
7. Run the gel at 50 W at a temperature of 60°C for 1.75 h.
8. Expose the gel after electrophoresis to an X-ray film using an intensifying screen at –70°C overnight.

4. Notes

1. The iron (II)- and the H_2O_2-solutions should be freshly made before use.
2. The solutions of ascorbate (20 mM), EDTA (4 mM), H_2O_2 (as a 30% stock solution), and the stop-mix are stable for months in a dark bottle at 4°C.
3. The solutions should be protected from light and kept at 4°C. Dilute buffers with bidistilled water.
4. It is advisable to keep the enzyme-to-DNA ratio <1 in order to assure proper selection conditions.
5. It is strongly recommended to check the quality of the labeled DNA before use on a sequencing gel. Nicks in the double strand, which could be derived from DNase activities during the preparation procedure, will appear as additional bands in the sequencing gel. This admixture of

bands would spoil the whole footprint, even when present in only small amounts. Furthermore it is recommended not to store the freshly labeled DNA longer than 2 wk because of the danger of radiation damage to the DNA.
6. The buffer conditions can be varied, for example, pH, but the ionic strength should not be too high (max 50 mM NaCl) in order to obtain sharp bands during the following electrophoresis. Many protein–DNA complexes are very stable at low ionic strength (e.g., complexes between RNA polymerase and promoters *[6]*). Therefore, in most cases the stability of the pH in the following electrophoresis limits the ionic strength.
7. The purpose of the dialysis is the removal of salt, whose presence would lower the quality of the electrophoresis pattern. As a rough approximation one can remove up to 80–90% of the salt present in the sample within 1 h of drop dialysis.
8. Longer heating or boiling creates additional cuts in the DNA.

References

1. Heumann, H., Metzger, W., and Niehörster, M. (1986) Visualization of intermediary transcription states in the complex between *Escherichia coli* DNA-dependent RNA polymerase and a promoter-carrying DNA fragment using the gel retardation method. *Eur. J. Biochem.* **158**, 575–579.
2. Chalepakis, G. and Beato, M. (1989) *Hydroxyl* radical interference: a new method for the study of protein-DNA interaction. *Nucleic Acids Res.* **17**, 1783.
3. Hayes, J. J. and Tullius, Th. D. (1989) The missing nucleoside experiment: a new technique to study recognition of DNA by protein. *Biochemistry* **28**, 9521–9527.
4. Schreck, R., Zorbas, H., Winnacker, E. L., and Baeuerle, P. A. (1990) The nf-kappa-b transcription factor induces DNA bending which is modulated by its 65-kd subunit. *Nucleic Acids Res.* **18**, 6497–6502.
5. Gartenberg, M. R., Ampe, C., Steitz, T. A., and Crothers, D. M. (1990) Molecular characterization of the GCN4–DNA complex. *Proc. Natl. Acad. Sci. USA* **87**, 6034–6038.
6. Werel, W., Schickor, P., and Heumann, H. (1991) Flexibility of the DNA enhances promoter affinity of *Escherichia coli* RNA polymerase. *EMBO J.* **10**, 2589–2594.
7. Schickor, P., Metzger, W., Werel, W., Lederer, H., and Heumann, H. (1990) Topography of intermediates in transcription initiation of *E. coli. EMBO J.* **9**, 2215–2220.
8. Maniatis, T., Fritsch, E. F., and Sambrock, J. (1982) *Molecular Cloning. A Laboratory Manual,* Cold Spring Harbor Laboratory, Cold Spring Harbor, NY.

CHAPTER 5

1,10-Phenanthroline-Copper Ion Nuclease Footprinting of DNA–Protein Complexes *in Situ* Following Mobility-Shift Electrophoresis Assays

Athanasios G. Papavassiliou

1. Introduction

The existence of cell-type specific promoter and enhancer elements has been known for several years. However, the mechanisms responsible for the remarkable specificity of such elements, in comparison to the ubiquitously active promoters and enhancers of "house-keeping" genes and DNA tumor viruses, have remained elusive until recently. Although transfection and mutagenesis experiments have taught us a great deal about the structure of cell type-specific *cis*-acting elements, the breakthrough in understanding the molecular basis for the differential activity of these elements has come from the analysis of their recognition by specific DNA-binding proteins.

Several techniques have been developed for the detection of cell type-specific DNA-binding activities and the identification of sequence-specific contacts or "footprints" of a protein on DNA. Many of these involve forming DNA–protein complexes (by incubating, under certain conditions, an asymmetrically labeled double-stranded DNA fragment containing the region of interest with crude or partly purified protein extracts), exposing them to enzymatic or to chemical reagents that can cleave or modify the DNA, and determining which bases are protected from attack when the protein is bound. The most widely used reagents are deoxyribonuclease I (DNase I, an endonuclease) and dimethyl sulfate (DMS). Footprinting experiments with DNase I are performed using parallel reactions with free DNA and with DNA–protein complexes, and the nuclease is allowed to digest DNA only to a limited extent *(1)*. The DNA is then purified and denatured, and the single-stranded end-labeled DNA fragments are resolved on a sequencing gel and autoradiographed. Comparing the digestion pattern of the DNA–protein complexes with that of free DNA shows a band-free region (footprint) where the bound protein has prevented access of the enzyme to DNA (*see* Chapter 1). In a similar analysis, the DNA is allowed to react mildly with DMS, which methylates primarily deoxyguanosine residues and renders their phosphodiester linkages labile under conditions of Maxam-Gilbert chemistry (*see* Chapter 6). The binding of a protein to a specific DNA region will result in a protection of the corresponding bases from chemical modification *(2)*.

The suitability of the above assays in determining the binding sequences of proteins on DNA is hindered by several disadvantages. First, the clarity of the footprint is highly dependent on the extent of occupancy of the binding site(s) (i.e., a "clear" footprint is observed only if all DNA molecules are involved in complexes). Unfortunately, this is not always easy to achieve, especially when the concentration and/or purity of the specific binding protein(s) is not satisfactory. Second, DNA–protein complexes formed in crude extracts may often be heterogeneous in terms of both binding specificity and kinetic stability. Therefore, direct footprinting in solution will not correspond to a single species, but instead reflect an "integral" of the multiple equilibria operating over the entire region of interest (i.e., the protection pattern will actually represent a composite of more than one complex, with complexes having a very low dissociation rate dominating the footprint). Finally, two different proteins that recognize the same sequence within the probe are most likely to yield indistinguishable footprints. These drawbacks may be overcome by coupling treatment with a footprinting reagent in solution with the mobility-shift electrophoresis (or gel retardation) assay *(3–5)*. In this approach, the protein and DNA molecules are incubated together, and the equilibrated reaction mixture is exposed to DNase I or DMS as before. DNA–protein complexes are subsequently isolated from the free probe by electrophoresis in a nondenaturing polyacrylamide gel. Although the negatively charged free DNA migrates rapidly toward the anode, once it is bound by a specific protein its mobility decreases *(3,4)*. Following the separation of the free and bound DNA species, the corresponding bands are cut out of the gel, and the DNA eluted and analyzed on a sequencing gel. The region(s) of protection evident in the DNA derived from the complexed fraction indicates the binding site *(5)*. Apparently, since the complexes are separated from contaminating unbound DNA fragments, the footprinting pattern will be free of background cutting, and thus considerably sharpened. Similar considerations apply when more than one complex can be formed on the fragment. As long as the DNA-binding proteins differ in their molecular masses and charges, they will cause altered electrophoretic mobilities of the corresponding complexes and, hence, different migration in the retardation gel. These complexes can be isolated and run in individual lanes on the sequencing gel. Thus, the exposure of the binding reaction to footprinting reagents, in combination with the fractionation offered by mobility-shift gels, permits identification of the regions of DNA bound by protein in different complexes, even if a low percentage of the initial DNA molecules has been complexed.

Although by employing the mobility-shift electrophoresis assay one can substantially increase the sensitivity of DNase I or DMS footprinting experiments in solution, several additional problems have still to be faced:

1. DNase I is a relatively bulky molecule (mol wt 30,400) that cannot cleave the DNA in the immediate vicinity of a bound protein because of steric hindrance. As a result, the region(s) protected from cutting extends beyond the actual protein-binding site.
2. The nonrandom nature of DNA cleavage by DNase I makes it impossible to assess the involvement in protein binding of nucleotides that lie in an area of the fragment resistant to the endonucleolytic activity of this enzyme (e.g., tracts of A and T residues, or TpA [as opposed to ApT] dinucleotide islands scattered within or adjacent to the binding site), so that binding sites or part of binding sites may not be detected.
3. The primary site of reaction of DMS with B-DNA is the N-7 atoms of guanine bases that are located in the major groove. Thus, those guanines in close proximity with the protein will be protected from methylation. However, if a protein primarily makes contact with a DNA sequence in the minor groove, or if there are no guanine residues in a major groove-binding site, DMS will not reveal these interactions.
4. In many instances, particularly when a complex has a relatively high "off" rate, the bound protein can dissociate from the protected DNA

fragment and reassociate to other DNA fragments that have already been nicked by DNase I, or modified by DMS. In this case, the DNA–cleavage pattern derived from the complexed fraction will closely resemble that of the uncomplexed DNA, rendering it difficult to observe a footprint. The limitations imposed by the size and the sequence, or base specificity of the aforementioned footprinting reagents, as well as the problem of protein exchange from the binding site(s) during treatment, are circumvented by merging the advantages inherent in the mobility-shift electrophoresis assay, with the *subsequent* exposure of the gel (hence of the resolved complexes while embedded in the polyacrylamide matrix) to a chemical DNA–scission reagent, namely 1,10-phenanthroline-copper ion (OP-Cu) *(6)*.

OP-Cu is an efficient chemical nuclease that cleaves the phosphodiester backbone of nucleic acids at physiological pHs and temperatures by oxidation of the deoxyribose (DNA) or ribose (RNA) moiety *(7)*. The kinetic scheme of the reaction is summarized in Fig. 1. The first step is the formation of the 1,10-phenanthroline-cupric ion coordination complex, under conditions that favor the 2:1 stoichiometry ($[OP]_2Cu^{2+}$). The DNA–scission process is initiated by adding a reducing agent, usually 3-mercaptopropionic acid, to the aerobic reaction mixture containing the target DNA. Under these conditions, the 2:1 cupric complex is reduced to the 2:1 cuprous complex ($[OP]_2Cu^+$) that is in turn oxidized by molecular oxygen to generate hydrogen peroxide. Hydrogen peroxide is an essential coreactant for the chemical nuclease activity and can be generated as indicated above or added exogenously *(8)*. The tetrahedral cuprous complex, present at the steady-state concentration defined by the experimental conditions (note the feedback mechanism in Fig. 1), then binds reversibly to the minor groove of DNA to form a central intermediate through which the reaction is funneled *(9)*. The DNA-bound cuprous complex is subjected *in situ* to one-electron oxidation by hydrogen peroxide to form a short-lived, highly reactive DNA-bound copper-oxo species that can be written either as a hydroxyl radical coordinated to the cupric ion or as a copper-oxene structure (Fig. 1). This species then reacts with H1'-deoxyribose protons of nucleotides that extend into the minor groove; this reaction initiates a series of reactions culminating in cleavage of the phosphodiester backbone. The products of the strand-scission event include the free base, DNA fragments bear-

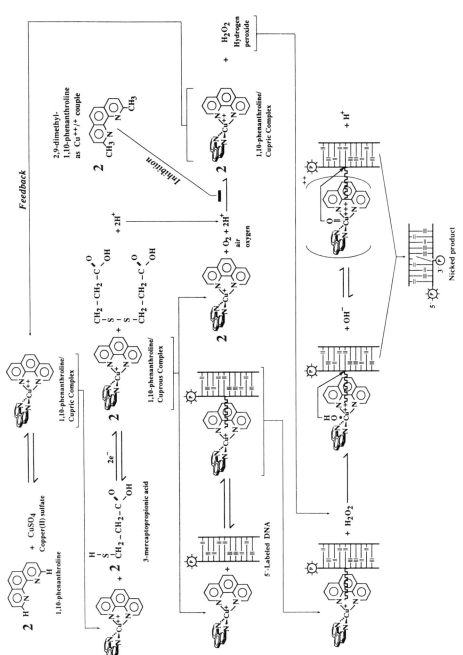

Fig. 1. Schematic representation of the kinetic mechanism for the nuclease activity of 1,10-phenanthroline-copper ion.

ing 5' and 3' phosphorylated termini, and the deoxyribose oxidation product 5-methylene-2-furanone *(10)*. The DNA–chain cleavage reaction can be efficiently quenched by adding to the mixture 2,9-dimethyl-1,10-phenanthroline. This phenanthroline derivative can also chelate copper ions to form a minor groove-associated cuprous complex (thus competing with [OP]$_2$Cu$^+$), but the reduction potential of the Cu^{2+}/Cu$^+$ couple is too positive to allow significant nuclease activity under normal assay conditions *(11)*.

In as much as the structural and functional properties of DNA are not altered by entrapment in a polyacrylamide gel matrix *(6)*, the small size and the ready diffusibility of all reaction components in solid supports permit the coupling of OP-Cu footprinting with the mobility-shift electrophoresis assay to study DNA–protein interactions *(12,13)*. In this method, the DNA-binding reaction is performed as usual, electrophoresed under established, nondenaturing conditions, and the entire retardation gel is immersed in a footprinting reaction mixture containing 1,10-phenanthroline, cupric ion, and 3-mercaptopropionic acid. Following the reaction quench with 2,9-dimethyl-1,10-phenanthroline, footprints are obtained after elution of the radioactive free and bound products from the mobility-shift gel and analysis on a sequencing gel (Fig. 2). Because the nuclease activity of (OP)$_2$Cu$^+$ produces 3'-phosphorylated and 5'-phosphorylated ends as cleavage products, sequencing gels can be accurately calibrated with the Maxam-Gilbert sequencing reactions.

The nuclease activity of (OP)$_2$Cu$^+$ bears several advantages as a footprinting reagent relative to protection analyses using DNase I or DMS as a probe. First, the (OP)$_2$Cu$^+$ chelate is a small molecule (compared to DNase I) that permits cleavage closer to the edge of the DNA sequence protected by protein binding, and therefore, a more precise definition of it. Second, since the scission chemistry involves attack on the deoxyribose moiety, (OP)$_2$Cu$^+$ is able to cut at all sequence positions regardless of base. However, the intensity of cutting (rate of cleavage) does depend on local sequence, with attack at adenines of TAT triplets being most preferred *(14; see also* Fig. 2). Interestingly, a preference for C-3',5'-G steps, rather than T-3',5'-A steps, is observed at a phenanthroline to copper ratio of 1:1, which strongly favors formation of the OPCu$^+$ complex *(15)*. Nevertheless, the cutting patterns obtained with $(OP)_2Cu^+$ are usually sufficiently well defined to identify protected regions, even though this endonucleolytic agent exhibits some degree of sequence specificity in its rate of cleavage of naked DNA. Third, because $(OP)_2Cu^+$ binds to the minor groove of DNA it will reveal minor groove interactions. Since the binding of the coordination complex should be restricted to three base pairs, the complex is more sensitive to local, protein-induced conformational changes than DNase I, which by possessing an extended minor groove-binding site may be unable to sense. In this context, the complex will also detect binding in the major groove when its approach to its minor groove-binding site is sterically blocked, or if the interaction of the protein in the major groove substantially alters the minor groove structure (both being frequent features of DNA–protein interactions). Furthermore, because of the difference in their respective mechanism of cleavage, DNase I and OP-Cu probe different aspects of the structure of a DNA–protein complex. DNase I cleavage relies on the accessibility of a particular phosphodiester bond, and thus protection is indicative of an interaction *on the outer face of the DNA helix*. In contrast, protection from $(OP)_2Cu^+$-mediated cleavage is most likely caused by the inhibition of its binding to the minor groove, and implies that *a portion of the protein occupies at least the minor groove*.

The major advantage, however, of the coupled OP-Cu footprinting procedure arises from the topography of treatment: Preformed DNA–protein complexes are exposed to the chemical nuclease *within the gel* (i.e., not prior but subsequent to a mobility-shift experiment). This characteristic of the technique makes it ideal for protection analysis of kinetically labile complexes *(16)*. At least three factors account for the latter. The first is that the background cleavage is greatly reduced by the separation of unbound DNA from the DNA–protein complex(es) pool. The second factor is the so-called "caging effect" *(3,4)*. The gel matrix forms "cage-like" compartments that prevent a dissociated protein from diffusing away from the DNA, so that by enhancing reassociation the apparent affinity constant will be higher than the true value. The protein could also interact with the gel matrix, thereby orienting its diffusion toward reassociation. Whatever the mechanism(s), the increase in stability of the complex contributed by

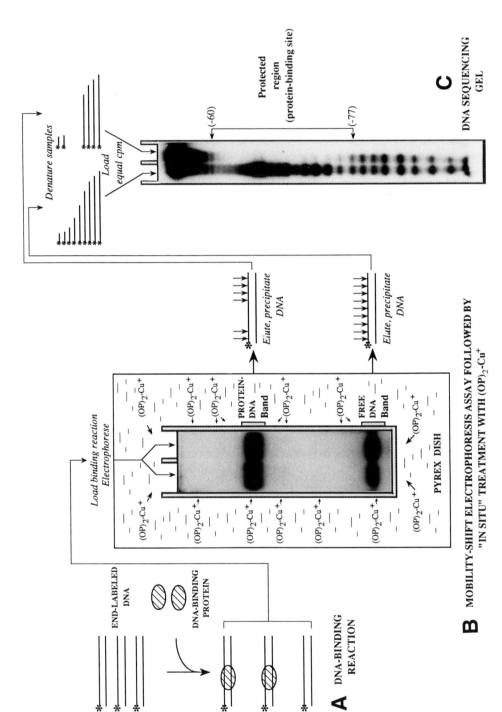

the gel leads to a more efficient blockage of the access of the $(OP)_2Cu^+$ chelate to the protein-binding DNA domain. The third factor comes from the nature and site of action of the cupryl intermediate through which the reaction is funneled, and acts synergistically with the previous one. Since this highly reactive oxidative species is generated near the surface of the DNA (*in situ*), diffusible radicals, if formed at all, will have a short or restricted diffusive path and, therefore, will be unable to achieve a fast equilibrium distribution along the DNA polymer. Consequently, protein-binding sites exposed during multiple dissociation events will most of the time escape the nucleolytic attack and hence remain intact.

In addition to the fact that discrete complexes with defined stoichiometries and a wide range of kinetic stabilities can be mapped simultaneously, the *in situ* OP-Cu footprinting procedure is superior to oligonucleotide binding competition assays in the analysis of multiple complexes frequently obtained in mobility-shift experiments employing unfractionated extract preparations. For example, numerous retarded bands can arise from protein–protein interactions between

---

Fig. 2. *(opposite page)* Outline of the combined mobility-shift/in gel OP-Cu footprinting assay. **A.** DNA restriction fragments containing a protein-binding site(s) are labeled with ^{32}P at a unique end and incubated with an extract containing the specific DNA-binding protein, under optimized binding conditions. **B.** After equilibration of the DNA-binding reaction, the free and bound DNA fragment populations are separated by electrophoresis through a nondenaturing polyacrylamide gel; the gel is then transferred into a buffer-containing Pyrex dish, and the retarded and unretarded DNA species are exposed *in situ* to the nuclease activity of $(OP)_2Cu^+$. The two DNA fractions are subsequently located by autoradiography of the wet gel, excised and eluted from the gel matrix, precipitated, and recovered in formimide buffer. **C.** Samples are heat-denatured, and equal amounts of radioactivity from the two fractions are electrophoresed on a denaturing polyacrylamide gel (DNA-sequencing gel) and autoradiographed. In the sample prepared from the free DNA band, bands will appear in the gel corresponding to positions of protein binding. For the sample(s) prepared from the protein–DNA band(s), bands will appear at all positions *except* those bound by the protein(s) (protected region). The particular example depicts the OP-Cu mapping of a DNA-protein complex formed between bacterially expressed LFB1 (a liver-specific transcription factor) and an oligonucleotide containing its binding site within the –95 to –54 region of the α1-antitrypsin promoter. Arrowheads connected by line demarcate the footprinted site. The enhanced cleavage observed within the protein-binding site in the free DNA sample is caused by the presence of repeated TA elements in this sequence (*see* Section 1.).

a transcription factor(s) and a specific DNA-binding protein, or from two proteins binding to distinct DNA sequences in a cooperative manner *(12)*. Although both complexes would be abolished by competition with oligonucleotides, these possibilities can be readily distinguished by direct footprinting within the gel. Finally, it is worth mentioning that OP-Cu footprinting *in situ* following gel retardation assays can also be applied to the mapping of RNA-protein complexes *(17)*.

2. Materials
2.1. Analytical Mobility-Shift Electrophoresis Assay
2.1.1. Solutions

1. A variety of concentrated binding and gel/running buffers commonly employed in mobility-shift electrophoresis assays (*see* for instance Chapter 21 in this volume).
2. 10% (w/v) ammonium persulfate: Weigh out 1 g of ammonium persulfate, and put it in a sterile plastic tube containing 10 mL of bidistilled sterile water. Vortex vigorously until the salt is dissolved. Store at 4°C after filtering; make fresh weekly.

2.1.2. Reagents / Special Equipment

1. Highly purified DNA probe labeled *exclusively* at one of its four ends (5' or 3'); use standard procedures for unique labeling (*see* Chapter 12 in vol. 4 of this series). All necessary precautions should be observed to minimize exposure to ionizing radiation during labeling and isolation of the probe.
2. All reagents and equipment for a typical mobility-shift electrophoresis assay (e.g., bulk carrier DNA, buffered extract containing DNA-binding protein), nondenaturing polyacrylamide gel electrophoresis, and autoradiography (*see below*, and Chapter 21 in this volume).

2.2. Preparative Mobility-Shift Electrophoresis Assay
2.2.1. Solutions

1. Binding and gel electrophoresis buffers employed in the optimized analytical assay.
2. Solution 2 in Section 2.1.1.
3. Dye-containing binding buffer: 0.05% (w/v) bromophenol blue in 1X optimized binding buffer (store at 4°C after filtering).

2.2.2. Reagents / Special Equipment

1. DNA probe used in the optimized analytical assay. Wear gloves and safety glasses during all manipulations involving the radioactive DNA, and work behind perspex or glass shields whenever possible.
2. All reagents employed in the optimized analytical binding reaction (*see* Section 2.1.).
3. 16–18 × 16–18 cm front and back glass plates: The plates must be absolutely clean before use. Wash them with warm soapy water; then holding them by the edges, rinse several times first in tap water and then in bidistilled water. Finally, rinse with ethanol, and let them air dry. Using a pad of Kimwipes, siliconize the inner side of the back plate with a 2% dimethyldichlorosilane solution in 1,1,1-trichloroethane in a fume hood (this product is harmful; wear gloves and handle with care).
4. 0.3-cm spacers.
5. Electroresistant plastic tape (e.g., 3M yellow electrical tape).
6. 0.45-µm filter (Millipore, Bedford, MA).
7. TEMED (N,N,N',N'-tetramethylethylenediamine).
8. 3-mm comb with 10-mm-wide teeth.
9. 10-mL syringe attached to a 18-gage needle.
10. 100–200-µL Hamilton syringe.
11. Additional reagents and equipment: solid acrylamide and *bis*-acrylamide (wear gloves and face mask when weighing these neurotoxic reagents); plenty of binder clamps (fold-back spring clips); razor blades; polyacrylamide gel electrophoresis apparatus; constant current power supply; peristaltic pump for recirculating electrophoresis buffer (if required); siliconized 1.5-mL Eppendorf microcentrifuge tube; spatula.

2.3. DNA Chemical Cleavage (Footprinting) Reactions within the Gel

2.3.1. Solutions

1. 10 mM Tris-HCl, pH 8.0 (store at room temperature after autoclaving).
2. MPA solution: Add 100 µL of neat 3-mercaptopropionic acid (Aldrich, Milwaukee, WI) to a sterile 50-mL conical tube containing 19.9 mL of bidistilled sterile water. Mix by vortexing. 3-mercaptopropionic acid is toxic and causes burns in contact with skin and eyes; wear gloves and handle accordingly. Dilute just prior to use (store the liquid reagent in a place protected from light).
3. OP solution: Weigh out 80 mg of 1,10-phenanthroline monohydrate (Aldrich), and dissolve (by vortexing and shaking vigorously for 2 min) in

10 mL of absolute ethanol in a sterile 50-mL conical tube. Wear gloves and dust mask when weighing this reagent. Prepare just prior to use (store the powdered reagent in a place protected from light).
4. Cu^{2+} solution: Weigh out 72 mg of anhydrous copper (II) sulfate (Aldrich), and dissolve (by vortexing for 1 min) in 50 mL of bidistilled sterile water in a sterile 50-mL conical tube. Powdered copper (II) sulfate is irritating to eyes, respiratory system, and skin; wear gloves and eye/face protection when weighing this chemical. Prepare just prior to use (store the powdered chemical sealed in a dry place).
5. $(OP)_2Cu^+$-STOP solution: Weigh out 127 mg of 2,9-dimethyl-1,10-phenanthroline (Neocuproine) monohydrate (Aldrich), and dissolve (by vortexing vigorously for 2 min) in 20 mL of absolute ethanol in a sterile 50-mL conical tube. Wear gloves and dust mask when weighing this reagent. Prepare just prior to use (store the powdered reagent in a place protected from light).

2.3.2. Reagents / Special Equipment

1. 20 × 20 cm Pyrex dish (available in most supermarkets); wash the dish with detergent, water, and then ethanol. Rinse with bidistilled sterile water and dry with tissues.
2. Sterile 50-mL conical tube.
3. Additional equipment: protective gloves; glass or plastic beaker; 20-mL glass pipet; vacuum aspirator.

2.4. Isolation of Free and Complexed DNA Fractions

2.4.1. Direct Elution from the Polyacrylamide Gel Matrix

2.4.1.1. SOLUTIONS

1. Gel elution buffer: $0.5M$ ammonium acetate, pH 7.5 (promotes diffusion of the DNA out of the gel matrix and is readily soluble in ethanol in the subsequent precipitation step), 1 mM EDTA, 0.1% SDS (w/v; effectively denatures any contaminating DNase activity). For improved recovery of DNA fragments smaller than 60 bp, the buffer should also include 10 mM magnesium chloride. This stock solution can be stored at room temperature protected from light for several months.
2. 25:24:1 (v/v) phenol/chloroform/isoamyl alcohol: Mix just prior to use (in a fume hood and wearing gloves) 25 vol of phenol (redistilled under nitrogen and equilibrated with 100 mM Tris-HCl, pH 8.0, 1 mM EDTA in the presence of 0.1% [w/v] 8-hydroxyquinoline) with 24 vol of chloroform and 1 vol of isoamyl alcohol. Avoid inhaling chloroform vapor.

Phenol can be stored at 4°C in dark (brown) bottles for up to 2 mo; it causes severe burns and should be handled with suitable precautions.
3. 24:1 (v/v) chloroform/isoamyl alcohol: Mix (in a fume hood and wearing gloves) 24 vol of chloroform with 1 vol of isoamyl alcohol. This organic mixture can be stored at room temperature in dark (brown) bottles indefinitely.
4. Low- and high-salt buffers required for the use of Elutip™-d minicolumns (*see* Chapter 6 in vol. 4 of this series).
5. 70% and 90% (v/v) ethanol.
6. Sequencing-gel loading buffer: 90% (v/v) deionized formamide (*see* Chapter 12 in vol. 4 of this series for a detailed protocol of how to deionize formamide), 1X TBE (*see* Section 2.4.2.), 0.025% (w/v) xylene cyanol FF, and 0.025% (w/v) bromophenol blue. Store at –20°C after filtering. Formamide is a teratogen; take all safety precautions to avoid contact during manipulations involving this reagent.

2.4.1.2. REAGENTS/SPECIAL EQUIPMENT

1. Plasticwrap, such as Saran Wrap® or Cling-film.
2. Old X-ray film covered with plasticwrap.
3. Glass stirring rod.
4. Small adhesive labels (or Scotch Tape®).
5. "Radioactive ink": Mix a small amount of ^{32}P with waterproof black drawing ink, to a concentration of ~200 cps (on a Geiger counter) per µL.
6. Fiber-tip pen.
7. Kodak X-Omat AR film.
8. Lightproof cardboard film holder.
9. Aluminum foil.
10. Lab marking pen.
11. Sharp scalpel.
12. Fine-tip waterproof marking pen.
13. Single-edged disposable razor blades.
14. 18-gage syringe needles or sterile forceps.
15. Sterile 3-mL syringes attached to a shortened 18-gage needle (broken with pliers).
16. Siliconized 1.5-mL Eppendorf microcentrifuge tubes.
17. 22-gage syringe needle.
18. Siliconized capless 0.5-mL Eppendorf microcentrifuge tubes.
19. Conformable self-sealing tape (e.g., Parafilm).
20. 1-mL sterile syringes (Becton Dickinson, Heidelberg, Germany).
21. 0.22-µm syringe filters (Millipore).
22. Ice-cold absolute ethanol.

23. Glycogen (from Boehringer, Indianapolis, IN, or Sigma, St. Louis, MO).
24. Drawn-out Pasteur pipet.
25. Elutip™-d mini-columns (Schleicher and Schuell, Dassel, Germany).
26. Additional equipment: Geiger counter; all equipment for autoradiography; low-speed centrifuge (Beckman J6B or Sorvall RC3); microcentrifuge (Eppendorf or equivalent); 37–42°C shaking air incubator; vacuum centrifuge (Savant, Hicksville, NY); scintillation counter or Bioscan Quick Count for ^{32}P; waterbath at 68°C.

2.4.2. Electrotransfer of the Entire Gel and Elution from NA-45 Membrane

2.4.2.1. Solutions

1. 10 mM EDTA, pH 7.6 (store at room temperature after autoclaving).
2. 0.5M NaOH.
3. 1X TBE buffer: 89 mM Tris base, 89 mM borate, 2.5 mM EDTA. To prepare 5 L of 5X TBE buffer, dissolve (under stirring for at least 1 h) 272.5 g of ultrapure Tris base, 139.1 g of boric acid, and 23.3 g of disodium EDTA dihydrate in 4.5 L of bidistilled water, and make up to a final volume of 5 L. It is not necessary to adjust the pH of this solution, which should be about 8.3. Store at room temperature; this stock solution is stable for many months, but it is susceptible to bacterial growth and should occasionally be inspected visually.
4. 20 mM Tris-HCl, pH 8.0, 0.1 mM EDTA (store at room temperature after autoclaving).
5. NA-45 membrane elution buffer: 1.0M NaCl, 20 mM Tris-HCl, pH 8.0, 0.1 mM EDTA (store at room temperature after autoclaving).
6. Solutions 2, 3, 5, and 6 in Section 2.4.1.1.

2.4.2.2. Reagents/Special Equipment

1. NA-45 membrane sheets (0.45-μm pore size; Schleicher and Schuell).
2. Filter paper (Whatman 3MM or equivalent).
3. Clean glass plate.
4. Items 1, 2, and 4–7 in Section 2.4.1.2.
5. Metal cassette.
6. Items 10–14, 16, and 22–24 in Section 2.4.1.2.
7. Additional equipment: protective gloves; electrotransfer unit (high current [2–3 A] power supply, e.g., Bio-Rad, Richmond, CA or Hoefer, San Francisco, CA); Kimwipes; Geiger counter; all equipment for autoradiography; microcentrifuge (Eppendorf or equivalent); 55°C waterbath with agitation; vacuum centrifuge (Savant); scintillation counter or Bioscan Quick Count for ^{32}P; waterbath at 68°C.

2.5. Preparation of the G + A Sequencing Ladder

2.5.1. Solutions

1. Carrier DNA stock solution: salmon sperm DNA extracted sequentially with phenol/chloroform/isoamyl alcohol (25:24:1 v/v) and chloroform/ isoamylalcohol (24:1 v/v), precipitated with ethanol, resuspended in bidistilled sterile water at 1 mg/mL, and sonicated to an average chain length of 200 bp.
2. 1.0M aqueous piperidine: Add 100 µL of concentrated piperidine (reagent grade, BDH) into 0.9 mL of bidistilled sterile water. Piperidine is somewhat hard to handle; wash out the micropipet used repeatedly, and mix well by vortexing when diluting. Make dilution in a fume hood just prior to use.
3. 1% SDS: Prepare a 10% (w/v) solution of SDS in bidistilled sterile water (wear dust mask when weighing solid SDS); heat at 68°C to assist dissolution (do not autoclave). Dilute 1:10 with bidistilled sterile water. Store at room temperature.
4. Solution 6 in Section 2.4.1.1.

2.5.2. Reagents / Special Equipment

1. Siliconized 1.5-mL Eppendorf microcentrifuge tube.
2. 20,000 cpm of the end-labeled DNA fragment used in the preparative mobility-shift assay.
3. 88% aqueous formic acid.
4. Conformable self-sealing tape (e.g., Parafilm).
5. 1-Butanol (n-butylalcohol).
6. Drawn-out Pasteur pipet.
7. Additional equipment: Wet ice; 37°C waterbath with agitation; thermostatted heating block at 90°C; lead weight; microcentrifuge (Eppendorf or equivalent); vacuum centrifuge (Savant); waterbath at 68°C.

2.6. Analysis of the Chemical Cleavage Products on a DNA Sequencing Gel

2.6.1. Solutions

1. 40% (19:1) acrylamide:*bis*-acrylamide solution (*see* Chapter 10 in vol. 4 of this series).
2. Solution 3 in Section 2.4.2.1.
3. Solution 2 in Section 2.1.1.
4. Fixing solution: 10% (v/v) acetic acid, 10% (v/v) methanol.

2.6.2. Reagents / Special Equipment

1. Urea (enzyme grade).
2. 34 × 40 cm front and back glass plates: treat the plates as described in Section 2.2.2.
3. 0.04-cm spacers.
4. 0.4-mm custom-ordered sample comb with 5-mm lanes spaced on 10-mm centers.
5. 10-mL syringe with a 22-gage needle.
6. Calibrated glass capillaries with finely drawn tips or disposable flat-capillary pipet tips (National Scientific Supply Company, Inc., San Rafael, CA).
7. 30-mL syringe with a bent 20-gage needle.
8. Backing paper (Whatman No. 1 or equivalent).
9. Plasticwrap, such as Saran Wrap or Cling-film.
10. Filter paper (Whatman 3MM or equivalent).
11. Kodak X-Omat AR film.
12. Large metal cassette.
13. Intensifying screen (DuPont [Wilmington, DE] Cronex Lightning Plus).
14. Additional reagents and equipment: TEMED (N,N,N',N'-tetramethylethylenediamine); plenty of binder clamps (fold-back spring clips); sequencing gel electrophoresis apparatus; power supply delivering high voltage (2500–3000 V) (e.g., Bio-Rad, LKB Turku, Finland; Pharmacia, Piscataway, NJ); dry-block heater at 90°C; wet ice; aluminum plate of the appropriate size; razor blade; plastic tank at gel dimensions; 10-mL glass pipet; Kimwipes; gel dryer; all equipment for autoradiography.

3. Methods

3.1. Analytical Mobility-Shift Electrophoresis Assay

1. Perform a preliminary mobility-shift electrophoresis assay (refer to Chapter 21 in this volume for a basic protocol for detecting DNA–protein interactions on native [nondenaturing] polyacrylamide gels) to obtain the complex(es) to map. Because no universal binding and/or gel system is likely to be found for the study of all DNA–protein interactions, it may be necessary to optimize conditions for formation and adequate resolution of the DNA–protein complex(es) of interest (*see* Note 1).
2. If crude or partly fractionated extracts are employed, ascertain the DNA-binding specificity of the resolved complex(es) by performing an analytical competition binding assay (*see* Chapter 21 in this volume and Note 2).
3. It is advisable, prior to proceeding to the more laborious preparative gel retardation/*in situ* footprinting assay, to perform an additional ana-

lytical experiment and obtain a qualitative estimation of the dissociation rates of preequilibrated protein–DNA complexes of interest (*see* Note 3). Although the $(OP)_2Cu^+$-mediated cleavage reactions in the gel are relatively insensitive to the kinetic stability of the DNA–protein complex(es) under investigation (*see* Section 1.), this information can be used in adjusting the exposure time to the chemical nuclease (*see* step 6 in Section 3.3. and Note 7), thereby enhancing the clarity of the expected footprint.

3.2. Preparative Mobility-Shift Electrophoresis Assay

1. Assemble 16–18 × 16–18 cm front and back glass plates and 0.3-cm spacers for casting a preparative mobility-shift polyacrylamide gel (3-mm-thick polyacrylamide gels are preferable, because they are easier to load and give sharper bands than 1.5-mm-thick gels). The plates must be scrupulously clean and free of grease spots to avoid trapping air bubbles while pouring the gel; it is highly recommended to use one glass plate (preferably the back) that has been siliconized on the inner side for ease of removal after electrophoresis is completed. Taking particular care (to prevent leakage), seal the entire length of the two sides and the bottom of the plates with electroresistant plastic tape.
2. Prepare, filter (through a 0.45-µm filter), and degas (by applying vacuum) 100 mL of the acrylamide gel solution found during optimization of the analytical assay (Section 3.1., step 1) to elicit abundant DNA–protein complex(es) of the highest resolution. Because of the ready permeability of the gel matrix by all reagent and quenching solutions used for the subsequent chemical treatment and the lack of diffusible radicals mediating the DNA–scission reaction, the $(OP)_2Cu^+$ *in situ* footprinting technique is compatible with a broad spectrum of gel porosities (ranging from 3.5 to 6% [w/v], with an acrylamide to *bis*-acrylamide molar ratio of 19:1 to 80:1) and gel/running-buffer compositions (from glycerol-containing/low-ionic-strength [pH 7.5–7.9] to high-ionic-strength TBE [pH 8.3], or Tris-glycine [pH 8.5] buffer systems).
3. Add to the solution 0.8 mL of 10% ammonium persulfate and 75 µL of TEMED, and swirl the mixture gently.
4. Slowly pour the acrylamide gel mix between the glass plates, and quickly insert a 3-mm comb bearing 10-mm-wide teeth. Allow the gel to polymerize (lying flat or nearly flat, to avoid undesirable hydrostatic pressure on the bottom) at room temperature for about 45 min.
5. After polymerization is complete, remove the electrical tape from the bottom of the gel (by cutting with a razor blade), and clamp the gel into place on the electrophoresis apparatus. Fill both chambers of the electro-

phoresis tank with the buffer used for preparation of the acrylamide gel mix (step 2), carefully remove the comb, and immediately rinse the sample wells with reservoir buffer using a 10-mL syringe with an 18-gage needle.

6. Prior to assembling the preparative binding reaction, preelectrophorese the gel for 60 min at 20 mA, with or without buffer recirculation between the two compartments, depending on the nature of the gel/running-buffer system used (low- or high-ionic strength). This removes any excess persulfate and unpolymerized acrylamide. Prerunning of the gel should be done at the temperature at which the binding reaction will be performed (known from step 1 in Section 3.1).

7. In a siliconized 1.5-mL Eppendorf tube, scale up the optimized analytical reaction 5- to 10-fold, depending on the relative proportion of the DNA–protein complex(es) obtained. If the detected specific DNA-binding activity(ies) (Section 3.1., step 1) represents <1% of the total label input, the amount of radioactive probe in the scaled reaction should be at least 250,000 cpm (*see also* Note 4).

8. Turn off the electric power. Using a 100–200-µL Hamilton syringe, load the preparative binding reaction onto one or two wells (depending on the total volume of the sample) in the middle of the gel. Raise the tip of the needle as the sample is loaded into the well. Do not attempt to expel all of the sample from the syringe since this almost always produces air bubbles that blow the sample out of the well. If the glycerol concentration in the binding buffer is low (<5%), it is important to load the well gently to prevent dilution (*see also* Note 5). Avoid adding bromophenol blue to the binding reaction prior to loading because this dye can rapidly disrupt some DNA–protein complexes. Instead, you may load just dye-containing binding buffer in one of the adjacent lanes to monitor the progress of electrophoresis.

9. Run the gel at 25–35 mA (it may be necessary to adjust the voltage occasionally if a constant power supply is not available) under the conditions described in step 6 for a time sufficient to allow migration of the free DNA probe to ~2 cm from the bottom of the gel. Provided the same plate size has been used in establishing the optimal electrophoresis conditions, this can be monitored by the migration of the tracking dye, in correlation with the position of the free probe on the autoradiogram obtained from the optimized analytical assay (step 1 in Section 3.1.). If electrophoresis is performed at room temperature, the glass plates should be allowed to become only slightly warm, since excess heating may perturb the equilibrated complexes, or even cause protein denaturation; decrease the current if the plates become any hotter.

10. Following electrophoresis, detach the glass plates from the gel apparatus, and using a spatula carefully remove the spacers and pry the glass plates

apart, taking extreme care not to distort or tear the gel, which should remain attached to only one of the plates (the nonsiliconized front plate).

3.3. DNA Chemical Cleavage (Footprinting) Reactions within the Gel

1. Wear protective gloves and wash your fingers extensively in a beaker containing bidistilled water to remove the talc powder. Immerse the whole gel, still attached to the lower plate (with the gel facing up), into a 20 × 20 cm scrupulously clean Pyrex dish (*never use a plastic tray*), containing 200 mL of 10 mM Tris-HCl, pH 8.0. Loosen it on its supporting glass plate (omit this step if using Tris-glycine-containing or low-percentage/low-ionic-strength polyacrylamide gels [e.g., a 3.5–4% gel, containing 6.7 mM Tris-HCl, pH 7.5, 3.3 mM sodium acetate, 1 mM EDTA, widely employed in mobility-shift assays], which are very sticky and extremely difficult to manipulate without fracturing).
2. Prepare the MPA (58 mM 3-mercaptopropionic acid), OP (40 mM 1,10-phenanthroline), and Cu^{2+} (9 mM $CuSO_4$) solutions. Use the recommended suppliers to obtain the liquid and powdered reagents (*see also* Note 6).
3. In a sterile 50-mL conical tube, transfer 1 mL of the freshly made OP solution. To this, add 1 mL of the freshly prepared Cu^{2+} solution, and wait 1 min while pipeting up and down (the mixture should turn light blue, indicating efficient formation of the $[OP]_2Cu^{2+}$ chelate). Add 18 mL of bidistilled sterile water and vortex the tube. This is the OP/Cu^{2+} solution (1,10-phenanthroline to copper ratio of ~4.5:1).
4. Add the OP/Cu^{2+} solution (20 mL) to the gel equilibrating in the 200-mL buffer, and shake the Pyrex dish while laying it on an even horizontal surface to distribute evenly.
5. Initiate the chemical nuclease reaction by adding the MPA solution (20 mL); distribute evenly by quickly shaking the Pyrex dish as before. The gel will turn brownish. The appearance of a dark brown precipitate indicates the presence of impurities in the original $CuSO_4$ solution, which will interfere with the cascade leading to DNA-strand scission. It is, therefore, crucial for the assay to use copper (II) sulfate of the best available analytical grade.
6. Incubate for a period of 8–30 min *without shaking* (do not disturb the equilibrated complexes; the small size and the ready diffusibility of all reaction components within the gel matrix are sufficient for a productive attack on the target DNAs). To obtain an intelligible and homogeneous cleavage pattern of all DNA species in the gel, the exact time of chemical treatment has to be adjusted for each particular case, based on the considerations discussed in Note 7.

7. During the last 5 min of the incubation period, prepare the $(OP)_2Cu^+$-STOP solution (28 mM 2,9-dimethyl-1,10-phenanthroline).
8. Quench the reaction by adding the $(OP)_2Cu^+$-STOP solution (20 mL), and wait 2 min while shaking the Pyrex dish. The gel will turn yellowish, which is diagnostic for the quality of 2,9-dimethyl-1,10-phenanthroline, and hence for efficient termination of the chemical nuclease action.
9. Using a 20-mL glass pipet, aspirate (staying away from the corners of the gel) all the liquid from the Pyrex dish, and carefully rinse the gel (still on the glass plate) with four changes of bidistilled water. Remove the plate with the gel on it from the Pyrex dish. It is not necessary to take any specific precautions in dispensing the original mixture and the washing material, because all reaction components are oxidatively destroyed. Immerse the Pyrex dish in household bleach for 1 h at room temperature followed by extensive washing down the drain with tap water.

3.4. Isolation of Free and Complexed DNA Fractions

3.4.1. Direct Elution from the Polyacrylamide Gel Matrix

1. Smoothly wrap the gel and plate with plasticwrap or, preferably, peel off the gel onto a suitable backing material (old X-ray film covered with plasticwrap is best) and wrap it with plasticwrap. Using a glass stirring rod as a rolling pin, remove any air bubbles trapped under the plasticwrap, being careful not to disturb the shape of the gel.
2. To aid accurate subsequent alignment of gel and film, trace three corners of the plasticwrap covering the gel with small adhesive labels (or pieces of Scotch Tape), marked with "radioactive ink" spots; use an almost equal amount of radioactivity in each ink dot to that in the DNA bound fraction(s) (this can be monitored by a Geiger counter). Use a fiber-tip pen to apply ink of the desired hotness to the sticky labels; let the ink dots dry completely before exposing to X-ray film.
3. In the darkroom, tape the sealed gel to a piece of Kodak X-Omat AR film. Enclose the assembly in a lightproof cardboard film holder, exerting an even gentle pressure on the "sandwich," and wrap the entire packet with aluminum foil to ensure a light-tight environment.
4. Expose the film at 4°C for 1–3 h (the length of exposure time depends on the relative abundance of the specific complex[es]) to assess the position of the retarded (bound) and unretarded (free) DNA fragments. The energy of the β particles produced by the ^{32}P decay is sufficient to penetrate several millimeters thickness of hydrated gels, without sig-

nificant absorption (quenching by the gel is <40%), thus allowing the direct autoradiographic detection of [^{32}P]-labeled DNA embedded in the gel matrix.

5. Develop the film, and using a lab marking pen, encircle the position of the complex(es) to be mapped as well as that of the unretarded probe. Any DNA released by dissociation during the run will trail just above the free DNA band as a smear; do not include this region in your marking.
6. Using a sharp scalpel, cut out from the X-ray film the marked rectangles containing the autoradiographic images of the free and bound probe.
7. Line up the "radioactive ink" spots on the film with the corresponding markings at the three corners of the plasticwrap. With a fine-tip waterproof marking pen, mark the position of the free and bound probe on the plasticwrap, using the periphery of the rectangular holes on the film as a template.
8. Remove the film, and cut through the marks on the plasticwrap with a disposable razor blade for each species. Separate the polyacrylamide slices from the rest of the gel (and from the plasticwrap), using either 18-gage single-use syringe needles or sterile forceps, and transfer them onto a piece of plasticwrap. It is desirable to keep the size of the polyacrylamide strips to a minimum (*see* Note 4).
9. Crush the gel slices by extruding them from a sterile 3-mL syringe barrel through a shortened 18-gage needle (broken with pliers) into a siliconized 1.5-mL Eppendorf tube by low-speed centrifugation (5 min at 2500*g*) in a swinging bucket rotor. Alternatively, punch a small hole by forcing a 22-gage sterile needle through the bottom of a siliconized capless 0.5-mL Eppendorf tube, place this tube into another siliconized capped 1.5-mL Eppendorf tube, put the gel slice in the upper tube, and spin at 12,000*g* in a microcentrifuge (minus rotor cover) for 1 min. The gel will be crushed through the hole into the lower tube.
10. To each tube add enough gel elution buffer to cover the gel paste, and mix well by vortexing. The volume of the buffer added depends on the size of gel slice, but as a guide, 0.5–0.6 mL is used for a slice $10 \times 3.5 \times 3$ mm.
11. Seal each tube with conformable self-sealing tape, and allow the DNA fragments to diffuse out by incubating at 37–42°C for 10–16 h in a shaking air incubator.
12. Vortex the tubes vigorously, and pellet the gel paste by centrifuging at room temperature for 1 min in a microcentrifuge (12,000*g*).
13. Using a micropipet, pipet off the supernatant solution, taking care to avoid polyacrylamide pieces, and transfer it to a 1-mL sterile syringe.
14. Remove any remaining tiny pieces of polyacrylamide by slowly passing the supernatant through a 0.22-μm syringe filter into a new sili-

conized Eppendorf tube (do not use polystyrene tubes to collect the filtrate, since they cannot withstand the subsequent organic extractions). The eluted yield of DNA fragments should be >90%.
15. Extract the filtered supernatant sequentially with an equal volume of phenol/chloroform/isoamyl alcohol (25:24:1 v/v) and chloroform/isoamylalcohol (24:1 v/v), to eliminate contaminating proteins that might distort DNA fragment migration during subsequent electrophoresis. In both steps, mix the contents of the tube thoroughly by vortexing for 30 s, and centrifuge at 12,000g (microcentrifuge) for 5 min at room temperature to separate the organic and aqueous phases (*see* Note 8).
16. With a micropipet, transfer the aqueous phase (no more than 0.55 mL) to a fresh siliconized Eppendorf tube. Add ~2 vol of ice-cold absolute ethanol (*no additional salt is required*), vortex well, and precipitate the radioactive DNA fragments by chilling the tube at −20°C for a minimum time of 60 min. Although it is generally not necessary to add carrier to aid precipitation (the small acrylamide polymers released from the crushed gel slice will suffice), it is recommended to precipitate the DNA in the presence of glycogen (10 µg/sample, added prior to ethanol) to improve even further the recovery of DNA.
17. Recover DNA by centrifugation at 12,000g for 30 min in a microcentrifuge at 4°C. Carefully aspirate off the ethanol supernatant with a drawn-out Pasteur pipet, taking care not to disturb the faintly visible radioactive pellet (its presence can be monitored by a Geiger counter, and its location identified from the position of the tube in the rotor). (*See also* Note 9).
18. Remove traces of salt trapped in the precipitate (which interfere with subsequent electrophoresis), by rinsing the pellet twice with 1 mL of 70 and 90% (v/v) ethanol, respectively, centrifuging each time at 12,000g (microcentrifuge) for 2 min at 4°C. For both washes, invert the tube gently several times; *do not vortex*.
19. Dry the pellet for 5 min in a vacuum centrifuge.
20. Measure each pellet by Cerenkov counting to determine radioactivity (2000 cpm is sufficient for an overnight exposure with intensifying screen).
21. Resuspend the pellets (by heating at 68°C for 2 min, vortexing vigorously, and repeatedly pipeting) in sequencing-gel loading buffer, so that 5 µL will contain equal Cerenkov cpm from the free and bound DNA fractions. It is important to equalize the number of cpm/µL in the two fractions in order to compare their cleavage patterns accurately. If the sequencing gel is not to be run immediately, you can store the DNA samples at −80°C.

3.4.2. Electrotransfer of the Entire Gel and Elution from NA-45 Membrane

If the mobility-shift electrophoresis assay was performed using Tris-glycine-containing or low percentage/low ionic-strength polyacrylamide gels, which behave like a poorly set Jello and are, therefore, extremely difficult both to manipulate for autoradiography and to handle as polyacrylamide strips in the subsequent DNA-elution steps, it is highly recommended to transfer the entire gel electrophoretically onto a sheet of NA-45 membrane (DEAE cellulose in membrane form). Following electroblotting, the NA-45 membrane is exposed to X-ray film, the bands corresponding to free and bound species are cut out, and DNA is eluted. The remaining steps in the procedure, beginning with organic extractions of the eluates, are identical to those described in Section 3.4.1., steps 15–21.

1. Cut a piece of NA-45 membrane and four pieces of filter paper to the exact size of the gel; cut the membrane between liner sheets wearing gloves.
2. To increase binding capacity, wash the membrane for 10 min in 10 mM EDTA, pH 7.6, and for 5 min in 0.5M NaOH, followed by several rapid washes in bidistilled water; let the membrane soak in 1X TBE buffer.
3. Remove the plate with gel from the Pyrex dish and place it on a flat surface. Carefully lay two pieces of prewet (in 1X TBE buffer) filter paper onto the surface of the gel, making sure no air bubbles are trapped between the filter paper and gel.
4. Slowly and with extreme care, lift the gel (adhered to the filter paper), and place it (with the gel facing up) on a clean glass plate. Wet the gel with a thin layer of 1X TBE buffer.
5. Wearing gloves, lay the wet membrane sheet over the gel, again being careful not to trap air bubbles beneath the membrane.
6. Complete the "sandwich" by placing the two remaining pieces of prewet (in 1X TBE buffer) filter paper on top of the membrane.
7. Insert the "sandwich" of filter paper/gel/membrane/filter paper into a gel-holder cassette, and load the assembly into one of the center slots in a transfer apparatus (any of the commercially available electroblot units are suitable), with the NA-45 membrane positioned between the gel and the anode (positive electrode).
8. Fill the transfer apparatus with 1X TBE buffer (precooled at 4°C), and transfer the chemically cleaved double-stranded DNA fragments electrophoretically from the gel to the NA-45 membrane. Electroblotting is

performed at 4°C for 3 h, at either 20 V (~1 V/cm), if small DNA fragments (40–90 bp) are being transferred, or at 35 V (~2 V/cm), if fragments >100 bp have been employed in the mobility-shift assay. The extra phosphate groups left on both the 3' and 5' end at the sites of chemical cleavage (Fig. 1) increase the electrophoretic mobility of the DNA molecules, thus facilitating transfer of the free and protein-bound species onto the membrane.

9. When transfer is completed, turn off the power, remove the gel "sandwich," lift the membrane sheet away from gel while wearing gloves, and rinse it in 20 mM Tris-HCl, pH 8.0, 0.1 mM EDTA to remove residual polyacrylamide. *Do not let the membrane dry!*
10. Place the wet membrane (with transferred DNA face up) on the surface of a used piece of X-ray film wrapped in plasticwrap, and cover it with a tightly drawn layer of plasticwrap. With a pad of Kimwipes, push out any trapped air bubbles under the plasticwrap. Efficient transfer can be monitored by checking with a Geiger counter.
11. Follow steps 2 and 3 in Section 3.4.1.; use a metal cassette instead of a cardboard film holder to expose the membrane to X-ray film.
12. Autoradiograph the membrane at 4°C for 15–45 min (about $1/4$ the time required for the wet gel).
13. Follow steps 5–8 in Section 3.4.1. Using sterile forceps, transfer the wet NA-45 membrane strips into siliconized 1.5-mL Eppendorf tubes.
14. Add to each NA-45 membrane strip 0.6 mL of NA-45 membrane elution buffer, and spin for a few seconds in a microcentrifuge to submerge the whole strip.
15. Incubate at 55°C for 2–3 h in a waterbath with agitation.
16. Vortex the tubes vigorously, and pellet the NA-45 membrane strips by centrifuging at room temperature for 10 s in a microcentrifuge.
17. Remove the buffer, and place it in a new siliconized 1.5-mL Eppendorf tube. Monitor paper for loss of radioactivity; typically, about 90% of the membrane-bound DNA is released with this technique.
18. Follow steps 15–21 in Section 3.4.1.

3.5. Preparation of the G + A Sequencing Ladder

Provided the sequence of the DNA probe is known, you may perform at this stage a Maxam-Gilbert guanine- and adenine-specific modification/cleavage reaction (G + A sequencing ladder) of the end-labeled DNA fragment used in the gel retardation assay. This reaction will be coelectrophoresed with the DNA samples eluted from the free and bound fractions, to identify nucleotides protected from chemical cleavage (protein-contact sites) in the final stage of the

Phenanthroline-Copper Footprinting

footprinting analysis (Section 3.6.). Below is a fast version (requiring only 1 h) of this otherwise time-consuming reaction.

1. In a siliconized 1.5-mL Eppendorf tube mix successively: 20,000 cpm of the end-labeled DNA fragment used in the mobility-shift assay, 1.5 µL of carrier DNA stock solution, and bidistilled sterile water to 10 µL.
2. Chill the tube on ice and add 1.5 µL of 88% aqueous formic acid.
3. Incubate at 37°C for 14 min in a waterbath.
4. Chill again on ice and add 150 µL of freshly prepared $1.0M$ aqueous piperidine. Close the tube and wrap the cap tightly with a conformable self-sealing tape.
5. Heat at 90°C for 30 min in a thermostatted heating block, with the wells filled with water. It is necessary to put a lead weight on top of the tube to prevent it from popping open as pressure builds up inside.
6. Cool the tube on ice. Remove the conformable tape and spin for a few seconds in a microcentrifuge; transfer to a new siliconized Eppendorf tube.
7. Add 1 mL of 1-butanol. Vortex vigorously until only one phase is obtained.
8. Mark the position of the tube in the rotor, and spin at $12,000g$ for 2 min in a microcentrifuge (room temperature).
9. Carefully remove and discard the supernatant using a drawn-out Pasteur pipet, taking care not to disturb the tiny pellet or the area of the tube where the pellet should be located.
10. Add 150 µL of 1% SDS and vortex the tube. Add 1 mL of 1-butanol. Mix well by repeatedly inverting the tube. This step removes remaining traces of piperidine trapped in the precipitate that interfere with the subsequent electrophoretic separation.
11. Resediment the precipitate by centrifuging at $12,000g$ for 2 min in a microcentrifuge, and remove the supernatant as in step 9.
12. Spin for a few seconds in a microcentrifuge to collect any traces of liquid at the bottom of the tube, and carefully remove it using a micropipet.
13. Dry the pelleted DNA for 5 min in a vacuum centrifuge.
14. Resuspend the samples (as in step 21 of Section 3.4.1.) in 5 µL of sequencing-gel loading buffer, and store at –80°C until ready to load onto the sequencing gel.

3.6. Analysis of the Chemical Cleavage Products on a DNA Sequencing Gel

Visualization of the length(s) on the DNA affected by protein(s) binding specifically to it (that is the protein-binding site[s], or footprint[s] left by the protein on the DNA) requires electrophoretic fractionation of the single-stranded fragments resulting from the chemical nuclease attack in a denaturing polyacrylamide gel of the type employed in DNA sequencing, followed by autoradiography. The location of the footprint(s) in the known DNA sequence is identified by including the sequencing marker G + A track prepared in Section 3.5.

1. Assemble and pour a 34 × 40 × 0.04 cm 6–15% (w/v) sequencing polyacrylamide gel, containing 1X TBE buffer and 8.3M urea. For a detailed description of how sequencing gels are prepared and cast refer to Chapter 10 in vol. 4 of this series. The percentage of acrylamide depends on the size of the DNA fragments to be separated as well as on the size and location (relative to the labeled end) of the suspected protein-binding site(s). As for the preparative mobility-shift assay, you should siliconize the inside surface of the back glass plate to aid pouring into gel mold and removal at the end of electrophoresis. To avoid dispersing of radioactivity across lanes (which might produce significant errors in a subsequent densitometric analysis of the free and bound DNA chemical cleavage patterns on the autoradiogram; see Note 11), it is recommended to use a 0.4-mm custom-ordered sample comb with 5-mm lanes and 5-mm spacing. You may wrap the polymerized gel in plasticwrap and keep it at room temperature until use (it can be stored as such for up to 36 h).
2. Attach the gel apparatus to the gel electrophoresis tank. Fill both the top and bottom electrode chambers with 1X TBE buffer, and remove the well-forming comb. Check that wells are free from "tails" of polyacrylamide adhering to sides, which may lead to uneven loading of samples and consequently band-shape distortion.
3. Preelectrophorese the gel for 45–60 min before loading the samples. This removes persulfate ions and heats the gel. Prerunning of the gel is performed at constant temperature (approx 55°C), which is most easily achieved by application of constant power (~50–70 W). If the surface temperature becomes too high (>65°C), the glass plates will crack.
4. Thaw the DNA samples (if frozen), heat-denature them (including the G + A sequencing ladder) at 90°C in a dry-block heater for 5 min, and quick-chill in wet ice.
5. Disconnect the power supply, and immediately prior to applying the samples, thoroughly rinse out (using a 10-mL syringe with a 22-gage needle) the wells of the gel with the upper reservoir TBE buffer; this prevents streaking of the DNA samples caused by urea that has diffused into the wells.
6. Using calibrated glass capillaries with finely drawn tips or, preferably, disposable flat-capillary pipet tips, load (as quickly as possible) 5 µL of each sample (plus the G + A sequencing ladder) onto the wells of the

sequencing gel, sweeping the sample evenly from side to side. An untreated naked DNA sample (not subjected to the gel retardation assay) should always be diluted in sequencing-gel loading buffer, heat-denatured, and coelectrophoresed with the treated samples to verify the integrity of the DNA, since single-strand nicks, for instance, might create artificial protein-binding sites.

7. Remove all bubbles from the bottom of the gel (they may prevent even migration of the samples) using a 30-mL syringe with a bent 20-gage needle.
8. Run the gel under preelectrophoresing conditions (constant power, ~50–70 W), taking care not to overheat the glass plates, which may then crack. Uneven migration of fragments ("smiling") caused by uneven distribution of the heat that builds up during electrophoresis can be avoided by clamping an aluminum plate to the front glass plate. It is customary to electrophorese the samples until the bromophenol blue marker dye is about 3–5 cm from the bottom of the gel, but longer electrophoresis times may be required to obtain single-band resolution in the area of the footprint(s). The location of this region(s) depends on the distance between the radioactive label and the suspected protein-binding site(s) as well as on the length of the DNA fragment. Make use of available tables in the literature (referring to the migration of oligodeoxynucleotides in sequencing gels in relation to marker dyes) to determine how long to run your gel in order to achieve the desired electrophoretic resolution in the region of the expected footprint(s); this will allow you to discern differences between the chemical cleavage patterns of the free DNA and that derived from the complexed fraction(s).
9. After completion of electrophoresis, remove the gel from the apparatus, and with the aid of a razor blade, slowly lift the siliconized plate. The thin polyacrylamide sheet will stick to the unsiliconized plate. Fix the gel for 15–20 min by gently immersing it (still attached to the lower plate) in a tank containing enough fixing solution; this removes excess urea that would otherwise crystallize out.
10. Carefully remove the plate bringing the gel on it from the tank, and lay it on a flat surface. Place a prewet (in fixing solution) sheet of backing paper (cut slightly bigger than gel dimensions) over the gel, press it gently down on the gel, roll out any air pockets (using a 10-mL glass pipet), and peel it off patiently and with extreme care together with the gel attached.
11. Cover the gel surface (but not the back of the filter paper) with a tightly drawn layer of plasticwrap. With a Kimwipe, push out any trapped air bubbles under the plasticwrap that might interfere with good uniform contact among the film, gel, and screen. Add two sheets of filter paper

next to the backing paper as a support pad, and put the "sandwich" into a gel dryer (paper pad closest to vacuum source).

12. Dry the gel under vacuum at 80°C for 45–60 min. Do not release the vacuum before the gel is completely dried (sequencing gels with acrylamide concentrations >10% are susceptible to fracturing).
13. In the darkroom, place a sheet of Kodak X-Omat AR film against the plastic-covered face of the gel. It is advisable to preflash the X-ray film, that is, exposing the film to a millisecond flash of light prior to placing it in contact with the sample to an optical density of 0.15 (A_{540}) above the absorbance of the unexposed film. This increases sensitivity (all of the time of exposure to the radioactivity-generated light produces blackening) and linearity of the film response (the degree of blackening above the background is proportional to the amount of radioactivity), which are both essential if a densitometric analysis of the chemical cleavage products is to be performed (see Note 11). Preflashing requires a photographic flash unit appropriately fitted with filters and adjusted as described by Laskey (18).
14. Autoradiograph in a metal cassette containing a single calcium tungstate intensifying screen (place the flashed face of film toward intensifying screen) at –70 to –80°C (to reduce scattering) for 12–16 h. Shorter or longer periods of time may also be required, since the clarity of the footprint depends on the intensity of the bands in the autoradiograph.
15. Immediately remove the film from the cassette and develop it, preferably in an automatic processing machine for X-Omat films. If reexposing the gel, it is necessary to let the cassette warm up before inserting a second film (cold cassettes will quickly collect moisture from the air).
16. Compare the chemical cleavage pattern of the naked DNA to that of the DNA-protein complex(es). The position of a band in the gel corresponds to the distance between the label and the point at which the DNA has been cleaved by the chemical reagent. Accordingly, bands at the bottom of the gel represent the smallest end-labeled DNA fragments, increasing in size as one reads up the gel until the pattern terminates abruptly in a strongly labeled band, corresponding to the uncleaved full-length probe (see Fig. 2). The protected region(s) (indicating sequence-specific protein binding) appears as an area resistant to cleavage (footprint), resulting in an almost complete absence of fragments (gap) arising from within the protein-binding site(s) in the chemical cleavage pattern of the complexed DNA. The nucleotides exhibiting protection are identified by aligning the bands in the cutting pattern of the free DNA with positions (bonds) in the sequence of the coelectrophoresed Maxam-Gilbert marker G + A track. In this comparison, it is necessary

to note that, regardless of the end of the DNA fragment labeled in the experiment (5' or 3'), the obtained set of products from the chemical cleavage reaction matches the mobilities of the G + A sequencing fragments exactly, as a consequence of the identical 3' and 5' ends generated by both chemistries at the cleavage points.

4. Notes

1. Optimization of the analytical mobility-shift electrophoresis assay is generally achieved by assessing binding-reaction parameters and gel electrophoresis conditions. Furthermore, success in interpretation of results from the coupled gel retardation/*in situ* footprinting assay depends critically on some properties of the DNA fragment used in the initial binding reaction.
   a. Properties of the DNA fragment used in the binding reaction. A mobility-shift electrophoresis assay employing crude extracts works best with short DNA fragments, since these reduce nonspecific interactions of proteins in the extract with sequences flanking the specific binding site(s) and are able to detect binding of large proteins more readily. Optimal sizes range between 100 and 150 bp, with the putative protein-binding site(s) located at approximately the center or at least 20–25 bp away from the radioactive labeled end. If a 20–25 bp synthetic oligonucleotide is to be used as a probe, it is advisable to design it in a way that it can be readily subcloned into the polylinker region of a suitable vector, and then labeled and released as a 40–45 bp restriction fragment, in order to obtain the desired single-base resolution within and around the expected footprint(s). Although for mobility-shift assays the DNA fragment may be labeled at all ends, the subsequent footprinting analysis requires the DNA fragment to be radioactively labeled (to a high specific activity) at the 3' or 5' end of *one* of the two strands. Klenow enzyme-labeled probes are preferable to kinased probes because some protein extracts contain substantial phosphatase activities. Finally, the labeled probe should be unnicked, because the resulting fragments may obscure the cleavage pattern obtained after the chemical attack in the footprinting analysis. Therefore, sufficient care should be taken to minimize nuclease activities during all steps of preparation, labeling, and isolation.
   b. Binding-reaction parameters include binding-buffer composition (pH, ionic strength, metal ion content, and presence or absence of nonionic detergents and/or stabilizer polycations), amount of crude or partly fractionated extract or purified protein, concentration of

labeled DNA probe, type and amount of bulk carrier DNA, and temperature and duration of incubation. Specifically, the following considerations should be evaluated: The optimal ratio of protein to DNA for the assay is best determined by titrating a fixed concentration of the radioactively labeled DNA fragment with increasing amounts of crude or partly fractionated extracts, or purified protein. Frequently, as protein concentration increases, binding passes through a maximum. Note, however, that increasing amounts of protein to a fixed concentration of DNA will not necessarily increase the yield of specific complex(es) seen. This is because of the fact that, whereas a given preparation of any DNA-binding protein(s) tends to be fully active in nonspecific binding, it is typically only fractionally active in site-specific binding activity (the apparent fractional activity varying from 5 to 75%, depending on the particular protein[s] and the individual sample). Therefore, too much protein, particularly with crude extract preparations, leads to occlusion of the binding site(s) by proteins interacting with DNA in a sequence-independent manner. This problem can be minimized by raising simultaneously the concentration of bulk carrier (competitor) DNA (typically of the order of 250- to 5000-fold [w/w] excess over binding-site DNA); this increases the occupancy of the binding site(s) by sequestering nonspecifically bound proteins, including nonspecific DNA-binding nucleases that may degrade the end-labeled DNA during the binding incubation. Bear in mind, however, that although some proteins are able to locate their target binding site(s) in the presence of vast excesses of nonspecific natural DNAs (sonicated salmon sperm or calf thymus DNA), other proteins cannot tolerate natural DNA carriers, but bind readily in the presence of an excess of synthetic polynucleotides, such as poly d(I-C) · d(I-C) or poly d(A-T) · d(A-T). It is noteworthy in the latter case that the efficacy of competition for nonspecific binding can vary among different batches from the same vendor. On the other hand, if too much carrier DNA is added, it will compete for the specific factor(s) of interest, and the level of complex(es) will decrease. Finally, provided an adequate resolution of DNA-bound species is obtained for a fixed concentration of competitor, increasing the amount of probe increases the fraction of DNA driven into complex(es), until the limit set by the binding constant(s) is reached.

c. Gel electrophoresis conditions. Gel parameters, such as percentage of acrylamide, degree of crosslinking, and pH and type of gel/running buffer system (high- or low-ionic strength) dramatically affect the size, aggregation state, and stability of DNA–protein complexes, hence their abundance and quality of separation. Accordingly, it may be necessary to try more than one gel fractionation/buffer system to obtain sufficient formation of the DNA–protein complex(es) of interest. The electrophoresis time has to be optimized for the complex(es) studied and the separation required (if more than one complex has to be mapped). The most promising conditions can then be applied in the subsequent preparative mobility-shift assay.

2. Binding competition analysis. If competitor DNA is identical to and in relatively large excess over the labeled DNA, >90% of the radioactive signal should be eliminated from complexes corresponding to protein(s) that interact with the binding-site DNA in a sequence-specific manner. Complexes unaffected or only slightly affected by the addition of competitor are thought to arise from nonspecific, low-affinity binding by abundant proteins that are present in excess to binding-site DNA; DNA derived from these complexes after the *in situ* chemical cleavage reactions can serve as a negative control in the footprinting analysis, because its cutting pattern will closely resemble that of the free (unbound) probe.

3. Assaying relative dissociation rates of DNA–protein complexes. To follow dissociation kinetics, complexes are allowed to form under optimal reaction conditions and, at time zero, exposed to a large excess of an agent that does not perturb their stability (commonly 100–250-fold mass excess of nonspecific competitor DNA or, preferably, of the same DNA fragment unlabeled). The "scavenger" molecules sequester the protein as it dissociates from its specific binding site, hence preventing it from rebinding. Analysis by the mobility-shift assay of aliquots at various times (from a few seconds to 2 h) after quenching the reaction with the sequestering agent, shows the amount of free DNA increasing while the level of protein-bound label diminishes. The experiment can be designed in a way that individual reactions can be started and quenched at different times, such that all reach the point at which they will be applied to the gel more or less simultaneously and electrophoresed for the same period of time. Since dissociation of typical DNA–protein complexes is a first-order process (i.e., independent of the concentration of complexes), the results of the analytical study are also applied to the subsequent preparative assay.

4. Trailing of the bands during electrophoresis. Glycerol-containing binding buffers tend to cause significant trailing at the edges of the bands during electrophoresis. If you noticed this trailing under the optimized conditions in the analytical assay, you may substitute glycerol with Ficoll (2.5% [w/v]; Type 400, Pharmacia) in the preparative binding buffer. The presence of Ficoll, although not interfering with the thermodynamic

and kinetic parameters of the binding reaction, gives rise to straight bands in the gel, thus minimizing the size of the polyacrylamide strips in the subsequent DNA–elution steps (Section 3.4.1.).
5. Loading the preparative polyacrylamide gel. If you have problems with the sample not sinking to the bottom of the well (which may be caused by substantial differences between the binding buffer and the electrophoresis buffer, and/or the large volume of the preparative reaction), you can preload the well(s) of the gel with binding buffer or, alternatively, load your sample with the power supply running at 10–15 mA (You must *take care* if you do this!).
6. You may try to footprint the complexes using a 1:1 ratio of 1,10-phenanthroline to copper, if your suspected protein-binding site(s) or the adjacent regions are particularly rich in 5'-CG-3' elements (*see* Section 1.). In this case, prepare the following solutions: 100 mM 3-mercaptopropionic acid, 5 mM 1,10-phenanthroline, and 5 mM CuSO$_4$.
7. Optimizing the time of *in situ* chemical treatment. The length of incubation period is determined by several factors, among which the following are of particular importance:
   a. Kinetic stability of the complex(es) (step 3 in Section 3.1.). In principle, the higher the dissociation rate of the complex(es), the lower the time of exposure to the chemical nuclease. However, because of the gel "caging effect" and the *in situ* (on the DNA surface) funneling of the reaction (*see* Section 1.), this rule of thumb is applicable only for DNA-protein complexes with half-lives either <<1 min or >>60 min; provided the complex is relatively abundant *(see below)*, incubation times <8 min and >30 min, respectively, should be used in these cases.
   b. Relative abundance of the complex(es). The aim of the reaction is to generate, on average, about one chemical cleavage event per DNA strand. Assuming that the cleavage process is governed by Poisson statistics, the product oligonucleotides will statistically be the result of a single cleavage (single-hit kinetics) when the concentration of the full-length labeled strand is ~50–70% of its original value (i.e., ~50–70% of the DNA fragments should be left uncleaved). Accordingly, when the abundance of the target complex(es) is very low (i.e., a bound DNA to free DNA ratio <<1), early termination of the reactions will be critical for high-"off"-rate complexes, beneficial for intermediate-stability complexes, and safe for extremely stable complexes.
   c. Temperature of incubation. If optimization of the DNA-binding reaction and, consequently, of the mobility-shift electrophoresis assay requires that both be performed at low temperature (4°C), you should

carry out the chemical nuclease treatment at low temperature as well. However, the amount of dissolved air oxygen in the reaction mixture under these conditions is considerably higher (increases with decreasing temperature); since molecular oxygen catalyzes a rate-limiting step that generates *in situ* hydrogen peroxide (an essential coreactant for the chemical nuclease activity; see Fig. 1), its presence in more than stoichiometric amounts will shift the subsequent equilibria toward the right side, leading to increased rates of DNA–strand scission. To compensate for this accelerated cleavage kinetics, you should decrease the time of exposure to the chemical nuclease. Inversely, incubation times should be longer than originally established (or you may add hydrogen peroxide exogenously) for chemical treatments performed at bench temperature during hot summer days in nonair-conditioned rooms.

d. Concentration of reagents. Increasing the concentration of 1,10-phenanthroline while holding the concentrations of copper (II) sulfate and 3-mercaptopropionic acid constant increases the overall rate of DNA–strand scission, without significantly affecting the sequence preferences of cleavage and the resulting fragment size distribution *(15)*. However, substituting 3-mercaptopropionic acid by the same concentration of ascorbic acid reduces the overall rate of DNA cleavage both at low (1:1) and high (>4.5:1) 1,10-phenanthroline/copper molar ratios *(15)*. These observations are of practical significance to estimating the time of chemical treatment, particularly when extremely labile or extraordinary stable complexes are being mapped. A higher concentration of 1,10-phenanthroline and a short incubation time might be used in the former case, whereas ascorbate (as the reducing agent) and prolonged treatments are recommended in the latter.

8. Phase inversion during organic extractions of the gel-, or membrane-eluted DNA samples. Because of the high-salt content of the elution buffers, the aqueous phase (which normally forms the upper layer) may sometimes be dense enough to form the lower layer. If this is the case, the aqueous phase can be easily identified by monitoring the eluted radioactivity with a Geiger counter or by following the strong yellow color of the organic phase (contributed by hydroxyquinoline that is added to phenol during equilibration as an antioxidant).

9. Excess acrylamide in the gel-eluted DNA samples. If the DNA pellet is highly contaminated with acrylamide monomers or other impurities from the gel matrix (a problem sometimes encountered when employing low-percentage mobility-shift gels and is usually apparent from the forma-

tion of an excessive turbidity during the ethanol-precipitation step), you can further purify it by passing through a preequilibrated Elutip™-d mini-column (an ion-exchange column) according to the manufacturer's instructions (or see Chapter 6 in vol. 4 of this series). Adsorption to and desorption from the column can be followed with a Geiger counter.

10. Appearance of the bands on the autoradiogram. If the band sharpness or shape in the chemical cleavage patterns on the autoradiogram suffers (e.g., narrowing of the bands toward the smallest end-labeled DNA fragments), the DNA samples contain residual proteins, salts, or acrylamide contaminants. These faults should be expected if uneven running and retardation of marker dyes is observed during electrophoresis, and can be avoided if careful attention is paid during steps 15-18 in Section 3.4.1. If necessary, the number of organic extractions (step 15 in Section 3.4.1.) and/or ethanol washes (step 18 in Section 3.4.1.) can be increased; acrylamide contaminants are efficiently removed by the use of Elutip™-d mini-columns (see Note 9). Note, however, that smeared or fuzzy bands without apparent electrophoretic problems may also arise from scattering during autoradiography; use a light-tight metal film cassette with a particularly effective closure mechanism, capable of firmly fixing filter/gel and film, and with intensifying screen in place (those produced by Picker International Health Care Products, Highland Heights, OH, are best in that).

11. Wondering about the footprint: scanning the autoradiogram. If the effects of protein binding on the cleavage rate of the chemical nuclease are not clearly discernible by eye, or when partial protection is obtained, you may analyze the ladders quantitatively by subtracting the cleavage pattern of the DNA derived from the bound fraction from that derived from the free fraction. This involves calculating the probability of cleavage at each bond (which is related to the amount of radioactivity, or intensity, in the corresponding band in the cutting pattern) and, finally, for each lane, the average number of cuts in the DNA strand, with the aid of automated laser densitometers linked to a computer (available from, for instance, Bio-Rad, LKB, or Pharmacia). Because quantification of band intensity is limited within the linear response of Kodak X-Omat AR films to radioactivity (bands with an optical density >0.15 and <1.8 absorbance units), it is essential not to use overexposed films for this type of analysis. Moreover, the film should be free of scratches, fingerprints, and dirt that will appear as optical signals indistinguishable from ^{32}P, thus interfering with the calculations. A difference probability plot along the entire length of the binding site(s) and surrounding

DNA can be obtained, revealing the protein-binding site(s) much more clearly and reliably than can be done by comparison by eye of the chemical cleavage patterns. If still in doubt, however, you may analyze the footprint on the other DNA strand.

Acknowledgments

This work has profited greatly from discussions with those who attended the 1991 EMBO Course on "DNA-Protein Interactions." It is dedicated to my sons George and Kostas.

Further Reading

Sigman, D. S. and Chen, C. B. (1990) Chemical nucleases: new reagents in molecular biology, in *Annu. Rev. Biochem.* vol. 59, Annual Reviews Inc., Palo Alto, CA.

References

1. Schmitz, A. and Galas, D. G. (1979) The interaction of RNA polymerase and *lac* repressor with the *lac* control region. *Nucleic Acids Res.* **6,** 111–137.
2. Humayun, Z., Kleid, D., and Ptashne, M. (1977) Sites of contact between λ operators and λ repressor. *Nucleic Acids Res.* **4,** 1595–1607.
3. Garner, M. M. and Revzin, A. (1981) A gel electrophoresis method for quantifying the binding of proteins to specific DNA regions: application to components of the *Escherichia coli* lactose operon regulatory system. *Nucleic Acids Res.* **9,** 3047–3060.
4. Fried, M. and Crothers, D. M. (1981) Equilibria and kinetics of *lac* repressor-operator interactions by polyacrylamide gel electrophoresis. *Nucleic Acids Res.* **9,** 6505–6525.
5. Topol, J., Ruden, D. M., and Parker, C. S. (1985) Sequences required for in vitro transcriptional activation of a Drosophila *hsp 70* gene. *Cell* **42,** 527–537.
6. Kuwabara, M. D. and Sigman, D. S. (1987) Footprinting DNA-protein complexes *in situ* following gel retardation assays using 1,10-phenanthroline-copper ion: *Escherichia coli* RNA polymerase-*lac* promoter complexes. *Biochemistry* **26,** 7234–7238.
7. Sigman, D. S., Graham, D. R., D'Aurora, V., and Stern, A. M. (1979) Oxygen-dependent cleavage of DNA by the 1,10-phenanthroline-cuprous complex. Inhibition of *E. coli* DNA polymerase I. *J. Biol. Chem.* **254,** 12,269–12,272.
8. Marshall, L. E., Graham, D. R., Reich, K. A., and Sigman, D. S. (1981) Cleavage of DNA by the 1,10-phenanthroline-cuprous complex. Hydrogen peroxide requirement: primary and secondary structure specificity. *Biochemistry* **20,** 244–250.
9. Sigman, D. S. (1990) Chemical nucleases. *Biochemistry* **29,** 9097–9105.
10. Pope, L. M., Reich, K. A., Graham, D. R., and Sigman, D. S. (1982) Products of DNA cleavage by the 1,10-phenanthroline copper complex. Identification of *E. coli* DNA polymerase I inhibitors. *J. Biol. Chem.* **257,** 12,121–12,128.

11. Tamilarasan, R., McMillin, D. R., and Liu, F. (1989) Excited-state modalities for studying the binding of copper phenanthrolines to DNA, in *Metal-DNA Chemistry* (Tullius, T. D., ed.), American Chemical Society Symposium Series **402**, pp. 48–58, Washington, DC.
12. Kakkis, E. and Calame, K. (1987) A plasmacytoma-specific factor binds the c-*myc* promoter region. *Proc. Natl. Acad. Sci. USA* **84**, 7031–7035.
13. Flanagan, W. M., Papavassiliou, A. G., Rice, M., Hecht, L. B., Silverstein, S., and Wagner, E. K. (1991) Analysis of the Herpes Simplex Virus type 1 promoter controlling the expression of U_L38, a true late gene involved in capsid assembly. *J. Virol.* **65**, 769–786.
14. Veal, J. M. and Rill, R. L. (1988) Sequence specificity of DNA cleavage by *bis*(1,10-phenanthroline)copper(I). *Biochemistry* **27**, 1822–1827.
15. Veal, J. M., Merchant, K., and Rill, R. L. (1991) The influence of reducing agent and 1,10-phenanthroline concentration on DNA cleavage by phenanthroline + copper. *Nucleic Acids Res.* **19**, 3383–3388.
16. Papavassiliou, A. G. and Silverstein, S. J. (1990) Interaction of cell and virus proteins with DNA sequences encompassing the promoter/regulatory and leader regions of the Herpes Simplex Virus thymidine kinase gene. *J. Biol. Chem.* **265**, 9402–9412.
17. Del Angel, R. M., Papavassiliou, A. G., Fernandez-Tomas, C., Silverstein, S. J., and Racaniello, V. R. (1989) Cell proteins bind to multiple sites within the 5' untranslated region of poliovirus RNA. *Proc. Natl. Acad. Sci. USA* **86**, 8299–8303.
18. Laskey, R. A. (1980) The use of intensifying screens or organic scintillators for visualizing radioactive molecules resolved by gel electrophoresis, in *Methods in Enzymology,* vol. 65 (Grossman, L. and Moldave, K., eds.), Academic, New York, pp. 363–371.

CHAPTER 6

Identification of Protein–DNA Contacts with Dimethyl Sulfate

Methylation Protection and Methylation Interference

Peter E. Shaw and A. Francis Stewart

1. Introduction

Dimethyl sulfate (DMS) is an effective and widely used probe for sequence-specific protein–DNA interactions. It is the only probe routinely used both for in vitro (methylation protection, methylation interference) and in vivo (DMS genomic footprinting) applications since it rapidly reacts with DNA at room temperature and readily penetrates intact cells *(1)*. DMS predominantly methylates the 7-nitrogen of guanine and 3-nitrogen of adenine. Thus, reactivity with G residues occurs in the major groove and with A residues in the minor groove. In standard Maxam and Gilbert protocols *(2)*, the methylated bases are subsequently converted to strand breaks and displayed on sequencing gels.

Methylation protection and interference are essentially combinations of the gel retardation assay *(3,4)* with the DMS reaction of the Maxam and Gilbert sequencing procedure. Protein–DNA interactions are reflected either as changes in DMS reactivities caused by bound protein (methylation protection) or as selective protein binding dictated by methylation (methylation interference).

In methylation protection, protein is first bound to DNA that is uniquely end-labeled, and the complex is reacted with DMS. DMS reactivities of specific residues are altered by bound protein either by exclusion, resulting in reduced methylation, increased local hydrophobicity, resulting in enhanced methylation, or by local DNA conformational changes, such as unwinding, resulting in altered reactivity profiles *(5–7)*. After the DMS reaction, free DNA is separated from protein-bound DNA by gel retardation, and both DNA fractions are recovered from the gel. Methylated residues are converted into strand scissions, and the free and bound DNA fractions are compared on a sequencing gel. A complete analysis requires examination of both strands. This is accomplished by preparing two DNA probes, each uniquely labeled at one end, and carrying both probes through the protocols. A binding site characterized by methylation protection will therefore appear as a cluster of altered DMS reactivities.

In methylation interference, DNA is first reacted with DMS, purified, and then presented to protein *(8,9)*. Under the reaction conditions used, methylation is partial, yielding approximately one methylated base per DNA molecule. Thus, the protein is presented with a mixture of DNA molecules that differ with respect to the positions of methyl groups. Some methyl groups will interfere with protein binding since they lie in or near the binding site. Gel retardation scparates the mixture into two fractions: free DNA, which, as long as DNA is in excess over binding activity, represents the total profile of methylation reactivity, and bound DNA, which will not contain any molecules with methyl groups incompatible with binding. Both free and bound DNA fractions are recovered, methylated residues are converted to strand scissions, and the fractions are compared on a sequencing gel. The binding site is observed as the absence of bands in the bound sample corresponding to the positions where methylation interferes with binding.

These two uses of DMS may not deliver identical results. For example, Fig. 1 presents a comparison obtained from experiments with the serum response element binding factors $p67^{SRF}$ $p62^{TCF}$ and their binding site in the human c-*fos* promoter (SRE). Since the use of DMS in vivo for genomic footprinting is limited to the equivalent of methylation protection, a direct comparison between in vivo and in vitro footprints requires the use of methylation protection rather than the more widely used methylation interference assay.

These two techniques are, however, very similar in practical terms and, thus, are presented together. Both techniques rely on preestablished conditions that permit a protein–DNA complex to be resolved

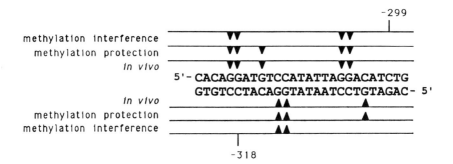

Fig. 1. Comparison of methylation interference and protection patterns formed by factors binding at the c-fos serum response element (SRE) in vitro and in vivo. G residues identified by methylation interference *(9)*, methylation protection, and in vivo genomic footprinting *(7)* are indicated. An additional G on both the upper and lower strands is implicated in the protein-DNA interaction by methylation protection.

in a gel retardation assay (the subject of Chapter 21) and on routinely used DNA sequencing methodologies, for which the reader will find a detailed treatment *(2)*.

2. Materials

1. Dimethyl sulfate (Merck, Rahway, NJ), analytical grade.
2. Piperidine (Sigma, St. Louis, MO), analytical grade: Use freshly made 1:10 dilution in double-distilled water.
3. Phenol/chloroform: 50% v/v, buffered with 50 mM Tris-HCl, pH 8.0.
4. NA45 paper (Schleicher and Schuell, Dassel, Germany).
5. Whatman 3MM paper or GB 002 paper (Schleicher and Schuell).
6. Gel retardation equipment (*see* Chapter 21).
7. Electro-blotting apparatus for Western transfer (e.g., Trans-Blot, Bio-Rad, Richmond, CA).
8. Standard DNA sequencing gel electrophoresis equipment.
9. Vacuum gel dryer (optional).
10. TBE buffer: 108 g Tris base, 55 g boric acid, 40 mL 0.5M EDTA, pH 8.3/L for 10X stock; working solution is 89 mM Tris base, 89 mM boric acid, 2 mM EDTA.
11. NA45 elution buffer: 10 mM Tris-HCl, pH 8.0, 1 mM EDTA, 1M NaCl.
12. Carrier DNA: Salmon testis DNA or calf thymus DNA, sheared (Sigma; dissolved at 3 mg/mL in 10 mM Tris-HCl, pH 8.0, 1 mM EDTA).
13. Sequencing gel loading buffer: 90% formamide, 10 mM EDTA, 0.1% w/v bromophenol blue, 0.1% w/v xylene cyanol blue.

14. Gel retardation loading buffer: 20% Ficoll, 20 mM EDTA, 0.1% w/v bromophenol blue.
15. 2X DMS buffer: 120 mM NaCl, 20 mM Tris-HCl, pH 8.0, 20 mM $MgCl_2$, 2 mM EDTA.
16. DMS stop buffer: 1.5M sodium acetate, pH 7.0, 1M 2-mercaptoethanol; store frozen.
17. X-ray film (e.g., Kodak X-Omat, Rochester, NY).
18. 0.25M Dithiothreitol (DTT).
19. Sequencing gel stock A: 1X TBE, 14.25% w/v acrylamide (Bio-Rad), 0.75% w/v *bis*-acrylamide, 8M urea. Sequencing gel stock B: As A except omit acrylamide and *bis*-acrylamide. Store in brown glass at 4°C.
20. 10% w/v ammonium persulfate made weekly.
21. *N,N,N',N'*-Tetramethylethylenediamine (TEMED; Bio-Rad).
22. 5% v/v dimethyldichlorosilane (Sigma) in chloroform.

3. Methods
3.1. Methylation Protection

1. Incubate 300,000 cpm of uniquely end-labeled DNA probe and a corresponding amount of protein together, under conditions previously optimized by gel retardation analysis (*see* Chapter 21 and Note 1), in a total vol of 100 µL.
2. Add 1 µL of DMS and incubate at room temperature (the incubation time depends on the length of the DNA probe and is empirical: As a guide 200-bp fragment $t = 1.5$ min; for 50-bp oligo duplex $t = 3.0$ min).
3. Add $1/_{10}$ vol of 250 mM DTT, mix gently, add 1/10 vol of gel retardation loading buffer, mix gently, load onto a 2-mm thick retardation gel in 1X TBE (or alternative buffer), and run as optimized for analytical retardation gels. However, the load should be spread over up to ten times more well area (*see* Note 2).
4. After electrophoresis, separate the glass plates carefully so that the gel adheres to one plate and cover the gel with clingwrap. Expose to X-ray film long enough to reveal complexes clearly (i.e., a few hours to overnight). The alignment of the film to the gel must be reliably marked.
5. Put the developed film on a lightbox. Remove the clingwrap from the gel and align the gel on the X-ray film. Cut pieces of NA45 paper sufficiently large to cover individual complexes in the gel, yet small enough to fit into 1.5-mL tubes when rolled up. Wet the paper pieces in retardation gel running buffer and, with the help of tweezers, position one over each complex of interest in the gel, as visualized from the underlying film. Also position a similar sized piece of paper over some of the uncomplexed DNA. NA45 paper can be labeled with pencil before wetting.

6. Carefully cover the gel and paper pieces with two sheets of 3MM paper wetted in 1X TBE (or alternative gel running buffer from step 3). Lay a scotchbrite pad from the electro-blotting apparatus on top of the paper and turn the gel over. Carefully remove the second glass plate, cover the other side of the gel with 3MM paper, scotchbrite as before, and insert the package into an electro-transfer apparatus as described in the manufacturer's instructions with the NA45 paper toward the anode. Transfer in 1X TBE (or alternative buffer) at 80 V for 1.5 h (*see* Note 3).
7. Stop the transfer, unpack the gel carefully with the NA45 paper on top, and transfer each piece to a labeled 1.5-mL tube containing 600 µL of elution buffer (check that the radioactivity has transferred to the paper using a hand monitor). Incubate at 70°C for 1 h.
8. Remove NA45 paper from each tube, check that at least half the radioactivity has eluted into the buffer (do not expect quantitative elution but at least 50% should come off; *see* Note 4), add 20 µg of carrier DNA, extract with 600 µL phenol/chloroform, and precipitate with 1 vol of isopropanol (*see* Note 5). Wash the precipitate once in 70% ethanol and dry under vacuum.
9. Dilute piperidine 1:10 in water and add 50 µL to each pellet (*see* Note 6). Vortex briefly, incubate at 90°C for 30 min (tubes must be clamped or weighted down to prevent the lids from opening), and then dry under vacuum. Take up the samples in 100 µL of water and repeat the drying process.
10. Measure the Cerenkov counts in each tube; then redissolve the samples in water (e.g., 10 cpm/µL) and transfer equivalent counts (1000 cpm is optimal; *see* Note 7) from each into fresh tubes. Dry down and redissolve the cleaved DNA in 5 µL sequencing gel loading buffer (*see* Note 8).
11. Prepare a standard sequencing gel. The percentage of the gel will depend on the DNA probe length, that is, for 30 mers or less, 10–15% gels should be used. For longer probes, lower percentage gels should be used (*see* Note 9). A standard sequencing gel is poured between two 40 × 20-cm glass plates separated by 0.4-mm spacers. One of the glass plates is cut to produce a 2-cm deep, 16-cm long U-shaped recess that leaves 2-cm wide ears on either side. This edge will form the top of the gel assembly. Clean the glass plates thoroughly, including a final wipe with ethanol, and then wipe one of the plates with 5% dimethylchlorosilane. This step must be performed in a fume hood and secure gloving must be worn. When this plate is dry, polish it with tissues (*see* Note 10). Seal the assembly of plates and spacers with waterproof tape along the two sides and bottom, and then clamp with bulldog clips. Make sure that the slot former fits in snugly. Mix sequencing gel stocks A and B to

produce 60 mL of the desired acrylamide concentration and warm up to room temperature. Add 60 µL TEMED, mix, then add 600 µL 10% ammonium persulfate, and pour immediately. Pouring is aided by tilting the gel assembly so that the solution is poured into the corner of the U-shaped recess and from this reservoir flows down one edge. Make sure that no air bubbles are trapped during pouring. If air bubbles are present, stand the assembly vertically and gently tap the glass next to the bubble. When the assembly is full, lay it down and insert the slot former, again ensuring that no air bubbles are trapped. Place clamps over the top end of the assembly so that the glass plates are squeezed onto the slot former. Polymerization of the remaining unpoured gel mix should occur within 5 min. Allow the gel at least 1 h to polymerize thoroughly.
12. After removing the sealing tape, clamp the gel assembly into a gel tank. Remove the slot former, top up the reservoirs with the running buffer (1X TBE), and clean the wells. Preelectrophorese at approx 50 W or 40 V/cm (*see* Note 11). Denature probes at 95°C for 5 min, snap cool in ice, and load onto the gel. Run the gel until optimal separation of sequence is achieved.
13. Stop electrophoresis, remove the gel from the tank, and lift off the glass plate that was silanized. Place the other glass plate with gel into 20% ethanol, 10% acetic acid for 10 min. Drain briefly; then overlay the gel with two sheets of 3MM paper and carefully peel it off the glass plate. Cover the gel surface with cling wrap and dry on a vacuum gel dryer (*see* Note 12). Expose the dry gel to X-ray film with intensifying screens as necessary (*see* Note 13).

3.2. Methylation Interference

1. Mix 300,000 cpm of end-labeled probe, 100 µL of 2X DMS buffer, and water to 200 µL. Add 2 µL of DMS and incubate at room temperature (the same guidelines as given in Section 3.1.2. apply for the reaction time). Stop the reaction by the addition of 50 µL cold DMS stop mix and precipitate with 850 µL cold ethanol. Redissolve the probe in 200 µL cold 0.3M sodium acetate, pH 7.0; add 700 µL cold ethanol and reprecipitate. Wash the precipitate twice in 80% ethanol, dry, and redissolve the probe in water or binding buffer to give about 20,000 cpm/µL.
2. Incubate the probe with protein for gel retardation as previously optimized for gel retardation analyses of the complexes in question (*see* Chapter 21) in a total vol of 100 µL.
3. Add $1/10$ vol of gel retardation loading buffer, mix gently, load onto a 2-mm thick retardation gel in 1X TBE (or alternative buffer), and run as optimized for analytical gels. However, the load should be spread over up to ten times more well area (*see* Note 2).

4. Continue with step 4 and all subsequent steps as described for methylation protection (*see* Section 3.1.).

4. Notes

1. To have sufficient counts to complete the procedure, proceed with at least 10 times the amount of material required for a simple gel retardation analysis (i.e., at least 300,000 cpm).
2. A common difficulty with these methods is the persistence of contaminants that accompany DNA after the preparative retardation gel. These contaminants interfere with the migration of DNA on the sequencing gel, producing blurred and distorted patterns. In order to minimize this problem, it is worth ascertaining the load limit of the retardation gel so that the protein–DNA complex will not smear, but will be well resolved and therefore concentrated in the gel before elution.
3. It is also possible to use a semidry electro-transfer apparatus (e.g., Bio-Rad Trans-blot SD) to transfer the DNA from the gel retardation gel onto NA45 paper. In this case, the transfer time and potential are both reduced.
4. In some instances, it may prove difficult to elute the DNA from the NA45 paper, in which case raising the salt concentration or the temperature may improve elution. (Extending the incubation time does not seem to help.) If not, the batch of NA45 may be to blame, or it is even conceivable that the DNA–protein complex in question is adsorbed too tightly onto the paper. It is not possible to phenol extract the NA45 paper in order to remove bound protein–DNA.
5. Retain the isopropanol supernatants until you are sure the samples have precipitated quantitatively. Add more carrier DNA if required.
6. The strand scission protocol described here should not convert methylated A residues into strand breaks. It is often observed, however, that breakages at As occur with reasonable efficiency. The following modification will produce efficient cleavage at both G and A residues. After the preparative retardation gel, resuspend the dried, purified DNA in 30 µL 10 mM sodium phosphate, pH 6.8, 1 mM EDTA. Incubate for 15 min at 92°C. Then add 3 µL 1M NaOH and incubate for 30 min at 92°C, followed by 320 µL 500 mM NaCl, 50 µg/mL carrier DNA, and 900 µL ethanol. Chill and centrifuge to pellet the radioactivity. Wash the pellet in 70% ethanol and dry. Proceed as above from step 10, Section 3.1.
7. As discussed in Note 2 above, it is similarly advisable to load as little material onto the sequencing gel as practicable. The practical lower limit for the sequencing gel is approx 1000 cpm/lane, but desperate individuals know no bounds.
8. Bromophenol blue and, to a lesser extent, xylene cyanol blue have been observed to cause exclusion distortions in the sequencing gels, and

either one can be excluded from the sequencing loading buffer if this occurs in a critical part of the gel. When dye(s) are omitted from the loading buffer, load the dye-containing loading buffer in a spare lane to act as a tracker.
9. The choice of percentage of the sequencing gel depends on the distance between the radioactive label and the site of the protein–DNA interaction. If the end-labeled DNA fragment is relatively long and multiple binding sites are to be resolved, a gradient or wedge sequencing gel can be used in step 11 of Section 3.1.
10. The polished, silanized glass plate can be baked in a 180°C oven. This fixes the silanization through repeated use so that the plate need not be silanized every time. It is better to silanize the plate with the U-shaped cut.
11. The exact settings for electrophoresis of sequencing gels vary greatly and should be determined empirically by assessing the temperature of the glass plates after running for about 1 h. Sequencing gels should be run so that their temperature remains stable between 45 and 55°C.
12. It is not essential to dry down the sequencing gel because, after one glass plate has been removed, it can be covered with cling wrap and exposed to X-ray film at –70°C with one screen. This alternative should only be considered if the signal is sufficiently strong or if a gel dryer is not available.
13. An appropriate complement for the final result is to perform the Maxam and Gilbert G + A reactions *(2)* on the end-labeled probe. On the sequencing gel, these reactions should provide unambiguous sequence information and, in case difficulties are encountered, clues as to the steps that are problematic.

References

1. Church, G. M. and Gilbert, W. (1984) Genomic sequencing. *Proc. Natl. Acad. Sci. USA* **81,** 1991–1995.
2. Maxam, A. and Gilbert, W. (1980) Sequencing end-labeled DNA with base-specific chemical cleavages. *Meth. Enzymol.* **65,** 499–560.
3. Fried, A. and Crothers, D. M. (1981) Equilibria and kinetics of lac repressor-operator interactions by polyacrylamide gel electrophoresis. *Nucleic Acids Res.* **9,** 6505–6525.
4. Garner, M. M. and Revzin, A. (1981) A gel electrophoresis method for quantifying the binding of protein to specific DNA regions: application to components of the *E. coli* lactose operon regulatory system. *Nucleic Acids Res.* **9,** 3047–3059.
5. Gilbert, W., Maxam, A., and Mirzabekov, A. (1976) Contacts between the LAC repressor and DNA revealed by methylation, in *Control of Ribosome Biosynthesis, Alfred Benzon Symposium IX* (Kjelgaard, N. O. and Maaloe, O., eds.), Academic, New York, pp. 139–148.

6. Johnsrud, L. (1978) Contacts between *Escherichia coli* RNA polymerase and a lac operon promoter. *Proc. Natl. Acad. Sci. USA* **75,** 5314–5318.
7. Herrera, R. E., Shaw, P. E., and Nordheim, A. (1989) Occupation of the c-fos serum response element in vivo by a multi-protein complex is unaltered by growth factor induction. *Nature* **340,** 68–70.
8. Siebenlist, U. and Gilbert, W. (1980) Contacts between *E. coli* RNA polymerase and an early promoter of phage T7. *Proc. Natl. Acad. Sci. USA* **77,** 122–126.
9. Shaw, P. E., Schröter, H., and Nordheim, A. (1989) The ability of a ternary complex to form over the serum response element correlates with serum inducibility of the human c-fos promoter. *Cell* **56,** 563–572.

CHAPTER 7

Diethyl Pyrocarbonate as a Probe of Protein–DNA Interactions

Michael J. McLean

1. Introduction

A rigorous study of the interactions between DNA and proteins or other molecules requires the accumulation of a wide range of information regarding the nature of such an interaction. Among the parameters on which it is desirable to gain information are: the size and number of the binding site(s) on the nucleic acid in question, the sequence specificities of the binding events, the strength of the interaction, the biochemical effects of the binding, and the structural consequences of such an interaction on all the molecules concerned.

Methods to obtain insight into most of the above parameters will be dealt with in other chapters of this volume. This chapter will focus on the use of a small molecule, diethyl pyrocarbonate (DEPC), to detect structural changes in a DNA as a consequence of its interaction with other molecules.

DEPC ($C_2H_5OCOCOOC_2H_5$) is a (relatively) water insoluble, electrophilic reagent that reacts with nucleophilic centers in nucleic acids. Although there is a weak reaction with cytosine residues, the chief reaction is with the purine nucleosides, which leads to scission of the glycosidic bond, generating an alkali-labile abasic site. The precise site of reaction of DEPC with purines (and thus the nature of any structural deformation detected using it) has been the subject of some controversy over recent years. In the early 1970s, Leonard and associates *(1,2)* showed that in free adenine, DEPC reacts at the N-1 and

From: *Methods in Molecular Biology, Vol. 30: DNA–Protein Interactions: Principles and Protocols*
Edited by: G. G. Kneale Copyright ©1994 Humana Press Inc., Totowa, NJ

N-3 positions, whereas N-9 substituted adenines react at N-7 and the exocyclic N-6. Thus, in deoxyadenosine, the most reactive center would appear to be the imidazole ring N-7. However, these investigations were performed under conditions in which the three remaining ring nitrogens were all equally accessible; consequently they do not afford reliable information as to the preferred reaction site(s) when one or more of the nitrogen atoms are impeded from reaction by steric blockage and/or hydrogen bonding. Thus, it was proposed *(3)* that in a Hoogsteen *(4)* base pairing configuration (in which the purine is in the syn conformation, N-7 is used in hydrogen bonding, and N-1 is exposed in the major groove), DEPC reaction with N-1 of purines would lead to chain cleavage.

In further experiments with 7-deaza-adenosine, it was shown that although reaction of DEPC does take place at N-1 or N-3, this does not lead to cleavage of the glycosidic bond, and thus chain scission does not occur. In "normal" B-form DNA, N-7 of purines, although projecting into the major groove, is relatively inaccessible to DEPC because of the stacking of neighboring bases. When this stacking is perturbed, reaction can take place, and this enhanced reactivity to DEPC has been used to map regions of altered secondary structure in naked RNA *(5)* and DNA *(6–8),* and in DNA–drug complexes *(9–12).*

Because "normal" DNA is relatively refractory to reaction with DEPC, this reagent is of little use in obtaining "footprints" of DNA–protein interactions since these are normally obtained from an inhibition of reaction/cleavage, thereby providing an estimate of binding site size. In the unusual case of Z-DNA, where purine N-7s are hyperreactive to DEPC, a diminution in reactivity can be used to measure specific protein binding *(13).*

This chapter will exemplify ways in which DEPC can be used to provide an indication of the structural perturbance caused by binding of agents to DNA. For the sake of simplicity, the example described herein was performed using echinomycin as the binding agent, but, given the appropriate buffer conditions, this approach can in principle be extended to work with any molecule interactive with DNA.

It should be stressed that any structural investigation relying solely on the evidence of one probing agent is dangerously blinkered and may give misleading results. Diethyl pyrocarbonate is no exception to this, and the most meaningful results are obtained when it is used in conjunction with other chemical and enzymatic probes, such as those discussed in other chapters of this volume.

However, DEPC does present at least two significant advantages over the use of most enzymatic and some chemical probes because most enzymes require stringent reaction conditions for their activity, whereas DEPC can be used over a range of conditions of pH, temperature, salt concentrations, and the like, enabling a study of the effects of environmental factors on protein–DNA interactions. In addition, DEPC does not cleave DNA by itself—it requires a subsequent treatment with base to effect cleavage, thus enabling a study of the effects of superhelical density on protein–DNA interactions.

2. Materials

1. Diethyl pyrocarbonate can be obtained from Sigma (St. Louis, MO), and should be stored at 4°C. Piperidine can be obtained from BDH (London, England) and should also be stored at 4°C. Both compounds are toxic and volatile, and should be handled with caution in a fume hood.
2. Eppendorf tubes should be silanized by briefly immersing the opened tubes in a beaker containing a 2% solution of dimethyldichlorosilane in 1,1,1-trichloroethane (BDH), pouring off excess solution and allowing the tubes to dry in air at room temperature in a fume hood.
3. The following stock solutions will be needed, and can be prepared in advance, autoclaved, and stored at 4°C until use:
   a. $1M$ Tris-HCl, pH 8.0.
   b. $1M$ Tris-HCl, pH 7.4.
   c. $0.5M$ EDTA, pH 8.3.
   d. $5M$ NaCl.
   e. $3M$ Sodium acetate.
   f. 10 mM Tris-HCl, 0.1 mM EDTA, pH 8.0.
   g. 500 mM NaCl, 100 mM Tris-HCl, pH 7.4.
   h. 0.1% Bromophenol blue, 0.1% xylene cyanol, 10 µM EDTA in 80% formamide. All the reagents necessary to make the above solutions are available from Sigma.
4. $0.3M$ Solutions of sodium acetate should be prepared fresh immediately prior to use. Over an extended period of time, this solution develops an extremely potent endonuclease activity—presumably because of the growth of an organism; $3M$ sodium acetate does not appear to sustain the growth of this life form. All water used to make solutions should be of the purest grade possible.

3. Methods

3.1. DEPC Experiments on Linearized Fragments

1. End-labeled DNA (either 5' or 3'), purified free of excess radiolabel (by electrophoresis/elution or by spin column), should be redissolved in 10 mM Tris-HCl, 0.1 mM EDTA, pH 8.0. An optimal concentration of DNA is between 3 and 10 pmol/µL and with an activity such that 1 µL of solution will register approx 20 cps on a Geiger counter (thus giving strong, clear bands on an autoradiograph after an overnight exposure).
2. Into each of two siliconized 1.5-mL Eppendorf tubes, carefully pipet 39 µL of water, 5 µL of an appropriate 10X buffer (e.g., 500 mM NaCl, 100 mM Tris-HCl, pH 7.4), and 1 µL of the solution of end-labeled DNA. Into one of the tubes, pipet 5 µL of a 125-µM solution of echinomycin in methanol. Into the other tube, pipet 5 µL of methanol. Mix the contents of the tubes by vortexing briefly and then incubate the tubes at 37°C for 15 min. (*See* Notes 1 and 2 for further suggested control experiments.) At the end of this period, place the tubes on ice and transfer the ice bucket to a fume hood.
3. Remove the DEPC from the refrigerator and transfer it to the fume hood. There may be a release of pressure on opening the bottle. Carefully pipet 1 µL of DEPC into each of the two Eppendorf tubes and cap them immediately. Return the DEPC to the refrigerator. Mix the contents of the tubes by tapping the side of the tube gently several times. This should be repeated every few minutes (*see* Note 3).
4. After 15 min, add 175 µL of water, 25 µL of 3M sodium acetate containing 5 mM tRNA (or sonicated calf thymus DNA), and 750 µL of ethanol to each tube. Mix the contents of the tubes by inverting them several times and incubate them at –70°C (preferably in a dry ice/acetone bath) for 15 min. Spin the tubes in an Eppendorf centrifuge for 15 min with the hinge of the tube perpendicular to the rim of the rotor. Carefully remove the tubes from the rotor and gently remove the supernatant with a Pasteur pipet held just under the surface of the liquid and held against the wall of the tube opposite the hinge. Resuspend the pellet in 250 µL of 0.3M sodium acetate and add 750 µL of ethanol. Mix as before and repeat the –70°C incubation/centrifugation procedure.
5. Prepare a 1 in 10 dilution of piperidine in water. To each of the two tubes add 100 µL of this solution, cap the tubes and place in a heating block at 90°C with a weight on the tubes to prevent the caps from opening. After 15 min, incubate the tubes briefly at –70°C (until the liquid is frozen), open the caps, and lyophilize the solutions. When all of the

liquid is apparently gone, add 10 µL of water to each tube and lyophilize, then add another 10 µL of water and lyophilize.

Resuspend the pellets in 2.5 µL of 80% formamide containing 0.1% bromophenol blue, 0.1% xylene cyanol, and 10 µM EDTA. The DNAs are now ready for analysis on sequencing gels by standard methods *(14)*. *See* Notes 4–7 for advice on the use of appropriate size markers and on interpretation of the results.

3.2. DEPC Experiments on Supercoiled DNAs

Since DEPC does not directly cleave DNA by its reaction but requires subsequent alkaline hydrolysis of abasic sites, it lends itself very well to an analysis of the effects of supercoiling of the DNA on the protein- or drug–DNA interaction.

1. To perform such an experiment, replace the end-labeled DNA in Section 3.1. with supercoiled, unlabeled DNA and perform steps 2–4 as before. In this procedure, exogenous DNA should not be used as a carrier.
2. Redissolve the pellets from step 4 of Section 3.1. in 50 µL of a 1X solution of a suitable restriction enzyme buffer and digest the DNA with appropriate restriction enzymes to liberate the fragment to be analyzed. End-label the fragment on one of the strands, and separate this from the main portion of the plasmid by gel electrophoresis, size exclusion chromatography, or some other suitable means.
3. The DNA thus obtained should be recovered by precipitation from a 70% ethanol solution, and then resuspended in 100 µL of 1M piperidine and treated exactly as in step 5 of Section 3.1.

4. Notes

1. Native DNA sometimes shows reproducible, enhanced reactivity to DEPC at defined sequences in the absence of any binding event *(10)*. It is therefore necessary to perform a control experiment as described herein, in which this reactivity (or lack of it) is identified.
2. Since DEPC also reacts with proteins, it is advisable to perform an additional control experiment in which the protein is incubated alone with DEPC. Excess, unreacted DEPC may be extracted into diethyl ether (the upper layer in an ether/aqueous mix). The DEPC-treated protein should then be incubated with DNA and some other functional test (e.g., DNase I footprinting; *see* Chapter 1) performed to ensure the presence of functional protein.
3. Sometimes the top of the Eppendorf tubes may pop open during the incubation with DEPC. This is caused by the production of carbon dioxide and is not a serious worry. Simply recap the tube.

4. Purines that are rendered hypersensitive to reaction with DEPC as a consequence of drug or protein binding to the DNA are visualized as a dark band on autoradiography of the gel, owing to cleavage of the DNA at the point where reaction has taken place. On its own, this information is of little use to the investigator since it does not reveal the nature of the sequence that has undergone reaction. Therefore, DEPC-treated samples should always be run adjacent to samples that have been sequenced according to Maxam and Gilbert method (*see* ref. *14* for a simple protocol), thus enabling one to determine the precise site of reaction at the base pair level.
5. The intensities of the bands (i.e., the darkness) mentioned above reflect the extent of reaction of a particular residue with DEPC and hence a measure of the distortion of the DNA duplex. In this context, it is worth noting that some DNA sequences appear to be inherently more deformable than others (and/or may be preferential binding sites for the ligand in question), and thus reactivity may be observed at lower relative concentrations of ligand at some sites than at others. Thus, to obtain the most information possible from one experiment, it is advisable to perform the DEPC experiment over a wide concentration range (3–4 orders of magnitude) of the ligand (*see* ref. *10* for an example).
6. Since DEPC reacts preferentially with purines, it is perfectly conceivable that a different pattern of reactivities will be observed with either strand of a given duplex. Therefore, to obtain the maximum information, it is advisable to repeat the experiment with the complementary strand radiolabeled.
7. Because the DNA is labeled at one of its termini, and because the DEPC reaction ultimately results in chain scission, it is imperative that no single DNA molecule reacts more than once with DEPC. If this occurs, only the shorter of the two reaction products will be observed on autoradiography since this is the fragment that bears the label. This will inevitably give rise to misleading results. In practical terms, this means that no more than one-third of the DNA should be allowed to react, that is, the majority of the radioactivity should be present as full-length material. If it is not, then the experiment should be repeated using less DEPC/more DNA, or a shorter reaction time.

References

1. Leonard, N. J., McDonald, J. J., and Reichmann, M. E. (1970) Reaction of diethyl pyrocarbonate with nucleic acid components, I: Adenine. *Proc. Natl. Acad. Sci. USA* **67**, 93–98.
2. Leonard, N. J., McDonald, J. J., Henderson, R. E. L., and Reichmann, M. E. (1971) Reaction of diethyl pyrocarbonate with nucleic acid components: adenosine. *Biochemistry* **10**, 3335–3342.

3. Mendel, D. and Dervan, P. B. (1987) Hoogsteen base pairs proximal and distal to echinomycin binding sites on DNA. *Proc. Natl. Acad. Sci. USA* **84,** 910–914.
4. Hoogsteen, K. (1959) The structure of crystals containing a hydrogen-bonded complex of 1-methylthymine and 9-methyladenine. *Acta Crystallogr.* **12,** 822,823.
5. Peattie, D. A. and Gilbert, W. (1980) Chemical probes for higher order structure in RNA. *Proc. Natl. Acad. Sci. USA* **77,** 4679–4682.
6. Herr, W. (1985) Diethylpyrocarbonate: a chemical probe for secondary structure in negatively supercoiled DNA. *Proc. Natl. Acad. Sci. USA* **82,** 8009–8013.
7. Collier, D. A, Griffin, J. A., and Wells, R. D. (1988) Non-B right-handed DNA conformations of homopurine-homopyrimidine sequences in the murine immunoglobulin switch region. *J. Biol. Chem.* **263,** 7397–7405.
8. McLean, M. J., Lee, J. W., and Wells, R. D. (1988) Characteristics of Z-DNA helices formed by imperfect (pur-pyr) sequences in plasmids. *J. Biol. Chem.* **263,** 7378–7385.
9. Portugal, J., Fox, K. R., McLean, M. J., Richenberg, J. L., and Waring, M. J. (1988) Diethylpyrocarbonate can detect the modified DNA structure induced by the binding of quinoxaline antibiotics. *Nucleic Acids Res.* **16,** 3655–3670.
10. McLean, M. J. and Waring, M. J. (1988) Chemical probes reveal no evidence of Hoogsteen base pairing in complexes formed between echinomycin and DNA in solution. *J. Mol. Recog.* **1,** 138–151.
11. Jeppesesen, C. and Nielsen, P. E. (1988) Detection of intercalation-induced changes in DNA structure by reaction with diethyl pyrocarbonate or potassium permanganate. Evidence against the induction of Hoogsteen base pairing by echinomycin. *FEBS Lett.* **231,** 172–176.
12. Fox, K. R. and Kentebe, E. (1990) Echinomycin binding to the sequence CG(AT)nCG alters the structure of the central AT region. *Nucleic Acids Res.* **18,** 1957–1963.
13. Runkel, L. and Nordheim, A. (1986) Chemical footprinting of the interaction between left-handed Z-DNA and Anti-Z-DNA antibodies by diethylpyrocarbonate carbethoxylation. *J. Mol. Biol.* **189,** 487–501.
14. Maniatis, T., Fritsch, E. F., and Sambrook, J. (1982) *Molecular Cloning: A Laboratory Manual.* Cold Spring Harbor Laboratory, Cold Spring Harbor, New York.

CHAPTER 8

Osmium Tetroxide Modification and the Study of DNA–Protein Interactions

James A. McClellan

1. Introduction

Osmium tetroxide is an evil-smelling chemical used as a fixative in electron microscopy that can also be used to modify thymidine residues within DNA *(1,2)*. The ability of osmium tetroxide to modify DNA is very sensitive to DNA conformation. In particular, osmium tetroxide will attack thymidines that are unstacked, either because they are in a single-stranded region or, for example, because the DNA is bent or because it is overwound *(3)*.

As a footprinting agent, osmium tetroxide has not been widely used. This is principally for historical reasons, and one of the motives for writing this chapter is to encourage experimentation with osmium tetroxide as a footprinting agent. However, it is true that the exquisite sequence specificity of osmium tetroxide is a disadvantage when one wishes to know *all* the contacts that a protein makes in a particular sequence. Additionally, osmium tetroxide has been shown to damage proteins, although this is probably a problem common to virtually all chemical footprinting agents, rather than a specific disadvantage of osmium tetroxide *(4)*. Nevertheless, osmium tetroxide has definite advantages in the study of certain DNA–protein interactions, specifically those interactions that result in a changed conformation of the DNA double helix. Examples of such interactions include the initiation of transcription *(5)* and the formation of nonstandard conforma-

From: *Methods in Molecular Biology, Vol. 30: DNA–Protein Interactions: Principles and Protocols*
Edited by: G. G. Kneale Copyright ©1994 Humana Press Inc., Totowa, NJ

tions in DNA that are known to be recombinogenic and to have the potential to modulate gene expression *(6–8)*.

The main advantages of osmium tetroxide are that it is very easy to work with and that the osmium modification reaction takes place under a very wide variety of environmental conditions *(9)*, including in vivo *(10–12)*. This makes it possible to study nucleic acid conformation as a function of environmental conditions, such as salt concentration, temperature, presence or absence of proteins, and polyamines. All that is required is that there should be susceptible Ts involved in the conformational change; since most sites where conformational changes occur in DNA are A + T rich, this is not a hard criterion to satisfy.

As an indication of the kinds of information that can be obtained using osmium tetroxide modification, we may cite the results of in vivo modification experiments on AT sequences *(12)*. Experiments like these provided the first direct evidence that cruciform structures may be seen in vivo. Our experiments also show that previously observed environmentally induced changes in plasmid linking numbers actually do result in changes in torsional stress in vivo, rather than being compensated for by the binding of extra histone-like proteins. However, the data also show that cruciforms are not a normal component of the bacterial genome; we do not see cruciforms if there are significant kinetic barriers to extrusion, and we do not see cruciforms unless we artificially raise the in vivo level of supercoiling by osmotically shocking the cells. Nevertheless, since environmentally induced changes in DNA supercoiling are thought to regulate the bacterial response to osmotic and other stresses, we cannot rule out the possibility that transient formation of unusual DNA structures could be instrumental in such responses *(13)*. It is particularly interesting that a sequence 5'-ATTATATATATATATATATATATAAT 3' is found around the promoter of a key pathogenic gene in *Haemophilus influenzae (14)*. This sequence could attain cruciform geometry at the levels of DNA torsional strain that we find to be operative inside environmentally stressed bacteria. Moreover, cruciform geometry at artificial promoters is thought to affect transcription *(15)*.

2. Materials

1. Microcentrifuge tubes.
2. Micropipets capable of dispensing 0.5–1000 µL.
3. 20 mM Osmium tetroxide, thawed (*see* Notes 1 and 2).

4. 6% (v/v) Aqueous pyridine or 20 mM aqueous 2, 2' bipyridyl (*see* Notes 2 and 3).
5. 10X Buffer: 50 mM Tris-HCl, 5 mM EDTA, pH 8.0.
6. 1–5 µg Plasmid DNA. Miniprep DNA is adequate; gradient-purified DNA is better.
7. Waterbath.
8. Absolute ethanol.
9. 3M Sodium acetate, pH 4.5.
10. Eppendorf microcentrifuge (e.g., 5415C).
11. Restriction enzymes.
12. Enzymes and radionucleotides for end-labeling DNA. In this protocol the enzyme used is the Klenow fragment of DNA polymerase I. One also needs an α-labeled dNTP at specific activity >6000 Ci/mmol and the other three dNTPs as nonradioactive 2 mM stocks.
13. Agarose gel with slots large enough to take >60 µL (they can be made by taping up smaller slots), powerpack, and gel tank.
14. 5X Loading dye: 10% Ficoll 400, 0.1M sodium EDTA, pH 8.0, 1% SDS, 0.25% bromophenol blue, 0.25% xylene cyanol.
15. Ethidium bromide (10 mg/mL stock).
16. Long-wave UV transilluminator.
17. Electroeluter. We use the IBI (Irvine, CA) model UEA.
18. Heating block capable of running at 90°C.
19. 1M Aqueous piperidine.
20. Liquid nitrogen.
21. High capacity vacuum pump with trap and dessicator.
22. Alkaline formamide dye; 200 µL of deionized formamide, 1 µL of 1M NaOH, and xylene cyanol + bromophenol blue (a needle dipped in the powdered dye and then tapped to remove excess will give quite enough).
23. Sequencing gel. The percent acrylamide and other details of the gel will depend on the exact system that is being studied; how far the site of modification is from the point of labeling, for example. As a rough guide, we have used 6% acrylamide (29 parts acrylamide to one part *bis*-acrylamide), 7M urea, 1X TBE gels run until the xylene cyanol (light blue dye) is on the foot of the gel to analyze modifications 170 bp from the site of labeling. To run such a gel takes about 2 h at a constant wattage of 75; this should correspond to about 1400 V and about 40 mA, but the important criterion is that the plates should be hot to the touch, but not so hot that they break during electrophoresis. If there is any doubt, turn the wattage down. To analyze modifications 40 bp from the site of labeling, we have used an 8% gel run until the bromophenol blue (dark blue dye) is at the foot of the gel. We do not deionize our acrylamide, but we do filter and degas the gel mixture (by attaching the

side-arm flask to a water pump and keeping a rubber bung in the neck of the flask until the fluid stops bubbling) prior to adding ammonium persulfate and TEMED (*in that order;* the order of addition is very important). We also prerun our gels for about 1 h until the plates are hot to the touch. In pouring the gel, the key points are to have clean unscratched plates (washed and scrubbed using ethanol and tissue paper) and cold acrylamide. With practice it is possible to get rid of virtually all of the bubbles, and the effects of small leaks can be minimized by keeping the gel plates as horizontal as possible during pouring and setting.
24. Sequencing apparatus. We have successfully used a number of different systems. Currently we are using a BRL (Bethesda, MD) model S2 gel stand and an IBI MBP 3000E power supply.
25. Whatmann 3MM paper.
26. Autoradiography cassette.
27. X-ray film and developing facilities.

3. Methods

3.1. DNA Modification and Detection of Adducts

This protocol describes the *in vitro* modification of plasmid DNA and the sequence-level detection of adducts by piperidine cleavage after specific 3' end-labeling. For alternative methods of detecting osmium adducts, *see* Note 4.

1. Combine the required volume of aqueous plasmid DNA with 5 µL of 10X modification buffer and add distilled water to a total volume of 45 µL.
2. In a separate tube, mix 2.5 µL of 20 mM osmium tetroxide with 2.5 µL of 6% pyridine or 20 mM bipyridine. This is the step at which the reactive osmium-ligand species is constituted.
3. Equilibrate the products of steps 1 and 2 at the desired reaction temperature for 5 min. To start with, try room temperature (20°C). If the reaction goes too far, it may be necessary to do it at a lower temperature, for example, on ice. You will be able to tell if the reaction has gone too far when you see the results; there will be a tendency for label-proximal modifications to be overrepresented.
4. Add the products of step 1 to the products of step 2, mix well, and incubate for the desired time. To start with, try 15 min at room temperature. If this is too vigorous, try 30 min on ice. If it is not vigorous enough, try 5 min at 37°C.
5. Meanwhile, make the stop solution; 180 µL of absolute ethanol and 5 µL of 3M sodium acetate, pH 4.5. Chill at –70°C.
6. Add stop solution to reaction, mix well, and chill at –70°C for 10 min.

7. Spin at maximum speed in Eppendorf microcentrifuge for 10 min. Note that non-Eppendorf microcentrifuges are not, in general, adequate substitutes. We do not know exactly why, but we suspect it has to do with heating of the rotor during spinning.
8. Pipet off the supernatant—do *not* pour—and discard as potentially carcinogenic waste. You have to be careful not to disturb the pellet.
9. Add 1 mL of absolute ethanol to the tube and spin in the microcentrifuge at maximum speed for 5 min.
10. Pipet off the supernatant and discard as in step 8.
11. Dry the pellet for 5 min in a vacuum dessicator.
12. Add 44 µL of distilled water.
13. Transfer to a fresh tube. This is important; sometimes residual osmium on the reaction tube can interfere with the subsequent steps.
14. Add 5 µL of 10X restriction enzyme buffer and 1 µL (about 10 U) of restriction enzyme. The enzyme chosen should give protruding 5' ends and ideally has a unique site about 40–200 bases from the site at which modification is expected.
15. Incubate at 37°C for about 1 h.
16. Meanwhile, dry down 5 µCi of a >6000 Ci/mmol α-[^{32}P] deoxynucleotide. This should be the first nucleotide to be incorporated by DNA polymerase; for example, if the enzyme used is *Eco*RI, it should be dATP.
17. To the dried-down radionucleotide, add 2 µL of 2 m*M* stocks of each of the cold nucleotides except for the hot one, that is, if the labeling nucleotide is dATP, add dGTP, dCTP, and dTTP. Mix well.
18. Add the hot and cold nucleotides to the restriction digest. Then add 1 µL (about 6 U) of the Klenow fragment of DNA polymerase. We find that Klenow polymerization works well in a variety of restriction enzyme buffers, especially those used for *Eco*RI and *Bam*HI.
19. Incubate for 1 h (not more) at 37°C. If you incubate for much longer than this, you get weird labeling artifacts.
20. Add 10 µL of 5X loading dye and electrophorese on a 1% agarose 1X TBE gel until the xylene cyanol (light blue) dye is about halfway down the gel, or in general, until the fragment of interest can be easily excised from the gel. Exactly which gel is chosen will depend on the particular system under study.
21. Stain the gel with 1 µg/mL ethidium bromide. Visualize by long-wave UV and excise the bands of interest from the gel.
22. Electroelute the labeled DNA into high salt (or use another method).
23. Precipitate and wash the DNA with ethanol. If precipitating from high salt, do not chill. You will bring down the salt, which is then hard to get rid of.

24. If it is desired to see only the signals on one strand, cut off one of the labeled ends (*see* Note 5). After restriction, the DNA should be ethanol precipitated, washed, and dried. Do this by repeating steps 5–11; it is not necessary to discard the waste as potentially mutagenic.
25. To the dried pellet add 100 µL of 1M piperidine. This is a strong base that enhances cleavage of the phosphate backbone next to modified bases. Heat at 90°C for 30 min.
26. Transfer to a new tube. Close the tube. Use a needle to punch holes in the cap.
27. Place the tube in liquid nitrogen for about 5 s.
28. Place the tube in a rack in a vacuum desiccator, attach a high-capacity vacuum pump with a trap, and turn it on. Do this quickly, so that the frozen sample does not have time to thaw.
29. Dry for about 2 h.
30. Add 50 µL of water to the tube and repeat lyophilization step 28 for about 1 h.
31. Repeat steps 29 and 30.
32. Resuspend the samples in at least 3 µL of alkaline formamide dye.
33. Heat to 90°C for 5 min and then chill on ice. This denatures any residual secondary structure in the DNA.
34. Load aliquots of about 50 counts/s on a sequencing gel and electrophorese. Exact details of the sequencing gel will depend on the system under study. For some suggestions, *see* item 23, Section 2.
35. Fix the gel in 10% acetic acid for 15 min, transfer to 3MM paper, dry, and autoradiograph.

3.2. Results of Osmium Tetroxide Modification

Figure 1 shows a time-course of osmium tetroxide modification on a 68-bp-long tract of alternating adenines and thymidines within a bacterial plasmid. This sequence is found in the first intron of a frog globin gene. The modification was done at 37°C in the absence of added salt, for the indicated times. A *Bam*HI-*Eco*RI fragment containing the tract was labeled at both ends by Klenow and α-[^{32}P]- dATP. The adducts were cleaved using hot piperidine and electrophoresed on a sequencing gel. The gel was fixed, dried onto paper, and autoradiographed.

The gel shows information from both strands. This is possible because the AT tract is asymmetrically placed on the *Bam*HI-*Eco*RI fragment and osmium modification is almost completely specific for the AT tract. As can be seen, the modifications are biased toward the label-proximal end of the tract on both strands, and this is intensified at

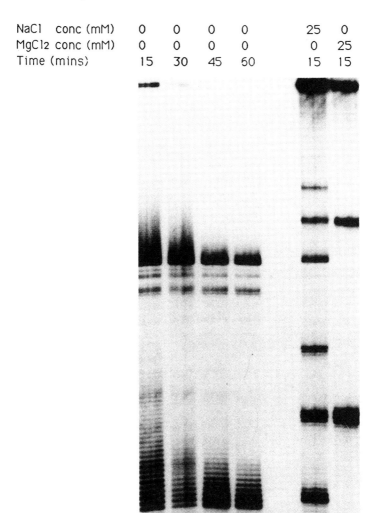

Fig. 1. Time-course of in vitro osmium tetroxide modification of $(AT)_{34}$ tract in plasmid pXG 540. An autoradiograph of end-labeled piperidine-cleaved *Eco*RI-*Bam*HI fragments from plasmid pXG 540 that had been treated with osmium tetroxide under the ambient conditions indicated above each lane (all reactions were done at 20°C). The fragments were separated on a 6% sequencing gel run hot to the touch at constant 70 W.

later time-points. In fact, this is also the pattern if the labeling is done on 5' ends (data not shown). Thus, the label-proximal bias of the signals indicates multiple modifications rather than an asymmetric structure.

Figure 1 also shows the result of modifying the AT tract in the presence of ions. Here, the patterns are quite different. In the presence of sodium ions, we see modifications at the center of the AT tract and at its ends. In the presence of magnesium ions, we see modifications at the center of the AT tract but not at its ends. These patterns are interpreted as indicating the presence of cruciform structures at the AT tract, either with tight scissor-shaped osmium resistant junctions (in the presence of magnesium), or with floppy square planar osmium sensitive junctions (in the presence of sodium) *(16)*. The central modifications are at what would be the loop of the cruciform, and the modifications at the end of the tract would be at the junction of the cruciform with flanking DNA (Fig. 2) *(see Note 6)*.

3.3. In Vivo Osmium Modification

It is possible to modify AT tracts in plasmids inside living bacteria *(10–12)*. Figure 3 shows the results of one such experiment, in which a number of plasmids with different lengths of AT were modified inside bacterial cells (*E. coli* HB101). The plasmid DNA was recovered by a modification of the Holmes-Quigley boiling method *(12)*; alkaline lysis methods are not used, so as to avoid premature alkaline cleavage of adducts. The DNA was then restricted and end-labeled, and the adducts were cleaved by hot piperidine before analysis on sequencing gels as described above.

In vivo modification requires a number of additional tricks if it is to be successful. One has to use bipyridine as the ligand. The number of cells is a critical parameter; the modification reaction should have cells at an A_{550} nm of 0.4. If the absorbance is even twice as high, the experiment is likely to fail. During the modification, the cells should turn a milk-chocolate brown; if they do not, discard the experiment and obtain fresh osmium tetroxide. We usually work on a 50 mL culture scale, and we usually wash the cells and do the modifications in 100 mM potassium phosphate buffer, pH 7.4. It is, however, possible to do the modifications in L-broth. After the cells have been boiled, the supernatant has to be collected, and it is very viscous indeed. We find that the best way to prepare the supernatant is to spin the lysates

Osmium Modification

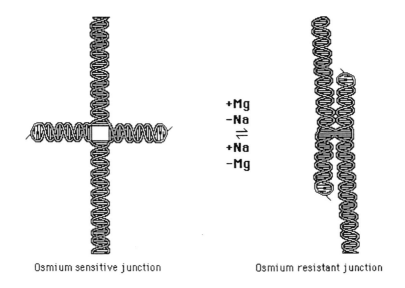

Osmium sensitive junction   Osmium resistant junction

Fig. 2. Two conformations of cruciforms. Two different cruciform conformations may be postulated in order to explain the different patterns of osmium modification seen at different ionic strengths.

at 30,000 RPM for 30 min in a Beckmann (Fullerton, CA) table-top ultracentrifuge (120 TS rotor). The supernatant can then be ethanol precipitated or made up as a CsCl/ethidium bromide gradient.

4. Notes

1. Health hazards of osmium tetroxide: Osmium tetroxide is a very powerful oxidizing and crosslinking reagent that formerly was used for tanning leather. It can react explosively with water. Fumes of osmium tetroxide can damage the cornea of the eye. It is thus a chemical to be treated with great respect. To our knowledge there is no evidence that it is a carcinogen, but this may be because the complex with heterocyclic activators has not been tested as such; this complex is certainly a very powerful and specific covalent modifier of exposed thymidines, and it would be surprising if it was not a carcinogen. It therefore seems prudent to dispose of osmium tetroxide waste as if it were carcinogenic.
2. Preparation and maintenance of osmium tetroxide stocks: Osmium tetroxide can be bought from various suppliers including Sigma (St. Louis, MO) and Johnson Matthey. The quality of the reagent varies widely, but in general it may be stated that the material bought from Johnson Matthey is superior. One usually buys a 250-mg aliquot that

| Salt Shock | + | + | + | + | | + | − |
| Marker (all Ts in AT 34 tract) | 34 | 22 | 15 | 12 | | 34 | 34 |

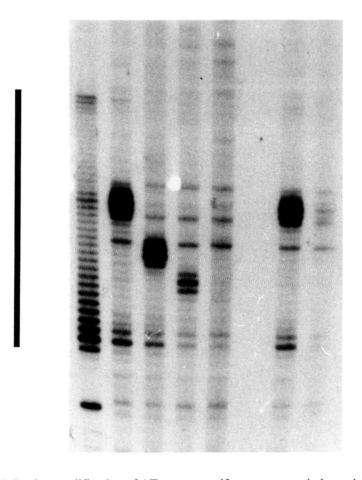

Fig. 3. In vivo modification of AT tracts: cruciform geometry in bacterial plasmids as a consequence of physiological salt shock. Bacterial cells containing isogenic plasmids with various lengths of AT were salt-shocked and treated with osmium tetroxide in vivo. Adducts were detected by preparing the DNA, end-labeling, and piperidine cleavage, followed by electrophoresis on thin 6% sequencing gels and autoradiography. Strong central modification of the AT tract shows that $(AT)_{34}$ adopts cruciform geometry in salt-shocked but not in control cells, and that $(AT)_{22}$ and $(AT)_{15,}$ but not $(AT)_{12,}$ also adopt cruciform geometry in salt-shocked cells.

should cost around £25–35 at current prices. The reagent is shipped inside a metal can. This contains a plastic tube, and inside the tube, wrapped in black paper, is a little sealed glass phial containing osmium tetroxide (yellow crystals). One makes up a stock 20-mM solution as follows:
  a. The glass tube is plunged into liquid nitrogen and kept there for about 15 s. This treatment prevents the chemical from reacting explosively with water, and it also makes the crystals less sticky and easier to remove from the glass phial.
  b. Inside the fume hood and over a washable tray, the glass phial is broken, ideally with the aid of a diamond knife, and the crystals are tipped into a glass beaker containing 49.2 mL of distilled water. The crystals take some time to dissolve—about 2–3 h.
  c. Once the crystals have dissolved, the reagent is aliquoted ready for use. Glass containers, such as Universals with screw tops, should be used, and storage should be at –70°C. It is best to keep the reagent as several separate aliquots and to use them one at a time. One problem with storage is that the glass containers often crack, which is obviously very dangerous. Volumetric flasks and non-Pyrex containers are particularly sensitive in this regard. The reagent should always be thawed with the container inside a glass beaker in the fume hood. If the reagent is blackish when thawed, discard it; the black color indicates lower oxidation states of osmium, including the metal.
3. Reactivity of osmium tetroxide complexes: By itself osmium tetroxide is not very reactive. The species that attacks DNA is a complex of osmium tetroxide with a heterocyclic compound, such as pyridine or bipyridine (Fig. 4). The attack is on the 5–6 double bond of thymidines *(3)*. Curiously, neither cytosines nor uracils show significant modifiability with osmium tetroxide, but guanine residues occasionally may.

Bipyridine is rather insoluble in water, apparently for kinetic reasons. To make a stock solution, it may be necessary to heat the bipyridine + water to 80°C for about 30 min. Once osmium tetroxide is added to pyridine, the mixed reagents should become a straw-yellow color; if they do not, discard them and obtain fresh osmium tetroxide.

One can change the reactivity of an osmium tetroxide preparation by changing the concentration of osmium and/or ligand, and the overall degree of reaction can be varied from single to multiple hits per molecule by changing the time and/or temperature of the modification reaction. Typical times and temperatures used in our laboratory to obtain single hits are as follows: 45 min at ice temperature; 15 min at 20°C; 5 min at 37°C; and 1 min at 40°C.

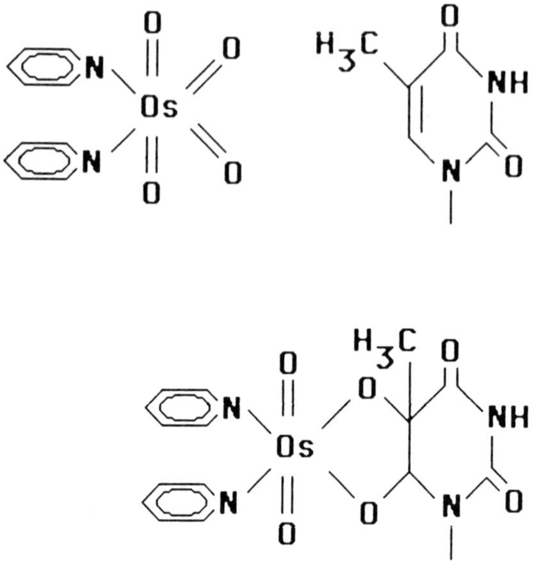

Fig. 4. Stereochemistry of OsO₄ attack on thymidines.

4. Detecting osmium tetroxide adducts: Various methods of detecting osmium tetroxide adducts exist, namely retardation of bands in agarose or acrylamide gels *(5)*, immunoprecipitation *(3)*, cleavage by S1 nuclease *(5)*, cleavage by hot piperidine *(6)*, and failure of primer extension *(11)*. In some cases, inhibition of restriction enzyme cleavage can be used *(10, 17)*.
5. Maxam-Gilbert single-strand sequencing without gel purification: A lot of people avoid Maxam and Gilbert sequencing because they think it involves too much gel purification. With a little ingenuity, however, the need for this can almost always be obviated, as in the following method. The basic strategy is to remove one of the two labeled ends on a little fragment that runs off the gel, so that signals from it do not interfere with those you want to see. To achieve this, first label both ends and then cut with a frequent cutter that has a site near to the end from which it is not required to see signals, and that does not have a site between the end from which it is desired to see signals and the tract where modifications are expected.

For example, the *Eco*RI site of pXG540 has a *Hae*III site 18 bp anticlockwise (in the direction of the Amp promoter). There is no *Hae*III site in the clockwise 170 bp up to the start of the AT tract. Thus, we often cut and label at *Eco*RI, and then do a limit digest with *Hae*III. This results in a long labeled fragment that has the information we want on it and a short fragment (18 bp) that we run off the gel.

6. Interpreting and overinterpreting chemical modification: Osmium tetroxide modification of DNA now has a substantial pedigree, having been used to modify a wide range of sequences in different unusual conformations in vitro, and a narrower range of sequences in different conformations in vivo. Figure 5 shows some of the actual or proposed structures that have been treated with osmium tetroxide. However, it is very important to be cautious in using osmium modification results, or any other kind of chemical or enzymatic probing, to deduce that a particular structure is forming. This is because osmium does not report on global conformation; it only tells you whether or not a particular T residue has an exposed 5–6 double bond. If a T is exposed, this could be for several reasons:

  a. The T might be in a single-stranded region of the DNA, such as a bubble, mismatch *(18)*, B-Z junction, or cruciform or H-structure loop.
  b. The T might be in a helix with a shallow or bulging major groove, for example, at a bend or within a GT/AC tract that was forming Z-DNA.
  c. The T might be in an overwound helix, with exposure of the 5–6 double bond resulting from a loss of base overlap, because of the sharp rotation of each base relative to its neighbors.

These considerations make it very difficult to conclude on the basis of osmium modification that, for example, Z-DNA is forming; osmium results do not distinguish in any simple way between what we have called the "U" structure *(9)*, Z-DNA, and a conformation (possibly the eightfold D-helix) that we have observed in locally positively stressed AT tracts *(19)*. Detecting the loops and junctions of cruciform or H-form DNA can sometimes be done in such a way as virtually to exclude alternative interpretations of the data, but even this is not always possible: For example, a GC tract with ATAT in the middle will react in the same way with osmium tetroxide whether it is in a Z-conformation or cruciform *(11)*, and there is a published interpretation of chemical modification at H-forming sequences that is radically at odds with the H-structure model *(20)*.

In addition to these difficulties, there is a need to be cautious about the effect of the probing chemical on the structural features deduced. Because osmium modification works under a wide range of conditions, it presents fewer such problems of interpretation than some other more fastidious chemicals. Nevertheless, we observe that the ligand used can have an effect on the result obtained; other things being equal, we find

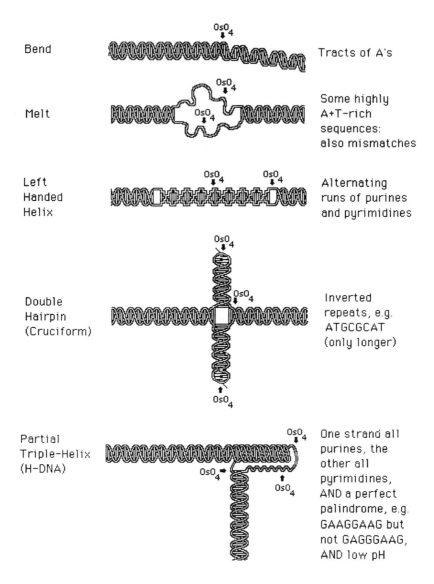

Fig. 5. Unusual DNA structures that react with osmium tetroxide.

that bipyridine is more likely to give a cruciform-like pattern of modification, and pyridine is more likely to give a U-structure-like pattern, when superhelically stressed AT tracts are probed in vitro. This probably has to do with different helix-unstacking potential of the two heterocycles; alternatively it could be an artifact of contaminating cations in the heterocycles or ion-heterocycle interactions.

Acknowledgments

This chapter is respectfully dedicated to Emil Palecek, who pioneered the use of osmium tetroxide as a probe of DNA structure, and whose group at the Biophysics Institute in Brno, Czechoslovakia, continues to lead the world in this area of research. In addition I wish to thank David Lilley and members of his research team at Dundee, Scotland, especially Fareed Aboul-Ela, Anamitra Bhattacharyya, Richard Bowater, and Alistair Murchie. I thank the Royal Society of London and the Czechoslovak Academy of Sciences for a Visiting Fellowship, and Pavla Boublikova and Alena Krejkova for making my stay in Brno such a happy one. Research in my laboratory is supported by the Smith-Kline Foundation, the Science Faculty Fund, and the Seedcorn Fund of Portsmouth University, Portsmouth, UK.

References

1. Beer, M., Stern, S., Carmalt, D., and Mohlenrich, K. H. (1966) Determination of base sequence in nucleic acids with the electron microscope. V. The thymine-specific reactions of osmium tetroxide with deoxyribonucleic acid and its components. *Biochemistry* **5**, 2283–2288.
2. Burton, K. and Riley, W. T. (1966) Selective degradation of thymidine and thymine deoxynucleotides. *Biochem. J.* **98**, 70–77.
3. Palecek, E. (1989) Local open DNA structures in vitro and in the cell as detected by chemical probes, in *Highlights of Modern Biochemistry* (Kotyk, A., Skoda, J., Paces, V., and Kostka, V., eds.), VSP International Sciences, Zeist, The Netherlands, pp. 53–71.
4. Behrman, E. J. (1988) The chemistry of the interactions of osmium tetroxide with DNA and proteins, in *Symposium on Local Changes in DNA Structure and Their Biological Implications*, Book of Abstracts, Brno, Czechoslovakia, p. 6.
5. Buckle, M., Spassky, A., Herbert, M., Lilley, D. M. J., and Buc, H. (1988) Chemical probing of single stranded regions of DNA formed in complexes between RNA polymerase and promoters, in *Symposium on Local Changes in DNA Structure and Their Biological Implications*, Book of Abstracts, Brno, Czechoslovakia, p. 11.
6. Lilley, D. M. J. and Palecek, E. (1984) The supercoil-stabilised cruciform of ColE1 is hyper-reactive to osmium tetroxide. *EMBO J.* **3**, 1187–1192.
7. Johnston, H. and Rich, A. (1985) Chemical probes of DNDA conformation: detection at nucleotide resolution. *Cell* **42**, 713–724.
8. Vojtiskova, M. and Palecek, E. (1987) Unusual protonated structure in the homopurine-homopyrimidine tract of supercoiled and linearised plasmids recognised by chemical probes. *J. Biomolec. Struct. Dyn.* **5**, 283–296.
9. McClellan, J. A. and Lilley, D. M. J. (1987) A two-state conformational equilibrium for alternating (A-T) n sequences in negatively supercoiled DNA. *J. Mol. Biol.* **197**, 707–721.

10. Palecek, E., Boublikova, P., and Karlovsky, P. (1987) Osmium tetroxide recognizes structural distortions at junctions between right- and left-handed DNA in a bacterial cell. *Gen. Physiol. Biophys.* **6,** 593–608.
11. Rahmouni, A. R. and Wells, R. D. (1989) Stabilisation of Z DNA in vivo by localized supercoiling. *Science* **246,** 358–363.
12. McClellan, J. A., Boublikova, P., Palecek, E., and Lilley, D. M. J. (1990) Superhelical torsion in cellular DNA responds directly to environmental and genetic factors. *Proc. Natl. Acad. Sci. USA* **87,** 8373–8377.
13. Dorman, C. J. (1991) DNA supercoiling and environmental regulation of gene expression in pathogenic bacteria. *Inf. Immun.* **59,** 745–749.
14. Langermann, S. and Wright, A. (1990) Molecular analysis of the *Haemophilus influenzae* type b pilin gene. *Molec. Microbiol.* **4,** 221–230.
15. Horwitz, M. S. Z. and Loeb, L. A. (1988) An *E. coli* promoter that regulates transcription by DNA superhelix-induced cruciform extrusion. *Science* **241,** 703–705.
16. Duckett, D. R., Murchie, A. I. H., Diekmann, S., von Kitzing, E., Kemper B., and Lilley, D. M. J. (1988) The structure of the Holliday junction, and its resolution. *Cell* **55,** 79–89.
17. Palecek, E., Boublikova, P., Galazka, G., and Klysik, J. (1987) Inhibition of restriction endonuclease cleavage due to site-specific chemical modification of the B-Z junction in supercoiled DNA. *Gen. Physiol. Biophys.* **6,** 327–341.
18. Cotton, R. G. H., Rodrigues, N. R., and Campbell, R. D. (1988) Reactivity of cytosine and thymine in single-base-pair mismatches with hydroxylamine and osmium tetroxide and its application to the study of mutations. *Proc. Natl. Acad. Sci. USA* **85,** 4397–4401.
19. McClellan, J. A. and Lilley, D. M. J. (1991) Structural alteration in alternating adenine-thymine sequences in positively supercoiled DNA. *J. Molec. Biol.* **219,** 145–149.
20. Pulleyblank, D. E., Haniford, D. B., and Morgan, A. R. (1985) A structural basis for S1 sensitivity of double stranded DNA. *Cell* **42,** 271–280.

CHAPTER 9

Diffusible Singlet Oxygen as a Probe of DNA Deformation

Malcolm Buckle and Andrew A. Travers

1. Introduction

The DNA double helix is highly malleable and when constrained, either as a small circle or by the action of a protein, can be readily distorted from its energetically favored conformation. Such distortions may be relatively moderate, as exemplified by smooth bending that maintains base stacking, or more extreme when this stacking can be disrupted. Deformations of this latter type include kinks, in which the direction of the double helical axis is changed abruptly at a single base step, and localized strand separation that may be a direct consequence of protein-induced unwinding or of high negative superhelicity in free DNA. Both kinks and localized unwinding can arise transiently during the enzymatic manipulation of DNA by recombinases and by protein complexes involved in the establishment of unwound regions during the initiation of transcription or of DNA replication.

The detection of localized lesions in the DNA double helix requires that the bases at the site of the lesion be accessible to a chemical reagent only when the DNA is distorted. Further, since protein-induced distortions are not necessarily sequence-dependent, it is desirable that any reagent used for detection possess minimal selectivity with respect to the base. Additionally the reagent itself should ideally be noninvasive, that is, it should not form a stable noncovalent complex with DNA and thereby possess the potential to perturb the local conformation of the double helix. Other highly desirable attributes for reagents

From: *Methods in Molecular Biology, Vol. 30: DNA–Protein Interactions: Principles and Protocols*
Edited by: G. G. Kneale Copyright ©1994 Humana Press Inc., Totowa, NJ

used for this purpose are that they should possess short half-lives and that they can be generated on demand *in situ*. These latter characteristics permit the study and detection of transient intermediates in the processes leading to the establishment of complexes competent to initiate replication or transcription or to catalyze recombination.

Chemical reagents so far described that specifically target bases in DNA that is locally deformed include dimethyl sulfate (*see* Chapter 6), diethyl pyrocarbonate (*see* Chapter 7), osmium tetroxide (*1; see* Chapter 8), and potassium permanganate *(2)*. However, all of these reagents react selectively with specific bases and are also relatively long-lived. Another reagent used extensively for the detection of locally unwound regions of DNA is copper-*o*-phenanthroline (*3; see* Chapter 5). This compound is a minor groove ligand that cleaves the sugar-phosphate backbone as a consequence of free-radical attack on a deoxyribose residue close to the site of binding.

One reagent that lacks these shortcomings is oxygen in the singlet state, of which there are two forms with energies of 155 and 92 kJ, respectively. The latter state has a much longer lifetime and can oxidize a variety of unsaturated organic substrates. Typically such a reaction may involve a Diels-Alder-like addition to a 1,3-diene. This highly reactive form of oxygen can be generated by the photochemical excitation of appropriate heterocyclic ring systems that can then promote the conversion of dissolved oxygen in the triplet state to a singlet form. In solution the singlet state generated in this way has a half-life of approx 4 µs and can react with accessible DNA bases to form an adduct across a double bond. Once formed such an adduct sensitizes the sugar-phosphate backbone to alkaline hydrolysis by piperidine, thus permitting the identification of the site of modification *(4)*.

1.1. The Reaction of Singlet Oxygen with DNA

The use of singlet oxygen as a reagent for analyzing DNA structure has been pioneered by the groups of Hélène and Austin, who have introduced two general methods of targeting the DNA *(4,5)*. In the first case, a DNA ligand is used as a sensitizer for singlet oxygen production in a manner analogous to the use of copper-*o*-phenanthroline for the generation of free radicals. Such ligands include methylene blue, which intercalates at sites where the DNA is relatively unwound *(4),* and also a porphyrin ring covalently linked to a defined DNA sequence designed to target a selected region of double helix *(5)*. In these examples, photochemical excitation produces singlet oxygen at the site of the bound ligand, and reaction is confined to the immediate proximity of the ligand. A second approach is to use singlet oxygen as a freely diffusible reagent. This use is similar in principle to that of a hydroxyl radical produced by the Fenton reaction *(6)*. In such experiments, the singlet oxygen is generated by the irradiation of a complex of eosin with Tris (Fig. 1) and is then free to diffuse. The eosin is irreversibly oxidized in the course of this reaction. However, because the half-life of the singlet oxygen is very short, the concentration of the eosin-Tris complex must be sufficiently high to ensure that singlet oxygen can access potentially reactive sites in the DNA before its reversion to the triplet state.

As with any chemical reagent reacting with a set of chemically distinct targets, the rate of reaction of singlet oxygen with the different bases varies. Notably guanine as the free base reacts up to 100-fold more rapidly than the other nucleic acid bases. However, the rate of reaction of diffusible singlet oxygen with duplex DNA appears not to be primarily determined by the nature of the bases at the target site but rather by their accessibility. In normal B-form DNA, the bases are generally tightly stacked so as to preclude the entry, and hence the reaction, of the reagent between the base pairs. In the structures of DNA oligomers, it is unusual for the average planes of adjacent base pairs to be separated by a roll angle of >10°. This tight structure is reflected in the relative lack of reactivity of DNA in solution toward singlet oxygen. By contrast, when bound by protein, the DNA can be locally unwound *(7)* or can be kinked so that adjacent base-pair planes can be inclined by up to 43° relative to each other *(8)*. This deformation of the DNA structure by bound protein would in principle be expected to increase the accessibility to singlet oxygen, as has been observed both for core nucleosome particles associated with DNA of mixed sequence *(4)* and for the ternary complex of RNA polymerase and CRP with the *lac* regulatory region *(9)*. The latter case is the only example for which information is so far available, and reaction is observed with all four DNA bases, although there are insufficient sites documented to preclude some base selectivity of the reagent. If the local structure of the DNA is the principal determinant of reactivity, the reagent should be able to access the bases through both the

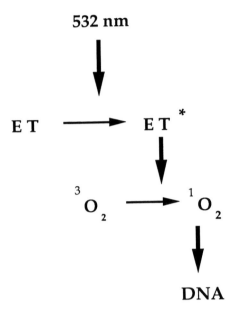

Fig. 1. Reaction cascade for the modification of DNA by singlet oxygen.

major and minor grooves as has been observed *(9)*. However, the precise range of DNA structures available for reaction with singlet oxygen remains to be established, as does the possible influence of bound protein on sensitizing or quenching the reactivity of the bases.

A major advantage of a photoactivated reaction is the ability to produce the reactive species under highly controlled conditions. Both porphyrins and methylene blue can be activated by a continuous laser beam. However, eosin absorbs maximally at 523.2 nm, a wavelength that is close to the 532-nm output from a neodymium-YAG laser. This fortuitous proximity is of particular utility because it permits the production of singlet oxygen either from an effectively continuous output or from a discrete number of pulses, each of approx 7-ns duration. This method of activation utilizes the high energy output of the Nd-YAG laser and also allows the kinetics of the protein-induced structural alterations in DNA to be followed with high precision. It should be noted that because the half-life of singlet oxygen in solution is only 4 µs *(10)*, the time available for reaction with the DNA is essentially limited by the time of irradiation. This short half-life also means that it is unnecessary to terminate any reaction by the addition of a quenching reagent.

2. Materials

1. A neodymium yttrium aluminum garnet (NdYAG) laser (Spectra-Physics [Les Ulis, France] DCR-11) set up as illustrated in Fig. 2 is used to generate a beam of polarized coherent light at a wavelength of 1064 nm. A doubling crystal correctly aligned in the beam path produces a mixture of light at 1064 and 532 nm. This mixture is subsequently separated by a dichroic mirror arranged so that the 532-nm beam is deflected down onto a thermostatted Eppendorf tube containing 20 µL of the sample to be irradiated. Alternatively, if the volume of the irradiated solution is small, the different wavelengths can be separated by an appropriate arrangement of prisms and the 532-nm beam directed into an Eppendorf tube held horizontally in a metal block maintained at the required temperature. It is essential to obtain an adequate separation of these two wavelengths since even a relatively low proportion of the primary emission at 1064 nm could result minimally in a rapid heating of the sample. The DCR-11 functions at a frequency of 10 Hz, each pulse of about 7-ns duration delivering 160 mJ of energy. Other NdYAG lasers are obtainable that can deliver up to twice this energy per pulse.
2. Eosin isothiocyanate obtained from Molecular Probes (Eugene, OR).
3. 10 mM Tris-HCl, pH 7.9.
4. DNA fragment containing the protein binding site. The fragment should be labeled at one end with ^{32}P using, for example, end-filling with either the Klenow fragment of DNA polymerase or reverse transcriptase (11). The concentration of the stock solution of fragment is typically in the region of 100 µg/mL.
5. An appropriate buffer that is suitable for the protein–DNA complexes under investigation. Avoid the use of reducing agents.
6. Bovine serum albumin: stock solution 10 mg/mL.
7. Phenol: equilibrated with an equal volume of 0.1M Tris-HCl, pH 8.0.
8. Absolute ethanol.
9. Piperidine: 0.1M piperidine is prepared by dilution from redistilled piperidine (10.1M).
10. Polyacrylamide gel: For a 200-bp DNA fragment, a 40 × 20 × 0.04 cm 8% denaturing polyacrylamide gel is used. The gel is prepared by mixing 10 mL 40% acrylamide solution (380 g DNA-sequencing grade acrylamide, 20 g N,N'-methylene bis-acrylamide in 1 L solution), 5 mL 10X TBE (108 g Tris base, 55 g boric acid, 40 mL 0.5M EDTA, pH 8.0, in 1 L solution), 23 g urea, deionized water to 50 mL. When the urea is fully dissolved, add 100 µL 10% ammonium persulfate, and mix with rapid stirring. Then add 60 µL TEMED ($N,N,N'N'$-tetramethylethylenediamine), and mix rapidly. Pour the gel solution between the sealed gel plates,

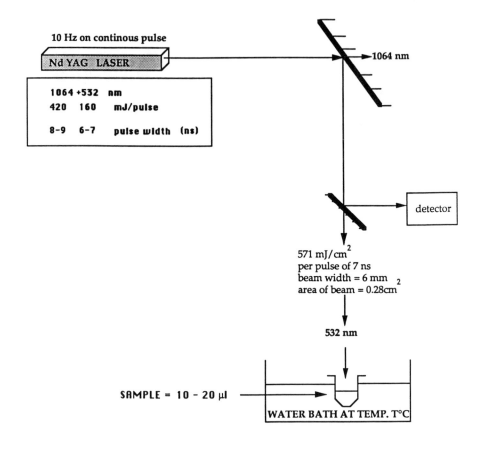

Fig. 2. Use of an Nd-YAG laser for irradiation of a 20-µL reaction mixture.

and insert a comb with 0.25-cm teeth into the gel solution. Allow to set, remove comb, and clean slots with gel buffer using a Hamilton syringe.
11. Gel running buffer: 90 mM Tris, 90 mM borate, pH 8.3, 10 mM ethylenediaminetetracetate (EDTA), pH 8.0.
12. Gel loading buffer: 95% formamide, 10 mM ethylenediaminetetracetate (EDTA), pH 8.0, 1 mg/mL xylene cyanol FF, 1 mg/mL bromophenol blue.
13. X-Ray film.
14. 10 mM dithiothreitol.

3. Methods
3.1. Preparation of Eosin-Tris Complex
1. A 10-mM stock solution of this complex is formed by incubating 10 mM eosin isothiocyanate in 10 mM Tris-HCl at pH 7.9 for 2 h at 37°C

taking care to avoid exposure to light. The eosin isocyanate is diluted from a freshly prepared 100-mM stock solution.
2. The concentration of eosin is estimated from the absorption of the solution at 525 nm ($\varepsilon_{523.2}$ = 25.6 for a 1M solution) and adjusted by the addition of the appropriate volume of double-distilled water.

3.2. Formation of Nucleoprotein Complexes

1. The first step in the detection of protein-induced deformation of DNA is the formation of a nucleoprotein complex. For example, RNA polymerase (100 nM) and end-labeled fragments of DNA containing the *lac* UV5 promoter (4 nM) are mixed in a buffer containing 100 mM KCl, 10 mM MgCl$_2$, 20 mM HEPES, pH 8.4, 3% glycerol, and 100 µg/mL bovine serum albumin in a total volume of 20 µL *(9)*.
2. This mixture is then incubated for 30 min at 37°C.

3.3. Irradiation of the Nucleoprotein Complex

1. Once the nucleoprotein complex has been formed, a fresh stock of the eosin-Tris complex is added at an appropriate concentration to 20 µL of the target solution in a small Eppendorf tube. Typically a final concentration of 50 µM is used (*see* Note 1).
2. Immediately after the addition of the eosin-Tris complex, the whole mixture is irradiated for 20 s at 10 Hz. This corresponds to a total energy dose of 115 J/cm^2 (*see* Note 2). Successful activation of the eosin-Tris complex is apparent by a detectable change in the color of the solution consequent on a shift in the absorption maximum from 523 to 514 nm on oxidation.
3. Although the half-life of singlet oxygen is sufficiently short to obviate the need to remove excess reagent, it is advisable to add a quenching agent, such as dithiothreitol, immediately on cessation of irradiation to minimize any secondary radical reactions. Typically 1 µL of 10 mM dithiothreitol is added to the irradiated solution.

3.4. Detection of Sites of Reaction with Singlet Oxygen

1. After irradiation and quenching, 30 µL double-distilled water are added to each sample.
2. 50 µL of buffered phenol freshly prepared by equilibration of melted phenol with an equal volume of 0.1M Tris-HCl are then added.
3. After mixing with a vortex mixer, the samples are centrifuged for 1 min in a benchtop microcentrifuge at 5000g to separate the aqueous and organic layers.

4. The upper aqueous layer is removed with an automatic pipet to a clean Eppendorf tube.
5. Add 3–4 vol of ethanol at 0°C, and place the samples in a dry ice/ethanol bath for 1 h.
6. Centrifuge the samples for 15 min at 5000g in a benchtop centrifuge.
7. Remove the ethanol using an automatic pipet.
8. Dry the samples in a centrifugal evaporator.
9. Resuspend in 100 µL of freshly prepared piperidine solution.
10. Heat at 90°C for 30 min (*see* Note 6).
11. Sites of cleavage are determined by separation on polyacrylamide gels (typically 40 cm, run at 60 W constant power until xylene cyanol FF marker has migrated 23 cm into gel) followed by autoradiography (typically 2–24 h exposure depending on the specific activity of the labeled DNA fragment). A typical result is shown in Fig. 3. Cleavage at a particular site results in the generation of a band of defined length. Standard Maxam and Gilbert sequencing reactions *(12)* can be performed and loaded on the same gel to identify the cleavage sites.

4. Notes

1. The half-life of singlet oxygen is short; therefore the average path length for diffusion is also short. Consequently, to ensure an adequate rate of reaction, the eosin-Tris concentration must be sufficiently high to ensure that all potential targets in the DNA are accessible to the reactive entity.
2. Although the energy dose used during irradiation may appear to be substantial, it should be kept in mind that even with a high intensity laser, each pulse delivers only 160 mJ and consequently full saturation of the system requires a considerable repetition of impulsion.
3. For optimum reactivity, it is essential that radical scavengers, such as mercapto-groups, should, as far as possible, be rigorously excluded from the reaction mixture, since they would effectively prevent any singlet oxygen from arriving at its target site. For the same reasons, the concentrations of alcohols, such as glycerol, should be kept as low as possible.
4. Ideally protein–DNA complexes should be insensitive to the presence of the unirradiated eosin-Tris complex. However, it has been observed that the half-life of certain complexes, in particular the binary CAP-DNA and RNA polymerase–DNA complexes, is reduced by approximately an order of magnitude with the sensitizer present *(9)*. For stable complexes in which the protein has a long residence time, this effect does not significantly interfere with the detection of DNA deformations, since the time of irradiation is short relative to the stability of the

Fig. 3. Reaction of singlet oxygen with a binary complex of *E. coli* RNA polymerase with the *lac*UV$_5$ promoter. The figure shows an autoradiograph of the pattern of reactivity on the transcribed strand in the presence and absence of RNA polymerase. Note that the bands visible in the DNA-only lane result from piperidine cleavage at (Py)$_3$ sequences, and their occurrence is independent of both irradiation and the presence of eosin-Tris.

complex. At this time it is unclear whether this effect is general or is restricted to particular complexes. Nevertheless it is essential to determine the stability of complexes under study under the precise conditions corresponding to those prevailing during irradiation. The method of choice is gel retardation.
5. The short time of irradiation allows the use of the singlet oxygen reaction in kinetic studies. Here again, to prevent any perturbation of an enzymatic manipulation of DNA, it would be necessary to add the eosin-Tris complex immediately prior to irradiation after the reaction under study had proceeded for the required time. For this purpose, a rapid mixing device would be necessary.
6. To obtain sharp bands on polyacrylamide gels, it is advisable for the samples to be transferred to clean Eppendorf tubes immediately prior to the evaporation of piperidine. Removal of piperidine in a centrifugal evaporator should also be carried out as rapidly as possible, and any form of heating should be avoided because this increases the nonspecific background cleavage of DNA by piperidine.

References

1. Lilley, D. M. J. and Palecek, C. (1984) The supercoil-stabilized cruciform of colE1 is hyper-reactive to osmium tetroxide. *EMBO J.* **3,** 1187–1195.
2. Sasse-Dwight, S. and Gralla, J. D. (1989) KMnO$_4$ as a probe for *lac* promoter DNA melting and mechanism *in vivo. J. Biol. Chem.* **264,** 8074–8081.
3. Sigman, D. S., Spassky, A., Rimsky, S., and Buc, H. (1985) Conformational analysis of *lac* promoters using the nuclease activity of 1,10-phenanthroline-copper ion. *Biopolymers* **24,** 183–197.
4. Hogan, M. E., Rooney, T. F., and Austin, R. H. (1987) Evidence for kinks in DNA folding in the nucleosome. *Nature* **328,** 554–557.
5. Le Doan, T., Perrouault, L., Hélène, C., Chassignol, M., and Thuong, N. T. (1986) Targeted cleavage of polynucleotides by complementary oligonucleotides covalently linked to iron-porphyrins. *Biochemistry* **25,** 6736–6739.
6. Tullius, T. D., Dombroski, B. A., Churchill, M. E. A., and Kam, L. (1987) Hydroxylradical footprinting: a high resolution method for mapping protein-DNA contacts. *Meth. Enzymol.* **155,** 537–558.
7. Ansari, A. Z., Chael, M. L., and O'Halloran, T. V. (1992) Allosteric underwinding of DNA is a critical step in positive control of transcription by Hg-MerR. *Nature* **355,** 87–89.
8. Schultz, S. C., Shields, S. C., and Steitz, T. A. (1991) Crystal structure of a CAP-DNA complex: the DNA is bent by 90°. *Science* **253,** 1001–1007.
9. Buckle, M., Buc, H., and Travers, A. A. (1992) DNA deformation in nucleoprotein complexes between RNA polymerase, cAMP receptor protein and the *lac* UV$_5$ promoter probed by singlet oxygen. *EMBO J.* **11,** 2619–2625.

10. Rougée, M. and Bensasson, R. V. (1986) Détermination des constantes de vitesse de désactivation de l'oxygène singulet ($^1\Delta_7$) en présence de biomolécules. *Comptes Rendues Acad. Sc. Paris* **302,** 1223–1226.
11. Travers, A. A., Lamond, A. I., Mace, H. A. F., and Berman, M. L. (1983) RNA polymerase interactions with the upstream region of the *E. coli tyrT* promoters. *Cell* **35,** 265–273.
12. Maxam, A. M. and Gilbert, W. (1980) Sequencing end-labeled DNA with base-specific chemical cleavages. *Meth. Enzymol.* **155,** 560–568.

CHAPTER 10

Ethylation Interference

Iain Manfield and Peter G. Stockley

1. Introduction

Structural studies of DNA–protein complexes have now made it clear that specific sequence recognition in these systems is accomplished in two ways, either directly by the formation of hydrogen bonds to base-pair edges from amino acid side chains located on a DNA-binding motif, such as a helix-turn-helix, or indirectly as a result of sequence-dependent distortions of the DNA conformation *(1)*. These contacts occur in the context of oriented complexes between macromolecules that juxtapose the specific recognition elements. As part of these processes, proteins make a large number of contacts to the phosphodiester backbone of DNA, as was predicted from biochemical assays of the ionic strength dependence of DNA binding.

Contacts to phosphate groups can be inferred by the ethylation interference technique *(2)*. Ethylnitrosourea (EtNU) reacts with DNA to form, principally, phosphotriester groups at the nonesterified oxygens of the otherwise phosphodiester backbone. Minor products are the result of the reactions of EtNU with oxygen atoms in the nucleotide bases themselves *(see* Note 1). Under alkaline conditions and at high temperature, the backbone can be cleaved at the site of the modification to form a population of molecules carrying either 3'-OH or 3' ethylphosphate groups.

The length of the ethyl group (approx 4.5Å) means that at a number of positions along a DNA molecule encompassing the binding site for a protein, complex formation will be inhibited by the presence of such a modification. At other sites, outside the binding site, no interference with protein binding at the specific site will be observed. Addition of the DNA-binding protein to a randomly ethylated DNA sample, followed by some procedure to separate the complexes formed from unbound DNA, will fractionate the DNA sample into those molecules able to bind protein with high affinity and those for which the ethylation has lowered the affinity (Fig. 1). In practice, modification at different sites produces molecules with a spectrum of affinities for the protein. It is therefore not possible to prove conclusively that a particular phosphate is contacted by the protein, but only that ethylation at that site interferes with complex formation.

Only occasionally are large amounts of pure protein readily available for in vitro biochemical assay of DNA-binding activity. Often only small amounts of crude nuclear extracts are available. In many commonly used assays, complex formation could not easily be detected in such situations. For example, using DNase I or hydroxyl radicals, a high level of binding site occupancy is required for a footprint to be detected. Fractional occupancy is readily detected by gel retardation of complexes but offers only limited characterization of the details of the protein–DNA interaction. Interference techniques, such as the ethylation and hydroxyl radical interference techniques *(3)*, do allow the molecular details of complex formation to be studied even when only small amounts of crude protein are available. Whatever the level of saturation, DNA fragments modified at sites reducing the affinity of protein for DNA are less likely to form complexes. The bound fraction on gel retardation assays will therefore always give an indication of the sites that do not inhibit complex formation when modified. The groups on the DNA recognized by the protein can then be inferred.

We have used the ethylation interference technique to probe the interaction of the *E. coli* methionine repressor, MetJ, with its binding site in vitro. Binding sites for MetJ consist of two or more immediately adjacent copies of an 8-bp site with the consensus sequence 5'-dAGACGTCT-3', which has been termed a "met box." X-ray crystallography has been used to determine the structure of the MetJ dimer, the complex with corepressor, *S*-adenosyl methionine (SAM), and a complex of the holorepressor with a 19-mer oligonucleotide containing two met boxes *(4,5)*. The structure of the protein–DNA

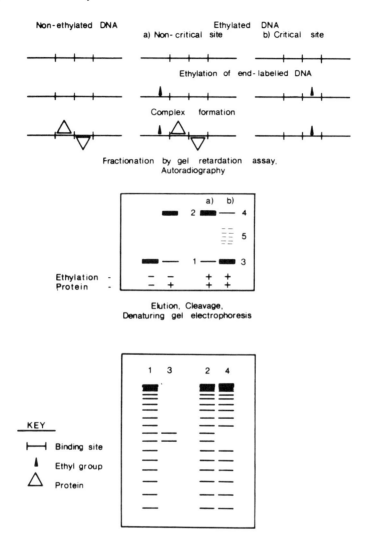

Fig. 1. Diagrammatic outline of the ethylation interference experiment. The upper section shows ethylation of end-labeled DNA (ethyl groups represented by small solid triangles) and complex formation with protein (represented by large open triangles). The expected mobility of each species on nondenaturing gels is shown in the middle section. In idealized form, the pattern of cleavage products that might be expected from recovery of each species after gel retardation assay is shown in the lower section. In practice samples represented by 1 + 3 and 2 + 4 migrate to the same position on the retardation gel and are therefore not separated. The resultant pattern is shown in Fig. 2.

complex in the crystal reveals two MetJ dimers (one per met box), binding to the DNA by insertion of a β-ribbon into the major groove. The general features of this model are corroborated by the results of the ethylation interference experiments and by data from a range of other footprinting techniques.

2. Materials

2.1. Preparation of Radioactively End-Labeled DNA

1. Plasmid DNA carrying the binding site for a DNA-binding protein on a convenient restriction fragment (usually <200 bp).
2. Restriction enzymes and the appropriate buffers as recommended by the suppliers.
3. Phenol: Redistilled phenol equilibrated with 100 mM Tris-HCl, pH 8.0.
4. Chloroform.
5. Solutions for ethanol precipitation of DNA: 4M NaCl and ethanol (absolute and 70% v/v).
6. Calf intestinal phosphatase (CIP).
7. CIP reaction buffer, 10X: 0.5M Tris-HCl, pH 9.0, 0.01M MgCl$_2$, 0.001M ZnCl$_2$.
8. TE buffer: 10 mM Tris-HCl, pH 8.0, 1 mM ethylenediaminetetra-acetic acid (EDTA).
9. Sodium dodecyl sulfate (SDS) 20% w/v.
10. EDTA (0.25M, pH 8.0).
11. T4 polynucleotide kinase (T4PNK).
12. T4PNK reaction buffer, 10X: 0.5M Tris-HCl, pH 7.6, 0.1M MgCl$_2$, 0.05M dithiothreitol.
13. Radioisotope: γ-[^{32}P]-ATP.
14. 30% w/v Acrylamide stock (29:1 acrylamide: N,N'-methylene-*bis*-acrylamide).
15. Polyacrylamide gel elution buffer: 0.3M sodium acetate, 0.2% w/v SDS, 2 mM EDTA.
16. Polymerization catalysts: Ammonium persulfate (10% w/v) and TEMED.

2.2. Ethylation Modification and Fractionation of DNA

1. End-labeled DNA in TE: 250,000 cpm are required per ethylation reaction (roughly 20 ng DNA fragment) to be performed. Standard safety

Ethylation Interference

procedures should be used when handling radiolabeled DNA (e.g., work behind Lucite shields).
2. Cacodylate buffer: 50 mM sodium cacodylate (used without adjustment of the pH, which is usually close to 8.0). Cacodylate buffer contains arsenic and therefore should be handled with caution. It is only necessary to prepare small volumes (≈10 mL) of this solution. Passing the solution through a 0.45-μm filter is the preferred method of sterilization.
3. Ethylnitrosourea (EtNU): This reagent is readily synthesized *(6)* or can be purchased from a commercial source (e.g., Sigma, St. Louis, MO, listed as *N*-nitroso-*N*-ethylurea, cat. no. N 8509). The solid material should be stored at −20°C and allowed to warm to room temperature before use. EtNU should be handled in a fume hood, and contaminated waste stored there until disposal. Wear two pairs of latex gloves when handling samples containing EtNU.
4. tRNA (1 mg/mL).
5. Solutions for ethanol precipitation of DNA: 4M NaCl and ethanol (absolute and 70% v/v).
6. Complex buffer: 10 mM Tris-HCl, pH 7.4, 150 mM NaCl, 1 mM *S*-adenosyl methionine *p*-toluene sulfonate salt (SAM). Add glycerol to this buffer to 10% v/v.
7. Purified DNA-binding protein or protein extract.
8. Non-denaturing gel acrylamide stock solution: 30% w/v (37:1 acrylamide:*N,N'*-methylene-*bis*-acrylamide).
9. Electrophoresis buffer stock solution: 1.0M Tris-HCl, pH 8.0, solid SAM.
10. Ammonium persulfate, APS (10% w/v).
11. TEMED: *N,N,N',N'*-tetramethylethylene diamine.
12. Glass plates: 150 × 150 × 1.5 mm.
13. Peristaltic pump capable of recirculating buffer at 5–10 mL/min.
14. X-ray film.
15. Autoradiography cassette.
16. X-ray film developer and fixer.
17. Plasticwrap.
18. Scalpel blade.
19. Syringe needle.
20. Polyacrylamide gel elution buffer: 0.3M sodium acetate, 0.1% w/v SDS, 1 mM EDTA.
21. 10 mM sodium phosphate, pH 7.0, 1 mM EDTA.
22. 1.0M NaOH (freshly prepared).
23. Acetic acid solutions: 1.0M and 10% v/v.

24. Sequencing gel loading buffer: 80% v/v formamide, 0.5X TBE, 0.1% w/v xylene cyanol, 0.1% w/v bromophenol blue.
25. Acrylamide stock solutions for sequencing gel: 19% w/v acrylamide, 1% w/v bis-acrylamide, 50% w/v urea in TBE.
26. TBE (1X): 89 mM Tris, 89 mM boric acid, 10 mM EDTA, pH 8.3.

2.3. Maxam-Gilbert Chemical DNA Sequencing Reactions

There is insufficient space to cover these methods in detail here, but extensive information, materials, methods, and trouble-shooting guides are readily available in published literature *(7,8)*.

3. Methods
3.1. Preparation of End-Labeled DNA

1. Digest the plasmid (e.g., 10 µg of the plasmid in a reaction volume of 200 µL) with one of the pair of restriction enzymes used to release a suitably sized DNA fragment (usually <200 bp). Extract the digest with an equal volume of buffered phenol, and then add 2.5 vol of ethanol to the aqueous layer to precipitate the DNA.
2. Add 50 µL 1X calf intestinal phosphatase (CIP) reaction buffer to the ethanol-precipitated DNA pellet (<50 µg). Add 1 U CIP, incubate at 37°C for 30 min, followed by addition of a further aliquot of enzyme, and incubate for a further 30 min. Terminate the reaction by adding SDS and EDTA to 0.1% and 20 mM, respectively, in a final volume of 200 µL, and incubate at 65°C for 15 min. Extract the digest with an equal volume of buffered phenol, then with 1:1 phenol:chloroform, and finally ethanol precipitate the DNA from the aqueous phase by addition of 2.5 vol of ethanol.
3. Redissolve the DNA pellet (≈2.5 µg) in 18 µL 1X T4PNK buffer. Add 20 µCi γ-[^{32}P]-ATP and 10 U T4PNK, and incubate at 37°C for 30 min. Terminate the reaction by phenol extraction (followed by ethanol precipitation and a second restriction enzyme digest) or by addition of nondenaturing gel loading buffer, and electrophoresis on a nondenaturing polyacrylamide gel. We use 12% (w/v) polyacrylamide gels (19:1, acrylamide:bis-acrylamide) with 1X TBE as the electrophoresis buffer.
4. After electrophoresis, locate the required DNA fragments by autoradiography of the wet gel. Excise slices of the gel containing the bands of interest using the autoradiograph as a guide. Elute the DNA into elution buffer overnight (at least) at 37°C. Ethanol precipitate the DNA, wash the pellet thoroughly with 70% v/v ethanol, dry under vacuum, and rehy-

drate in a small volume of TE buffer (e.g., 50 µL). Determine the radioactivity of the sample by liquid scintillation counting of a 1-µL aliquot.

3.2. Ethylation Modification and Fractionation of DNA

3.2.1. Ethylation Reaction

1. Dispense the required volume containing ≈250,000 cpm of radiolabeled DNA solution into an Eppendorf tube, add cacodylate buffer to a final vol of 100 µL, and heat solution to 50°C in a heating block. Prepare the minimum volume of EtNU-saturated ethanol (at 50°C) for the required reactions, add 100 µL of this to the DNA, mix, and incubate at 50°C for 60 min (see Notes 1 and 2).
2. Add to the sample 5 µL of 4M NaCl, 2 µg of tRNA, and 150 µL of ethanol. Mix and place at –70°C for 60 min or in a dry ice/ethanol bath for 15 min. Pellet the DNA by centrifugation in a microfuge for 15 min, and remove the supernatant (store separately to be destroyed by incineration). Add 500 µL of 70% ethanol, mix thoroughly, recentrifuge, and remove the supernatant. Dry the pellet briefly under vacuum.

3.2.2. Fractionation of DNA by Gel Retardation (see Note 3)

The following procedure allows the separation of free and bound DNA by means of gel retardation. The precise conditions will depend on the protein under investigation (see Notes 3–5).

1. Mix 10 mL of nondenaturing acrylamide stock solution, 1.5 mL of 1.0M Tris-HCl, pH 8.0, 29.5 mL of distilled water, 0.2 mL of APS, 15 µL of TEMED, and 1.5 mg of SAM, and pour into gel frame. Insert the well-former and leave to polymerize for 1–2 h.
2. When polymerized, insert the gel into the tank and connect peristaltic pump tubing such that buffer is pumped in both directions (i.e., top to bottom and bottom to top reservoirs). Preelectrophorese gel for 30 min at 100 V.
3. Redissolve the pellet of ethylated DNA in complex buffer plus glycerol (see Note 5). Set aside 10% of the ethylated DNA sample, which will be used as an unfractionated control to indicate the variation in level of modification at each residue. Dry this sample under vacuum and store at –20°C until step 1 of Section 3.2.4. Add DNA-binding protein to a concentration that would saturate unmodified DNA and incubate at 37°C for 15 min to allow complex formation (**N.B.** exact conditions will vary depending on the protein being studied.

4. Load the DNA–protein complex solution onto the gel and electrophorese into the gel at 250 V for 2 min. Reduce the voltage to 100 V and continue electrophoresis until a small amount of bromophenol blue dye loaded into an unused lane has reached the bottom of the gel.

3.2.3. DNA Recovery

1. After electrophoresis, remove one plate, and wrap the gel and remaining plate in clear plasticwrap film (we use Saran Wrap™). Cut a piece of X-ray film large enough to cover the lanes used on the gel and fix it to the gel firmly with masking tape. Using a syringe needle, make a number of holes through the film and gel that will serve to orient the developed film with respect to the gel.
2. Place the assembly in an autoradiography cassette and leave at 4°C overnight.
3. Take the film off the gel and develop as usual. When dry, align the film and gel using the holes created previously. Using a syringe needle, make a series of holes into the gel around the fragments of interest using the bands on the film as a guide. Remove the film and excise the marked regions of polyacrylamide. Place gel fragments (10 × 5 mm) in Eppendorf tubes, add 600 µL of gel elution buffer, and incubate at 37°C overnight (at least).
4. Transfer 400 µL of the eluate to a fresh tube, add 2 µg of tRNA and 1 mL of ethanol, mix, and place at –70°C for 60 min. Pellet DNA by centrifugation in a microfuge for 15 min. Discard the supernatant (check for absence of radioactivity), wash the pellet with 500 µL of 70% ethanol, recentrifuge, and discard the supernatant. Dry the pellet briefly under vacuum. If the DNA does not pellet readily, incubate the sample at –70°C for longer or recentrifuge at 4°C.

3.2.4. Phosphotriester Cleavage and DNA Sequencing

1. Redissolve each pellet in 15 µL sodium phosphate buffer and add 2.5 µL of 1.0M NaOH. Seal tube with plastic film (e.g., Parafilm) and incubate at 90°C for 30 min. Centrifuge samples briefly to collect any condensation. Add 2.5 µL of 1.0M acetic acid, 2 µL of 4M NaCl, 1 µg of tRNA, and 70 µL of ethanol. Leave samples at –70°C for 60 min. Pellet DNA as described in step 4 of Section 3.2.3.
2. Redissolve the pellet in 4 µL of sequencing gel loading buffer. Heat to 90°C for 2 min and load samples onto a 12% w/v polyacrylamide sequencing gel alongside Maxam and Gilbert sequencing reaction markers. Electrophorese at a voltage that will warm the plates to around 50°C. After electrophoresis, fix the gel in 1 L of 10% (v/v) acetic acid for 15

min. Transfer the gel to 3MM paper and dry under vacuum at 80°C for 60 min. Autoradiograph the gel at −70°C with an intensifying screen.
3. Compare lanes corresponding to bound, free, and control DNA for differences in intensity of bands at each position. A dark band in the "free fraction" (and a corresponding reduction in the intensity of the band in the "bound fraction") indicates a site where ethylation interferes with complex formation. This is interpreted as meaning that this residue is contacted by the protein or a portion of the protein comes close to the DNA at this point. For 5' end-labeled DNA, the ethylation reaction products migrate slightly more slowly than the Maxam-Gilbert chemical sequencing products (see Note 7 and 8, and also Fig. 1).

3.3. Results and Discussion

The result of an ethylation interference experiment with MetJ is shown in Fig. 2 along with densitometer traces showing quantitative comparisons of the distribution of products in bound and free fractions (Fig. 3) *(10)*. Visual examination of the autoradiograph shows that ethylation at 5'-pG2 results in total exclusion of such fragments from protein–DNA complexes. Densitometry indicates that ethylation at other sites inhibits complex formation to varying degrees and that there are more sites in the 3' half-site than in the 5' half-site for which ethylation inhibits complex formation. These data can be interpreted in terms of the MetJ-DNA crystal structure *(5)*. The site at which ethylation completely inhibits complex formation corresponds to the phosphate 5' to the guanine at position 2 of each met box. The crystal structure shows a contact to this phosphate from the N-terminal dipole of the repressor B-helix. Indeed, at the center of the operator site, sequence-dependent distortions of the oligonucleotide fragment away from B-DNA result in displacement of this 5' G2 phosphate by up to 2 Å in the direction of the protein. Thus, this site of complete inhibition of complex formation corresponds to an important contact between the DNA and a secondary structural element in the protein, which presumably is unable to adjust to the presence of a bulky ethyl group. The other sites of ethylation interference effects are contacted by amino acid side chains and peptide backbone groups in extended loops of the repressor. It might be expected that side chains and loops could be flexible enough to reorient in order to reduce steric hindrance between the protein and the ethyl group, thus explaining the partial interference effects. Similar good correlations between the contacts

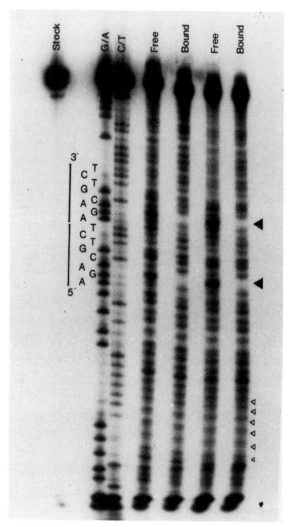

Fig. 2. Ethylation interference of MetJ:DNA interaction. Samples for denaturing gel electrophoresis were prepared following the methods given here. "Stock" indicates nonethylated DNA that has been through the cleavage reaction. "G/A" and "C/T" are the products of the purine- and pyrimidine-specific Maxam-Gilbert chemical cleavage reactions, respectively. "Free" and "Bound" are the DNA fractions that can and cannot form complexes, respectively. The two sets of data represent results obtained with DNA ethylated for 30 min (left-hand lanes) or 60 min (right-hand lanes). The sequence of the MetJ binding site is indicated along the side of the autoradiograph. The phosphate, ethylation of which interferes most strongly with complex formation, is indicated by a large solid triangle. Small, open triangles indicate the minor reaction products of the cleavage reaction, which, for small fragments, are resolved on these gels.

Ethylation Interference

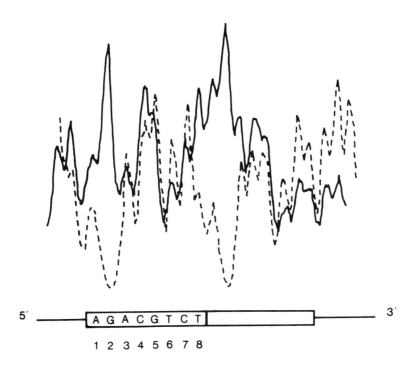

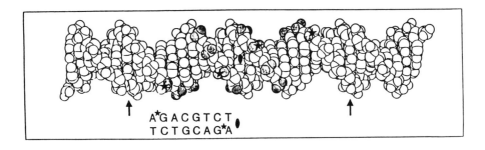

Fig. 3. Summary of the results. **Top:** Densitometer traces of ethylated DNA from the bound and free fractions. DNA from the free fraction is indicated by the trace with solid lines and DNA from the bound fraction by the trace with dashed lines. The position of each binding site is indicated at the bottom of the trace. The sequence within each box is 5'-AGACGTCT-3'. Note that the greatest inhibitory effect is at the second residue (G2) of each box and that the number of positions, ethylation of which inhibits complex formation, is greater in the 3' box than in the 5' box. **Bottom:** Space filling representation of the operator site showing the positions where ethylation results in complete inhibition of complex formation (starred phosphates).

identified by ethylation interference experiments and those seen in crystals have been demonstrated in other systems, such as phage 434 repressor *(11)* and phage Lambda repressor *(12).*

4. Notes

1. Modification at secondary (nonphosphate) sites: Early work on the reaction of alkylating agents with DNA showed that a number of products are obtained. With EtNU, phosphotriester groups comprise 60–65% of the reaction products *(6)*. The remaining products are the result of reactions at base oxygen groups with relative abundance in the order thymine O2 = guanine O6 > thymine O4 >> cytosine O2 for double-stranded DNA *(9)*. To our knowledge, the effects of such modifications on DNA binding by proteins have not been addressed in the literature on ethylation interference experiments. However, some prediction of the effects might be made based on a knowledge of the structure of DNA. Thymine O2 and cytosine O2 are in the minor groove, and guanine O6 and thymine O4 are in the major groove. For sequence-specific DNA-binding proteins interacting with DNA via the major groove, it would be expected that the presence of an ethyl group would inhibit complex formation, but that a modification in the minor groove would be less inhibitory.
2. Level of modification of DNA by EtNU: A similar intensity of each band (subject to the position-dependent variation of reactivity observed with EtNU) is the required level of modification. We have used a single batch of "home-produced" EtNU for all our ethylation interference experiments. The conditions of time and temperature used with this batch give an appropriate level of probe modification. There may be variations in the reactivity of EtNU from different sources leading to under- or overmodification. The correct modification conditions can readily be determined by performing a test ethylation on a small amount of DNA, such as 20,000 cpm, followed by alkaline cleavage (omitting the gel fractionation step), sequencing gel electrophoresis, and autoradiography. Overmodification will produce a bias toward short fragments.
3. Choice of fractionation method: The original report of the use of the phosphate ethylation reaction used the filter binding assay to fractionate protein-bound and protein-free DNA *(2)*. Filter-bound DNA is then eluted by washing the filter in a high salt buffer containing SDS. For other experiments, we find that TE + 0.1% SDS efficiently elutes DNA from nitrocellulose filters. The rapid recovery of DNA from filters is an advantage of using the filter binding assay compared with the gel retardation assay. Despite this, fractionation by gel retardation assay has proven to be by far the most popular method in the literature. The

advantage of the gel retardation assay is that the presence and amounts of multiple complexes can be determined, which is not possible by filter binding. Parallel binding reactions with ethylated and unethylated DNA and separation on nondenaturing polyacrylamide gels readily demonstrate any differences in the mobility of complexes formed with each DNA sample

4. Cofactor requirements: High affinity DNA binding in our system (MetJ) is dependent on the presence of S-adenosyl methionine at millimolar concentrations. This cofactor is present in binding reactions and is included in the gel mix, but it would be prohibitively expensive also to include it in the electrophoresis buffer. This does not seem to affect the results obtained by this technique. Electrophoresis for extended times does deplete the lower region of the gel of corepressor leading to some complex dissociation. Another feature specific to this system is the hydrolysis of the corepressor to presumably inactive products. For this reason, gels were not left to polymerize for more than 2 h.

In other systems where a cofactor is not required or in which the cofactor is cheap and/or stable over extended periods, the conditions used for the gel retardation assay fractionation step should be optimized by a consideration of the specific features of the system under study. For further details, consult other chapters of this book.

5. Effects of salt precipitates: During ethanol precipitation of DNA, salt is often also precipitated. This white crystalline precipitate is readily distinguished from the almost clear nucleic acid pellets. The interaction of DNA-binding proteins with their sites is strongly ionic strength-dependent, and therefore the presence of a high concentration of salt following ethanol precipitations will inhibit complex formation in addition to any ethylation interference effects. After the cleavage reaction at modified sites, another ethanol precipitation is performed. The presence of a large amount of salt at this stage will prevent complete dissolution of the pellet and will interfere with subsequent electrophoresis. A dark background between each band was often observed on autoradiographs. It is believed that this is caused by the presence of salt in the sample. Reprecipitation as described below helps to reduce this problem.

To remove any salt precipitate, the pellet can be dissolved in a small volume of TE (e.g., 100 µL) and precipitated by addition of 2 vol of ethanol without addition of further amounts of salt. The DNA can be pelleted as described in step 4 in Section 3.2.3.

6. Recommended controls for the gel retardation assay fractionation step: With a premodification reaction such as this, it is important to perform appropriate control reactions especially for the gel retardation step. For

such binding reactions, we use 20,000 cpm unmodified DNA in the presence and absence of MetJ at the concentration used in the binding reaction with ethylated DNA samples. The specific activity of protein samples may vary from batch to batch. The control reaction outlined above will confirm that the protein sample used is active for DNA binding. In gel retardation assays, the mobility of complexes is a function of a number of properties of the system, such as charge and molecular weight of the protein, stoichiometry of the complex, and bending of the DNA induced by binding of the protein. Demonstration that the mobility of complexes formed with ethylated and unethylated DNA is the same is probably good evidence that there are no significant differences between the complexes.

The protein concentration used in the fractionation step dictates how many interfering sites are reported. Because the ethyl groups at different sites affect protein binding to differing degrees, increasing the protein concentration can mask any weak interference effects such that only the most strongly interfering sites will be detected. The control reaction with unmodified DNA will show the level of binding site saturation, and therefore indicate the level of discrimination between strongly and weakly interfering sites that can be expected. Using a range of protein concentrations in the binding reaction with aliquots of the ethylated DNA should therefore allow the strength of the inhibitory effect at each site to be placed in rank order. This is valuable structural information, since it might be expected that the strongest effects will be observed at sites that are in closest contact to the protein in the complex.

7. The presence of multiple cleavage products at each phosphate: The products of the cleavage reaction at phosphotriester groups carry either 3'-OH or 3'-ethylphosphate groups. For large fragments on low percentage polyacrylamide gels, these two species are not resolved. However, for short fragments on high percentage gels, two bands are observed at each residue. In practice, this does not produce problems with data analysis.

8. Trouble shooting: We have experienced few problems with this technique. Most problems have been associated with the specific properties of the proteins we have studied. However, it is possible to envisage a number of potential problems and explanations for these, and remedies are presented here.

Of the available structures of DNA-binding proteins complexed to DNA fragments, there are none in which the protein does not make some contacts to the phosphodiester backbone. Thus, it is expected that because of the size of the ethyl group, an interference effect will always be observed. In the event that no inhibition of complex formation is observed, it should be confirmed that ethylation has occurred by performing a titration of the ethylation reaction as discussed in Note 1.

It is possible that an interference effect is observed but that no cleavage products are observed on the sequencing gel, although the full length DNA is present. This could be caused by an error with the buffer used to resuspend the eluted DNA pellet, the NaOH solution used to catalyze strand scission, or the temperature of the reaction, all of which can be readily checked.

References

1. Otwinowski, Z., Schevitz, R. W., Zhang, R.-G., Lawson, C. L., Joachimiak, A., Marmorstein, R. Q., Luisi, B. F., and Sigler, P. B. (1988) Crystal structure of the *trp* repressor/operator complex at atomic resolution. *Nature* **335,** 321–329.
2. Siebenlist, U. and Gilbert, W. (1980) Contacts between *Escherichia coli* RNA polymerase and an early promoter of phage T7. *Proc. Natl. Acad. Sci. USA* **77,** 122–126.
3. Hayes, J. J. and Tullius, T. D. (1989) The missing nucleoside experiment: a new technique to study recognition of DNA by protein. *Biochemistry* **28,** 9521–9527.
4. Rafferty, J. B., Somers, W. S., Saint-Girons, I., and Phillips, S. E. V. (1989) Three-dimensional crystal structures of *Escherichia coli met* repressor with and without corepressor. *Nature* **341,** 705–710.
5. Somers, W. S. (1990) Crystal structures of methionine repressor of *E. coli* and its complex with operator. Ph.D. Thesis, University of Leeds, Leeds, England.
6. Jensen, D. E. and Reed, D. J. (1978) Reaction of DNA with alkylating agents. Quantitation of alkylation by ethylnitrosourea of oxygen and nitrogen sites on poly [dA-dT] including phosphotriester formation. *Biochemistry* **17,** 5098–5107.
7. Maxam, A. M. and Gilbert, W. (1977) A new method for sequencing DNA. *Proc. Natl. Acad. Sci. USA* **74,** 560–564.
8. Maxam, A. M. and Gilbert, W. (1980) Sequencing end-labeled DNA with base-specific chemical cleavages. *Meth. Enzymol.* **65,** 499–560.
9. Singer, B. (1976) All oxygens in nucleic acids react with carcinogenic ethylating agents. *Nature* **264,** 333–339.
10. Phillips, S. E. V., Manfield, I., Parsons, I., Davidson, B. E., Rafferty, J. B., Somers, W. S., Margarita, D., Cohen, G. N., Saint-Girons, I., and Stockley, P. G. (1989) Cooperative tandem binding of Met repressor from *Escherichia coli*. *Nature* **341,** 711–715.
11. Bushman, F. D., Anderson, J. E., Harrison, S. C., and Ptashne, M. (1985) Ethylation interference and X-ray crystallography identify similar interactions between 434 repressor and operator. *Nature* **316,** 651–653.
12. Ptashne, M. (1987) *A Genetic Switch: Gene Control and Phage Lambda.* Blackwell Scientific, Boston, MA.

CHAPTER 11

Uranyl Photofootprinting of DNA–Protein Complexes

Peter E. Nielsen

1. Introduction

It has long been known that the uranyl(VI) ion (UO_2^{2+}) forms strong complexes with various inorganic and organic anions, including phosphates, and that the photochemically excited state of this ion is a very strong oxidant *(1)*. For instance, uranyl-mediated photooxidation of alcohols has been studied in detail *(2,3)*. It is also widely recognized that uranyl chemistry and photophysics/photochemistry are very complex. Thus monomeric UO_2^{2+} is only present at low pH (pH ~ 2.0), whereas polynuclear species and various "hydroxides," which often precipitate, form at higher pH *(4)*.

In spite of this complexity, we have found that uranyl-mediated photocleavage of DNA can be used to probe for accessibility of the phosphates in the DNA backbone *(5–8)*. Thus, uranyl is a sensitive probe for protein–DNA phosphate contacts *(5,6)* as well as for DNA conformation in terms of DNA minor groove width *(7,8)*.

The systems that we have analyzed so far by uranyl-mediated DNA photocleavage include the λ-repressor/O_R1 operator complex *(5)*, *E. coli* RNA polymerase/deoP1 promoter transcription initiation open complex *(6)*, transcription factor IIIA (TFIIIA)/*Xenopus* 5S internal control region (ICR) complex *(9)*, catabolite regulatory protein (CRP)/operator DNA complex and the CRP/RNA polymerase/deoP_2 promoter initiation complex *(23)*, bent kinetoplast DNA *(7)*, and the 5S-gene ICR-DNA *(8)*. Furthermore, we have found that drug-DNA complexes (exempli-

From: *Methods in Molecular Biology, Vol. 30: DNA–Protein Interactions: Principles and Protocols*
Edited by: G. G. Kneale

fied by mitramycin *[10]* and distamycin) *(21)* may also be studied by uranyl photofootprinting.

The molecular mechanism for uranyl-mediated photocleavage of DNA is not fully understood, but we have shown that uranyl binds to the phosphates of DNA and oxidizes the proximal deoxyriboses. The main products are 3'- and 5'-phosphate termini in the DNA, and the free nucleobases are liberated in the process *(22)*. Since uranyl binding is to the phosphate groups of the DNA, very little sequence dependence of the photocleavage is seen.

2. Materials

1. Uranyl nitrate ($UO_2[NO_3]_2$), analytical grade: 100 mM stock solution in H_2O. This solution was found to be stable for photofootprinting purposes for at least 12 mo and was diluted to working concentrations immediately prior to use (*see* Note 1).
2. ^{32}P end-labeled DNA restriction fragments (*see* Note 2).
3. Buffer for formation of protein-DNA complex (*see* Note 3).
4. 0.5M Sodium acetate, pH 4.5.
5. Ethanol 96%.
6. Ethanol 70%.
7. Calf thymus DNA, 2 mg/mL.
8. Gel loading buffer: 80% formamide in TBE buffer, 0.05% bromphenol blue, 0.05% xylene cyanol.
9. TBE buffer: 90 mM Tris-borate, 1 mM EDTA, pH 8.3.
10. Polyacrylamide gel: 8% acrylamide, 0.3% *bis*-acrylamide, 7M urea, TBE buffer. Size: 0.2 mm × 60 × 20 cm.
11. λ-Repressor: 1 µg/µL (*see* Note 4).
12. Buffer for λ-repressor footprinting: 40 mM Tris-HCl, pH 7.0, 2.5 mM $MgCl_2$, 1 mM $CaCl_2$, 0.1 mM EDTA, 200 mM KCl.
13. DNase I: 1 mg/mL in 10 mM Tris-HCl, pH 7.4, 1 mM $MgCl_2$.
14. X-ray film: Agfa Curix RP1.
15. Philips TL 40 W/03 fluorescent light tube that fits into standard (20 W) fluorescent light tube sockets if the transformer is changed to 40 W (*see also* Note 5).

3. Methods

3.1. Uranyl Photofootprinting

1. Form the complex to be analyzed by mixing the [^{32}P]-end-labeled DNA fragment (> 20,000 cpm/sample) (*see* Note 2) with the DNA binding ligand in 90 µL of the buffer (*see* Note 3) (containing 0.5 µg calf thymus DNA carrier) in a 1.5-mL polypropylene microfuge tube at the desired temperature.

2. Dilute the 100 mM uranyl nitrate stock solution to 10 mM in H_2O.
3. Add 10 μL of this to the sample and mix well (see Note 6).
4. Place the sample in a thermostatted heating/cooling block if the temperature is critical.
5. Irradiate the sample for 30 min at 420 nm by placing the open microfuge tube directly under the fluorescent light tube at a distance of 10 cm (see Note 5).
6. Add 20 μL of 0.5M sodium acetate, pH 4.5, to prevent coprecipitation of uranyl salts (which will interfere with subsequent gel analysis) and precipitate the DNA by addition of 250 μL 96% ethanol.
7. Place the sample on dry ice for 15 min (or overnight at –20°C) and centrifuge for 30 min at 20,000g.
8. Wash the pellet with 100 μL 70% ethanol, dry in vacuo, and redissolve in 4–10 μL of 80% formamide gel loading buffer.
9. Heat the sample at 90°C for 5 min.
10. Load 10,000 cpm on a polyacrylamide sequencing gel (0.2–0.4 mm thick, 60 cm long) and run the gel at 2000 V. A sequence ladder (e.g., A + G) is run in parallel (see Fig. 1).
11. Visualize radioactive bands by autoradiography overnight (or longer) at –70°C using amplifying screens.
12. Quantitate the results by densitometric scanning of the autoradiograms, if desired (see Note 7).

3.2. Example

Figure 1 shows a footprinting experiment of the complex between λ-repressor and the O_R1 operator DNA using uranyl and DNase I. Quantitative analysis of these results by densitometric scanning and displaying the results on the DNA sequence (Fig. 2A) reveal that mainly four regions of the O_R1 operator are protected from photocleavage by uranyl. Furthermore, the results show that these regions coincide with those protected from attack by hydroxyl radicals as well as the phosphates indicated by ethylation interference and X-ray crystallography to be involved in protein binding (15). However, the footprinting patterns are not identical, as discussed below. Displaying of uranyl footprinting results on a DNA double-helix model is very informative and, in the case of the λ-repressor/O_R1 complex, shows (Fig. 2B) that the repressor binds to one face of the helix and that the phosphate contacts are positioned mainly in regions across the major groove of the DNA, thus in full accord with binding of the $α_3$-recognition protein helix of each repressor subunit in the major groove.

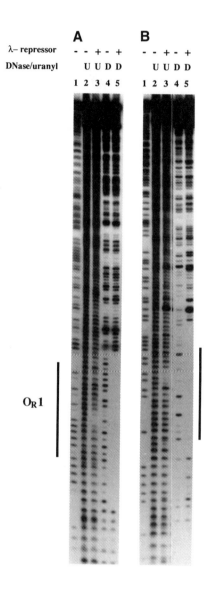

Fig. 1. Uranyl photofootprint and DNase I footprint of the λ-repressor/O_R1 operator complex (see ref. 5 for details). The O_R1 operator sequence was cloned into the Bam/HindIII site of pUC19, and the 225-bp EcoRI/PvuIII fragment labeled with [^{32}P] at the 3'- (A) or 5'-end (B) of the EcoRI site was used in the experiments. Lanes 2 and 4 are controls without added λ-repressor (0.7 µg/sample). Lanes 2 and 3 are uranyl photofootprints, whereas lanes 4 and 5 are DNase I (0.5 µg/mL, 5 min at room temperature) footprints. Lanes 1 are A + G sequence reactions obtained by treating the DNA with 60% formic acid for 5 min at room temperature and subsequent piperidine treatment. The samples were analyzed on an 8% polyacrylamide gel and run at 2500 V. The gel was subjected to autoradiography for 16 h at –70°C using intensifying screens.

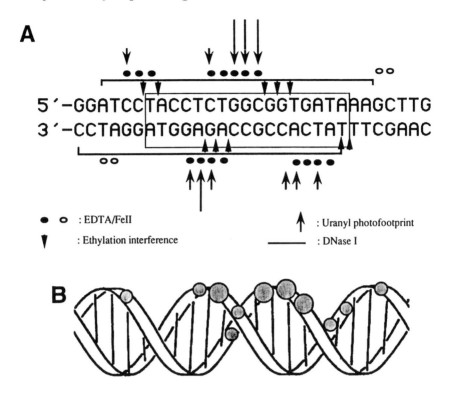

Fig. 2. **A.** O_R1-operator sequence (box) showing uranyl photofootprint (arrows), EDTA/FeII footprint (dots *[16]*) ethylation interference (this is for O_L1) (arrowheads *[15]*), and DNase I footprint (brackets). **B.** Display of the uranyl photofootprint on O_R1-helix model. The size of the dots signifies degree of protection.

3.3. Comparison of Uranyl Photoprobing to Other Techniques

The results obtained by uranyl photofootprinting are comparable to those obtained by hydroxyl radical (EDTA[FeII]) footprinting *(16,17)* and ethylation interference experiments *(18)*.

Since the uranyl ion binds to the phosphates of the DNA, a uranyl photofootprinting experiment reports on phosphates of the DNA backbone that are accessible to the uranyl and that are therefore not involved in contacts with the bound ligand. Hydroxyl radical footprinting reports accessibility of the deoxyriboses of the DNA backbone, and in the cases studied with both uranyl and hydroxyl radical probing (λ-repressor *[5,16]*, RNA polymerase *[6,19]*, and TFIIIA

[9,20]), the footprint obtained by uranyl covers fewer bases than that obtained by EDTA(FeII).

Interference probing by phosphate ethylation using ethylnitrosourea also reports the involvement of individual phosphate groups in protein-DNA interaction. However, this is an interference technique and therefore only phosphate groups that are indispensable for complex formation are detected. Thus for small complexes, such as the λ-repressor, O_R1 complex ethylation interference and uranyl results are virtually equivalent, whereas for larger complexes, such as RNA polymerase-promoter-complexes, only part of the contacts detected by uranyl photofootprinting are picked up by ethylation interference. In conclusion, ethylation interference, hydroxyl radical, and uranyl footprinting complement each other. Hydroxyl radicals report on the accessibility of individual deoxyriboses, uranyl footprinting reports on the accessibility of individual phosphates, and ethylation interference reports on phosphates that are indispensable for complex formation. Finally, both hydroxyl radicals and uranyl are able to report variations in DNA conformation, and for both probes, groove width has been implicated as the determinant parameter, although the features of the DNA structure probed are not identical.

4. Notes

1. Uranyl acetate ($UO_2[CH_3COO]_2$) gives identical results, but the 100-mM stock solution in this case has to be made 50 mM in HCl in order to be stable.
2. [^{32}P]-end-labeled DNA fragments of ~50–300 bp in length are prepared by standard techniques *(11):* Typically, the plasmid containing the protein binding site is opened by a restriction enzyme that cleaves at a distance of 20–50 bp from the binding site. This distance is important since the best resolution is obtained in the 20–70 bases interval and the bands of uranyl-cleaved DNA fragments become "fuzzy" above ~100 bases. The plasmid is labeled either at the 3'-end with α-[^{32}P]-dNTP and the Klenow fragment of DNA polymerase, or at the 5'-end (after dephosphorylation with alkaline phosphatase) with γ-[^{32}P]-ATP and polynucleotide kinase. The plasmid is then treated with a second restriction enzyme cutting 50–300 bp from the labeling site and the DNA fragment containing the protein binding site is purified by gel electrophoresis on a 5% polyacrylamide, in TBE buffer. The DNA fragment is extracted from the excised gel slice with 0.5M ammonium-acetate, 1

mM EDTA (16 h, room temperature), and precipitated by addition of 2 vol of 96% ethanol. The pellet is washed with 70% ethanol and dried.
3. Choice of buffer: The right choice of buffer for a uranyl photofootprinting experiment is crucial for a successful result. In particular, the pH of the medium is important. The uranyl-mediated photocleavage of DNA is extremely dependent on pH being most efficent at ~6.0, less efficent at 5.0 and 7.0, and virtually absent at 8.0. Furthermore, as the pH is lowered, a strong modulation of the sequence dependence of the cleavage is observed. In fact, this modulation reflects the conformation of the DNA. Thus, for photofootprinting where an even cleavage is warranted, buffers of pH 6.5–7.0 are advantageous, whereas buffers of pH 6.0–6.5 should be chosen for studies of DNA structure.

The composition of the buffer and the buffer capacity are also of importance. Since the uranyl solution is acidic, it should be checked if addition of uranyl changes the pH of the medium. Furthermore, uranyl-mediated photocleavage of DNA is most efficient in acetate or formate buffers, less efficent in Tris-HCl, very inefficient in HEPES or PIPES buffer, and virtually absent in phosphate buffers (uranyl phosphate precipitates). The ionic strength (as Na^+) is of minor importance, and the cleavage is not affected by the presence of Mg^{2+} or DTT either. Finally, the uranyl photoreaction is not influenced by the temperature (0–70°C) *(22)*. Within these constraints, a buffer that allows protein-DNA binding must be chosen.
4. λ-Repressor was prepared according to ref. *12* using an overproducer plasmid, pAE305 in *E. coli.*
5. Light source: Any light source emitting at 300–420 nm can be used. This could be the Philips TL 40 W/03 tube emitting at 420 nm ± 30 nm. Alternative fluorescent light tubes are Philips TL 20 W/12 (300 nm) or TL 20 W/09N (365 nm). Lamps emitting below 300 nm are not recommended because of absorption by the DNA bases at these wavelengths. The fluorescent light tubes suggested in this chapter are not very powerful, but quite sufficent for footprinting, and they are inexpensive and do not require sophisticated power supplies. However, if shorter irradiation times are required, uranyl photofootprinting experiments are quite adequately performed with Pyrex-filtered light from high pressure Hg-lamps, Xenon lamps, or lasers of the appropriate wavelength (300–420 nm).
6. Order of mixing: It is important that the uranyl be added last since the uranyl-DNA complex is very stable *(K_a is estimated to 10^{10} M^{-1} [22])* and uranyl-DNA aggregates that precipitate often form without this adversely affecting the outcome of the footprinting reaction. Conversely,

if uranyl is added prior to the DNA binding ligand, the ligand will only have limited access to the DNA. It is also extremely important that dilution of the uranyl stock solution be performed immediately prior to use, since uranyl solutions are not stable at pH ≥ 2.0.

7. Examples of densitometric scanning and quantification of footprinting experiments can be found in refs. *8, 13,* and *14.*

References

1. Burrows, H. D. and Kemp, T. J. (1974) The photochemistry of the uranyl ion. *Chem. Soc. Rev.* **3,** 138–165.
2. Azenha, M. E. D. G., Burrows, H. D., Furmosinho, S. J., and Miguel, M. G. M. (1989) Photophysics of the excited uranyl ion in aqueous solutions. Part 6. Quenching effects of aliphatic alcohols. *J. Chem. Soc. Faraday Trans.* **85,** 2625–2634.
3. Cunningham, J. and Srijaranai, S. (1990) Sensitized photo-oxidations of dissolved alcohols in homogeneous and heterogenous systems. 1. Homogeneous photosensitization by uranyl ions. *J. Photochem. Photobiol. A. Chem.* **55,** 219–232.
4. Greenwood, N. N. and Earnshaw, A. (1984) in *Chemistry of the Elements,* Oxford, Pergamon, p. 1478.
5. Nielsen, P. E., Jeppesen, C., and Buchardt, O. (1988) Uranyl salts as photochemical agents for cleavage of DNA and probing of protein-DNA contacts. *FEBS Lett.* **235,** 122–124.
6. Jeppesen, C. and Nielsen, P. (1989) Uranyl-mediated photofootprinting reveals strong *E. coli* RNA polymerase-DNA backbone contacts in the +10 region of the deoP1 promoter open complex. *Nucleic Acids Res.* **17,** 4947–4956.
7. Nielsen, P. E., Møllegaard, N. E., and Jeppesen, C. (1990) Uranyl photoprobing of conformational changes in DNA induced by drug binding. *Anti-Cancer Drug Design* **5,** 105–110.
8. Nielsen, P. E., Møllegaard, N. E., and Jeppesen, C. (1990) DNA conformational analysis in solution by uranyl-mediated photocleavage. *Nucleic Acids Res.* **18,** 3847–3851.
9. Nielsen, P. E. and Jeppesen, C. (1990) Photochemical probing of DNA complexes. *Trends Photochem. Photobiol.* **1,** 39–47.
10. Nielsen, P. E., Cons, B. M. G., Fox, K. R., and Sommer, V. B. (1990) Uranyl photofootprinting. DNA structural changes upon binding of mithramycin, in *Molecular Basis of Specificity in Nucleic Acid Drug Interactions,* vol. 23 (Pullman, B. and Jortner, J., eds.) Jerusalem Symposium on Quantum Chemistry and Biochemistry, Kluwer Academic, Dordrecht, pp. 423–432.
11. Maniatis, T., Fritsch, E. F., and Sambrook, J. (1982) *Molecular Cloning. A Laboratory Manual.* Cold Spring Harbor Laboratory, Cold Spring Harbor, NY.
12. Amann, E., Brosins, J., and Ptasne, M. (1983) Vectors bearing a hybrid trp-lac promoter useful for regulated expression of cloned genes in *Escherichia coli. Gene* **25,** 167–178.
13. Jeppesen, C. and Nielsen, P. E. (1989) Photofootprinting of drug-binding sites on DNA using diazo- and azido-9-aminoacridine derivatives. *Eur. J. Biochem.* **182,** 437–444.

14. Dabrowiak, J. C., Kissinger, K., and Goodisman, J. (1989) Quantitative footprinting analysis of drug-DNA interactions: Fe(lII)methidium-propyl-EDTA as a probe. *Electrophoresis* **10,** 404–412.
15. Schultz, S. C., Shields, G. C., and Steitz, T. A. (1991) Crystal structure of a CAP-DNA complex: the DNA is bent by 90°. *Science* **253,** 1001–1007.
16. Tullius, T. D. and Dombroski, B. A. (1986) Hydroxyl radical "footprinting": high-resolution information about DNA protein contacts and application to lambda repressor and Cro protein. *Proc. Natl. Acad. Sci. USA* **83,** 5469–5473.
17. Burkhoff, A. M. and Tullius, T. D. (1987) The unusual conformation adopted by the adenine tracts in kinetoplast DNA. *Cell* **48,** 935–943.
18. Siebenlist, U., Simpson, R. B., and Gilbert, W. (1980) *E. coli* RNA polymerase interacts homologously with two different promoters. *Cell* **20,** 269–281.
19. O'Halloran, T. V., Frantz, B., Shin, M. K., Ralston, D. M., and Wright, J. G. (1989) TheMerR heavy metal receptor mediates positive activation in a topologically novel transcription complex. *Cell* **56,** 119–129.
20. Vrana, K. E., Churchill, M. E., Tullius, T. D., and Brown, D. D. (1988) Mapping functional regions of transcription factor TFIIIA. *Mol. Cell. Biol.* **8,** 1684–1696.
21. Møllegaard, N. E. (1993) Uranyl probing of DNA structures and protein DNA interactions. Thesis, University of Copenhagen.
22. Nielsen, P. E., Hiort, C., Buchardt, O., Dahl, O., Sönnichsen, S. H., and Nordèn, B. (1992) DNA binding and photocleavage by uranyl(VI) ($UO2^{2+}$) salts. *J. Amer. Chem. Soc.* **114,** 4967–4975.
23. Møllegaard, N. E., Rasmussen, P. B., Valentin-Hansen, P., and Nielsen, P. E. Characterization of promoter recognition complexes formed by CRP and Cy + R for repression and by CRP and RNA polymerase for activation of transcription on the *E. coli deo*Cp_2 promoter. *J. Biol. Chem.* (in press).

CHAPTER 12

Nitration of Tyrosine Residues in Protein–Nucleic Acid Complexes

Simon E. Plyte

1. Introduction

Chemical modification is a powerful tool for investigating the accessibility and function of specific amino acids within folded proteins. It has provided significant information regarding the role of different amino acids at the binding sites of numerous enzymes and DNA binding proteins. This information has frequently been used to plan subsequent site-directed mutagenesis experiments. Additionally, the data from chemical modification experiments complements that from crystallographic and NMR data in elucidating the residues located at the DNA binding site.

Reagents exist to modify cysteine, methionine, histidine, lysine, arginine, tyrosine, and carboxyl groups selectively. However, in this chapter we are only concerned with the selective modification of tyrosine residues (for reagents and conditions for the modification of the other amino acids, *see* ref. *1*). The side chain of tyrosine can react with several compounds, the most commonly used being *N*-acetylimidazole and tetranitromethane (TNM). *N*-acetylimidazole will *O*-acetylate tyrosine residues in solution *(2)*, and this reagent has been used to modify numerous proteins including the Fd gene 5 protein *(3)*. However, this reagent can also *N*-acetylate primary amines, and in the study on the Fd gene 5 protein *(3)*, in addition to acetylation of three tyrosine residues, all five lysine residues were found to be modified. Tetranitromethane is a reagent highly specific for tyrosine resi-

From: *Methods in Molecular Biology, Vol. 30: DNA–Protein Interactions: Principles and Protocols*
Edited by: G. G. Kneale

dues and reacts under mild conditions to form the substitution product 3-nitrotyrosine *(4)*. The modified tyrosine has a characteristic absorption maximum at 428 nm, and this can be made use of to quantitate the number of tyrosine residues modified *(4)*. However, under harsher conditions, there have been some reports of modification of sulfhydryl groups and limited cases of reaction with histidine and tryptophan *(5)*.

1.1. Strategies

1.1.1. Tyrosine Accessibility

The general strategy employed in chemical modification experiments is to determine the accessibility of the target residues within the native protein and the extent of protection offered by the bound substrate. Peptide mapping of the labeled protein then allows the roles of the individual residues to be assessed. First, the free protein is nitrated and then digested into fragments by proteolysis. These peptides are subsequently separated to enable identification of the modified residue(s). The nucleoprotein complex is then nitrated and the modified residues identified in a similar way. From these results, the extent of protection at each site can be established.

For peptide mapping, a protease should be chosen that, on digestion of the target protein, will place each tyrosine in a separate peptide. However, this is not essential if the modified residues are identified by N-terminal sequencing. It is possible that tyrosine modification may affect the efficiency of α-chymotrypsin digestion, and this should be allowed for. The peptides can be separated by reverse- phase HPLC, and those containing tyrosine purified for further analysis. The tyrosine containing peptides can be easily identified directly after HPLC purification by their characteristic fluorescence emission maximum at 305 nm (when excited at 278 nm). A particular tyrosine residue can then be identified by N-terminal sequence analysis.

The identification of nitrated tyrosine residues in the free protein provides information concerning the solvent accessibility of these residues in the protein and indicates which residues are likely to be buried within the protein. DNA protection studies will indicate which residues may be involved in protein–DNA interactions. However, the protection from nitration by bound DNA is only an indication of a functional role for a particular residue, since the bound DNA may confer protection to a residue several angstroms away. Consequently, functional studies need to be performed to determine further the role of the protected residue(s). The situation is analogous to the two types of analysis frequently used in the investigation of the DNA bases involved in complexes—"footprinting" and "interference" techniques. The data obtained from chemical modification and protection studies can then be used to design site-directed mutagenesis experiments to look at the function of an individual residue by observing the effects of its replacement with other amino acid residues.

1.1.2. Functional Studies

A protocol for functional studies will not be described in this chapter, but some general considerations will be mentioned here. One should nitrate the free protein and determine whether the modified protein still binds to DNA. This information should indicate whether the residues protected in the nucleoprotein complex are implicated in DNA binding. However, with proteins that bind cooperatively to DNA, a reduction in DNA binding affinity may result from disruption of protein–protein rather than protein–DNA interactions. A possible way to resolve this ambiguity is to bind the native and modified protein to short oligonucleotides where the cooperativity factor is negligible. Modification of residues involved in protein–protein interactions should not significantly affect the intrinsic binding of the modified protein to DNA, when compared to the native protein.

Tyrosine residues can interact with DNA either by hydrophobic interactions via stacking with DNA bases or by hydrogen bonding with the nucleotide through the phenolic OH group *(6)*. Nitrotyrosine has a pK_a of 8.0, which may disrupt H-bonding as well as base stacking interactions. However, the addition of sodium dithionate reduces 3-nitrotyrosine to 3-amino tyrosine (which has a pK_a similar to that of native tyrosine) and may restore H-bonding interactions *(7)*. Reduction with this reagent may provide further information concerning the nature of the tyrosine–nucleic acid interaction.

1.1.3. Rates of Modification

Nitration of a protein will initially report on the accessibility of specific tyrosine residues in the presence and absence of DNA. However, if the modified tyrosine residues can be analyzed individually, one can look at the nitration rates of the tyrosines to determine the degree of accessibility of each residue. This is achieved by removing aliquots of protein (at various time intervals) from a nitration experiment and determining the percentage nitration of each tyrosine for a given time-point. This can be done by quantitating the nitrated and unnitrated products after digestion, either by measuring the peak areas (recorded at 214 nm) taken directly from the HPLC profile *(8)*, or by amino acid analysis of the purified peptides.

2. Reagents

1. All chemicals should be of AnalaR grade or higher, and dissolved in double-distilled water.
2. Trifluoroacetic acid (TFA), water, and acetonitrile should be of HPLC grade.
3. Buffers for HPLC should be filtered (0.2 μm) and degassed before use.
4. Tetranitromethane stock solution: a 300-mM stock solution of TNM in ethanol. Store in the dark at 4°C. Note that TNM can cause irritation to the skin and lungs, and the solution should be made up in the fume hood. Additionally, TNM can be explosive in the presence of organic solvents, such as toluene.
5. Nitration buffer: 150 mM NaCl, 10 mM Tris-HCl, pH 8.0.
6. Desalting column: Disposable "10DG" Econo columns (Bio-Rad, Richmond, CA) are preferred.
7. μBondapak C18 HPLC column (Waters Associates, Milford, MA) or similar reverse phase column.
8. Trypsin (TPCK treated).
9. Standard SDS PAGE equipment with a DC power supply capable of 150 V.
10. SDS polyacrylamide gel stock solutions:
    Solution A: 152 g acrylamide, 4 g *bis*-acrylamide. Make up to 500 mL.
    Solution B: 2 g SDS, 91 g Tris base, pH 8.8. Make up to 500 mL.
    Solution C: 2 g SDS, 30 g Tris base, pH 6.8. Make up to 500 mL.
    When making up these three solutions, they should all be degassed and filtered using a Buchner filter. They should be stored in light-proof bottles and will keep for many months.
11. 10% Ammonium persulfate (APS): Dissolve 0.1 mg in 1 mL of dH_2O.
12. 15% SDS polyacrylamide gel: Mix together 8.0 mL of solution A, 4.0 mL of solution B, and 3.9 mL of dH_2O. Add 150 μL of 10% APS and 20 μL of TEMED. Mix well and then pour between plates. Immediately place a layer of dH_2O on top of the gel to create smooth interface with the stacking gel. When the resolving gel has set, pour off the water and prepare the stacking gel. This is done by adding 750 μL of solution

Nitration of Tyrosine Residues

A and 1.25 mL of solution C to 3.0 mL of dH_2O. Finally, add 40 µL of APS and 10 µL of TEMED, pour on the stacking gel, and insert the comb. Remove the comb as soon as the gel has set to avoid the gel sticking to the comb.
13. 10X SDS running buffer: 10 g SDS, 33.4 g Tris base, 144 g glycine, made up to 1 L.
14. High methanol protein stain: Technical grade methanol 500 mL, 100 mL glacial acetic acid, 0.3 g PAGE 83 stain (Coomassie blue), made up to 1 L.
15. Destain solution: 100 mL methanol, 100 mL glacial acetic acid, made up to 1 L.
16. 2X SDS PAGE loading buffer: 4% (w/v) SDS, 60 mM Tris-HCl, pH 6.8, 20% glycerol, 0.04% (w/v) bromophenol blue, and 1% (v/v) β-mercaptoethanol.

3. Methods

The method is a fairly general one for protein nucleic acid complexes. However, precise details of the conditions for dissociation and peptide mapping will vary with the system under investigation. As an example of the technique, nitration of the Pf1 gene 5 protein and nucleoprotein complex will be described *(8)*.

3.1. Nitration

1. Desalt the protein or nucleoprotein complex into nitration buffer to a concentration between 0.5 and 5 mg/mL (*see* Note 1). For initial determination of nitrated residues, 0.5 mg of protein should be sufficient. However, if a time-course experiment is performed, larger amounts of protein are required.
2. To 1 mL of sample, add a 10-fold molar excess of 300 mM TNM (in ethanol) and incubate at room temperature for 1 h, stirring gently (*see* Note 2). The reaction is stopped by the addition of acid (add HCl to pH 2.0) or by rapid desalting into 10 mM Tris-HCl, pH 8.0 (*see* Note 3). Run an aliquot of the modified protein/nucleoprotein complex on an SDS gel, together with native protein, to determine whether there has been any TNM induced crosslinking (*see* Note 4). If analyzing the free protein, proceed to step 4; if modifying the nucleoprotein complex, proceed to step 3.
3. Dissociate the nucleoprotein complex by the addition of salt (*see* Note 5). Large DNA fragments can be removed by ultracentrifugation, whereas smaller fragments can be either digested with nucleases or removed by gel filtration. The protein is then dialyzed or desalted into the appropriate protease digestion buffer.

4. Digest the protein to completion with the desired protease(s) and then lyophilize the peptides for separation by HPLC. The peptides may be stored at –20°C.

In the example provided, the Pf1 gene 5 protein and nucleoprotein complex were incubated at room temperature in the presence of a 64-fold molar excess of TNM (300 mM in ethanol) for 3 h. The reaction was stopped by desalting the protein (and nucleoprotein complex) into 10 mM Tris-HCl, pH 8.0. The nucleoprotein complex was dissociated by the addition of $MgCl_2$ to 1M and the phage genomic DNA was then removed by ultracentrifugation at 221,000g (in a Beckman L8 ultracentrifuge; 70.1 Ti rotor) for 2.5 h. The protein was desalted into 10 mM Tris-HCl, pH 8.0, for proteolysis and digested with trypsin (Sigma [St. Louis, MO], TPCK treated) at an enzyme:substrate ratio of 1:25 (w/w) for 3 h at 37°C. Phenylmethane sulfonyl fluoride was added to a final concentration of 1 mM and the sample lyophilized overnight. This procedure results in the complete separation of the three tyrosine-containing tryptic peptides.

3.2. Peptide Mapping

1. Peptides can usually be separated by reverse-phase HPLC on a C18 column. Generally the peptides are applied to the column in 8M urea, 2% β-mercaptoethanol, and separated in an acetonitrile gradient in the presence of 0.05–0.1% trifluoroacetic acid (TFA). The acetonitrile gradient profile must be determined empirically for each particular protein.
2. Determine separation conditions for peptides from the native protein (see Note 6) and identify tyrosine-containing peptides (see Note 7).
3. Apply peptides from the nitrated protein and initially elute under the same conditions that were used for the native protein (see Note 8). If necessary, change the acetonitrile gradient to achieve separation of tyrosine-containing peptides and their nitrated counterparts.

In the example provided, the tryptic peptides from native gene 5 protein were resuspended in 200 µL 8M urea/2% β-mercaptoethanol and clarified prior to HPLC analysis (Fig. 1A). Tyrosine-containing peptides were initially detected by their fluorescence properties (see Note 7) and then identified by automated Edman degradation on an Applied Biosystems 477A pulsed liquid amino acid sequencer. Nitrated peptides were applied to the C18 column and separated under identical conditions (Fig. 1B). The nitrated peptides were initially detected by their altered retention times and by virtue of their yellow color in

Nitration of Tyrosine Residues

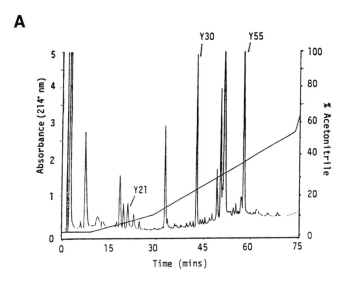

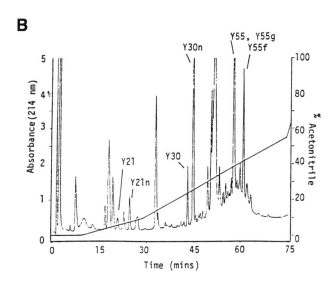

Fig. 1. HPLC elution profile of tryptic peptides of the Pf1 gene 5 protein for (**A**) native protein and (**B**) nitrated protein. Peaks Y21, Y30, and Y55 correspond to tryptic peptides containing tyrosine 21, 30, and 55, respectively. (n) Denotes a nitrated peptide. Aliquots (100 µL) were applied to a µ-Bondapak C18 HPLC column (Waters Associates) (300 × 4.6 mm id) fitted with a C18 guard column. The HPLC buffers for this experiment were: Buffer A: 0.05% TFA/H_2O; Buffer B: 0.05% TFA/acetonitrile. Peptides were separated in the following gradient at a flow rate of 2 mL/min: 0% B for 5 min; 0–10% B in 20 min; 10–55% B in 45 min; 55–90% B in 5 min, 90% B for 5 min, and 90–0% B in 5 min.

10 mM Tris-HCl, pH 8.0. The identity of nitrated peptides was subsequently confirmed by N-terminal sequencing (*see* Note 9).

3.3. Functional Studies

As discussed in Section 1., one should check whether nitration of the protein impairs the ability to bind DNA (other chapters in this volume can be consulted for details). The protein isolated from the nitrated nucleoprotein complex should also be checked for DNA binding. Since the target amino acid residues in contact with the DNA should have been protected from modification, the protein from the nitrated complex would be expected to retain DNA binding ability.

4. Notes

1. As an alternative to desalting, the protein can be dialyzed into nitration buffer.
2. The molar excess of TNM can be increased to ensure maximal modification (e.g., Pf1 gene 5 protein was nitrated in a 64-fold molar excess of TNM in the example provided). Note, however, that at high concentrations, protein insolubility can become a problem.
3. One can desalt the protein into a buffer appropriate for proteolysis or dissociation of the nucleoprotein complex at this stage, as required.
4. TNM induced crosslinking has been widely reported, and an SDS gel should be run to check for the appearance of adducts. Reducing the concentration of the protein and molar excess of TNM may help to limit adduct formation. Gel filtration is another way of removing the aggregates prior to peptide mapping (*see also* Note 10).
5. Usually protein-nucleic interactions can be disrupted by the addition of NaCl or MgCl$_2$ to 1–2M. The conditions required to effect separation will vary with the nature of the complex.
6. It is not essential to have complete separation of all fragments, only separation of the tyrosine-containing peptides and their nitrated counterparts; the HPLC conditions should be adjusted to achieve this.
7. Tyrosine residues have a characteristic fluorescence emission maximum at 303 nm when excited at 278 nm. This phenomenon can be used initially to determine which peptides contain a tyrosine residue (this may not be possible, however, if there is a tryptophan residue present in the same peptide because of energy transfer). If on-line fluorescence detection is not available, the fractions can be taken directly from the HPLC and analyzed in a fluorimeter. The peptides should then be unambiguously identified by either N-terminal sequencing or amino acid analysis.

8. The addition of a nitrate group to the tyrosine should alter the hydrophobicity, and hence retention time of that particular peptide in an acetonitrile gradient. This should allow immediate identification of the nitrated peptides. However, it is possible that a nitrated peptide comigrates with another unmodified peptide. Therefore, freeze-dry all peptides from HPLC and resuspend in 10 mM Tris-HCl, pH 8.0: The nitrated peptides will have a faint yellow color (absorbance maximum at 428 nm).
9. For peptides sequenced on an Applied Biosystems 477A pulse liquid amino acid sequencer (fitted with a 120A separation system for the analysis of PTH derivatized amino acids), PTC-3-nitrotyrosine elutes just after DTPU.
10. TNM induced oligomerization has been observed on the nitration of numerous proteins including several DNA binding proteins *(3,9,10)*. This is usually considered undesirable, and steps are often taken to reduce the crosslinking and remove adducts before analysis (e.g., by gel filtration). However, advantage can be taken of this crosslinking ability; Martinson and McCarthy *(11)* used TNM as a reagent to crosslink histones specifically. On nitration with TNM, we have also shown that the Pf1 gene 5 protein forms an SDS-stable dimer *(10)*. Initial analysis of the peptide adduct in this case suggested that tyrosine 55 from one monomer was crosslinked to Phenylalanine 76 from the other monomer (forming an intradimer crosslink rather than an interdimer crosslink). The adducts are thought to form via a free radical mechanism resulting in zero length crosslinks between residues in close proximity *(12,13)*. Thus, if adduct formation is limited to one or two species, additional structural information can be obtained from the experiment. The crosslinked proteins should be digested and the peptide adduct purified by HPLC. N-terminal sequencing, amino acid analysis, and G. C. mass spectroscopy (of the hydrolyzed peptide) should enable unambiguous identification of the two residues participating in the crosslink and provide structural information concerning the relative positions of these residues in the protein.

References

1. Lundblad, R. and Noyes, M. (1984) in *Chemical Reagents for Protein Modification I and II,* CRC, Boca Raton, FL.
2. Riordan, J., Sokolovsky, M., and Vallee, B. (1967) Environmentally sensitive tyrosine residues. Nitration with tetranitromethane. *Biochemistry* **6,** 358.
3. Anderson, R., Nakashima, Y., and Coleman, J. (1975) Chemical modification of functional residues of the Fd gene 5 DNA-binding protein. *Biochemistry* **14,** 907–917.

4. Sokolovsky, M., Riordan, J., and Vallee, B. (1966) Tetranitromethane. A reagent for the nitration of tyrosyl residues in proteins. *Biochemistry* **5,** 3582–3589.
5. Sokolovsky, M., Harell, G., and Riordan, J. (1969) Reaction of tetranitromethane with sulphydryl groups in proteins. *Biochemistry* **8,** 4740–4745.
6. Dimicoli, J. and Helene, C. (1974) Interaction of aromatic residues of proteins with nucleic acids I and II. *Biochemistry* **13,** 714–730.
7. Sokolovsky, M., Riordan, J., and Vallee, B. (1967) Conversion of 3-nitrotyrosine to 3-aminotyrosine in peptides and proteins. *Biochem. Biophys. Res. Commun.* **27,** 20.
8. Plyte, S. E. and Kneale, G. G. (1991) Mapping the DNA binding site of the Pf1 gene 5 protein. *Protein Eng.* **4(5),** 553–560.
9. Anderson, R. and Coleman, J. (1975) Physiochemical properties of DNA-binding proteins: Gene 32 protein of T4 and *Escherichia coli* unwinding protein. *Biochemistry* **14,** 5485–5491.
10. Plyte, S. (1990) Ph.D. Thesis. The biochemical and biophysical characterization of the PF1 gene 5 protein and its complex with nucleic acids. Portsmouth Polytechnic, Portsmouth, UK.
11. Martinson, H. and McCarthy, B. (1975) Histone-histone associations within chromatin. Cross-linking studies using tetranitromethane. *Biochemistry* **14,** 1073–1078.
12. Williams, J. and Lowe, J. (1971) The crosslinking of tyrosine with tetranitromethane. *Biochem. J.* **121,** 203–209.
13. Bruice, T., Gregory, M., and Walters, S. (1968) Reactions of tetranitromethane (I). Kinetics and mechanism of nitration of phenols by tetranitromethane. *J. Am. Chem. Soc.* **90,** 1612–1619.

CHAPTER 13

Limited Proteolysis of Protein–Nucleic Acid Complexes

Simon E. Plyte and G. Geoff Kneale

1. Introduction

Limited proteolysis is a useful structural probe for investigating the globular nature of proteins by preferentially digesting the more accessible regions often found between domains. Generally, proteases require a small region of polypeptide chain possessing conformational flexibility for accommodation in the active site *(1)*. The regions of a protein possessing conformational flexibility are often found between tightly folded domains and are therefore preferential sites for proteolysis. In practice, limited proteolysis is achieved by dilution of the enzyme sufficiently so that it will only digest the most accessible regions, leaving the domains intact. Digestion of protein–nucleic acid complexes is often advantageous in that the DNA may provide steric protection of the DNA binding domain not afforded by the free protein. The generation of domains by limited proteolysis relies directly on the tertiary structure of the protein under investigation and provides much firmer evidence for their existence than that provided by sequence homology.

An increasing number of nucleic acid binding proteins are known in which regions of their polypeptide chain are folded separately into compact globular domains, each possessing a distinctive function. For example, digestion of the A1 heterogenous nuclear ribonucleoprotein (A1 hnRNP) with *Staphylococcus aureus* V8 protease produces two discrete domains, both capable of binding single-stranded nucleic acids *(2,3)*. Similarly, digestion of the Pf1 gene 5 nucleoprotein complex results in the production of a 12 kDa domain that retains much of the single-stranded DNA binding ability of the intact protein *(4)*. Limited tryptic digestion of the *Escherichia coli* DNA gyrase A protein produces two distinct fragments: an N-terminal DNA supercoiling domain and a C-terminal DNA binding domain *(5,6)*. In addition to its use for the analysis of domain structure, limited proteolysis is a technique that can be used preparatively. The isolation of two proteolytic fragments from *Escherichia coli* DNA gyrase B, for example, permitted the successful crystallization and structure determination of one of its domains *(7)*.

1.1. Strategy

The strategy adopted for the limited proteolysis of nucleoprotein complexes can be considered in four parts: optimization of proteolysis, characterization of proteolyzed complex, purification of DNA binding domains, and sequence characterization of the fragment(s).

1.1.1. Proteolysis of Nucleoprotein Complex

The nucleoprotein complex should be digested with various proteases to establish which conditions are optimal for generating a protease resistant domain. We routinely vary two parameters (enzyme/substrate ratio and duration of digestion) when determining the best conditions for limited proteolysis. However, other parameters, such as temperature, ionic strength, and pH may also be varied. To determine the appropriate enzyme/substrate ratio for a particular protease the nucleoprotein complex is digested at several enzyme/substrate ratios, removing samples at regular time intervals for SDS-PAGE analysis. The appearance of a discrete domain, resistant to further degradation (even if only transiently), is evidence for the existence of a domain, although not necessarily one that binds DNA. Choice of protease is often critical (*see* Table 1). Initially, it is best to try a relatively nonspecific enzyme (e.g., papain) because this decreases the likelihood of activity being dependent on primary sequence rather than tertiary structure.

1.1.2. Preliminary Characterization of DNA Binding Properties of the Proteolyzed Nucleoprotein Complex

An initial indication of DNA binding can be found during the proteolysis experiment by removing two aliquots for gel analysis that

Limited Proteolysis

Table 1
Useful Enzymes for Limited Proteolysis

Enzyme	Substrate Specificity	Inhibitors
α Chymotrypsin	Preferentially cuts C-terminally to aromatic amino acids	Aprotinin, PMSF, DFP, TPCK, Chymostatin
Elastase	Cuts C-terminally to aliphatic noncharged amino acids (e.g., A, V, L, I, G, S)	PMSF, DFP
Endoproteinase Arg-C	Cuts C-terminally to arginine residues	DFP, TLCK
Endoproteinase Lys-C	Cuts C-terminally to lysine residues	Aprotinin, DFP, TLCK
Papain	Nonspecific protease, but shows some preference for bonds involving Arg, Lys, Gln, His, Gly, and Tyr	PSMSF, TPCK, TLCK, lupeptin, heavy metal ions
Pepsin	Nonspecific protease	Pepstatin
Subtilisin	Nonspecific protease	DFP, PMSF
Trypsin	Cuts C-terminally to lysine and arginine residues	DFP, PMSF, TLCK
Endoprotinase Glu-C (V8 protease)	Cuts C-terminally to glutamic acid and/or aspartic acid residues (*A)	DFP

*A. Will cut C-terminally to glutamic acid residues in ammonium bicarbonate, pH 8.0, or ammonium acetate, pH 4.0. Will cut C-terminally to glutamic and aspartic acid in phosphate buffer, pH 7.8.

Abbreviations used: DFP, diisopropyl fluorophosphate (extremely toxic!); PMSF, phenglmethyl sulfonyl fluoride; TPCK, N-tosyl-L phenylalanine chloromethyl ketone; TLCK, Nα-p-tosyl-L-lysine chloromethyl ketone.

can be run on polyacrylamide or agarose gels appropriate for the size of the complex in the presence and absence of the denaturant SDS. A retardation in the mobility of the DNA (seen under UV light) in the absence of SDS implies that the fragment is still associated with DNA and constitutes a DNA binding domain. However, this does not prove that the proteolyzed fragment is a discrete DNA binding domain; it is possible that the nucleoprotein complex has only been "nicked" by the protease and maintains its native tertiary structure by noncovalent interactions. Therefore, it is necessary to purify the domain and fully characterize its DNA binding properties.

1.1.3. Purification of the DNA Binding Domain

Purification of the fragment can make use of the fact that it will still be associated with DNA. Ultracentrifugation of the proteolyzed nucleoprotein complex (if large fragments of DNA are used) concentrates the domain and removes residual protease and small proteolytic fragments. The proteolyzed nucleoprotein complex can then be dissociated and the domain further purified if necessary. Alternatively, the DNA binding fragment can be purified by affinity chromatography on DNA agarose. Several techniques are available to determine whether the purified domain binds DNA (*see* other chapters in this book) and include gel retardation assay, a variety of footprinting techniques, fluorescence spectroscopy, and circular dichroism.

1.1.4. Determination of the Amino Acid Sequence of the Domain

N-terminal sequencing and amino acid analysis of the purified DNA binding domain should be sufficient to establish the sequence of the domain, if the native amino acid sequence is known. Mass spectroscopy should also assist in identification of the domain. If certain proteases have been used (e.g., trypsin, α-chymotrypsin, endoprotinase Arg-c, and so forth) the C-terminal amino acid may also be known. If there are still ambiguities, carboxypeptidase digestion of the fragment can also be used to help identify the C-terminal residues, although this is not always reliable. If this still does not yield an unambiguous result, one must resort to amino acid sequencing of the entire fragment.

2. Materials

1. Spectra-por dialysis membrane washed thoroughly in double-distilled water.
2. All proteases should be of the highest grade available and treated for contaminating protease activity, if necessary. A list of useful enzymes and their inhibitors is given in Table 1.
3. Buffers should be of AnalaR grade or higher and made up in double-distilled water.
4. Standard SDS PAGE equipment with a DC power supply capable of 150 V.
5. SDS polyacrylamide gel stock solutions:
   Solution A; 152 g acrylamide, 4 g *bis*-acrylamide. Make up to 500 mL.
   Solution B; 2 g SDS, 91 g Tris base, pH 8.8. Make up to 500 mL.
   Solution C; 2 g SDS, 30 g Tris base, pH 6.8. Make up to 500 mL.

When making up these three solutions they should all be degassed and filtered using a Buchner filter. They should be stored in lightproof bottles and will keep for many months.
6. 10% ammonium persulfate (APS): Dissolve 0.1 mg in 1 mL of dH_2O.
7. 15% SDS polyacrylamide gel: Mix together 8.0 mL of solution A, 4.0 mL of solution B, and 3.9 mL of dH_2O. Add 150 µL of 10% APS and 20 µL of TEMED. Mix well and then pour between plates. Immediately place a layer of dH_2O on top of the gel to create smooth interface with the stacking gel. When the resolving gel has set, pour off the water and prepare the stacking gel. This is done by adding 750 µL of solution A and 1.25 mL of solution C to 3.0 mL of dH_2O. Finally, add 40 µL of APS and 10 µL of TEMED, pour on the stacking gel, and insert the comb. Remove the comb as soon as the gel has set to avoid the gel sticking to the comb.
8. 10X SDS running buffer: 10 g SDS, 33.4 g Tris base, 144 g glycine, made up to 1 L.
9. High methanol protein stain: 500 mL technical grade methanol, 100 mL glacial acetic acid, 0.3 g PAGE 83 Stain (Coomassie blue), made up to 1 L.
10. Destain solution: 100 mL methanol, 100 mL glacial acetic acid, made up to 1 L.
11. 2X SDS PAGE loading buffer: 4% (w/v) SDS, 60 mM Tris-HCl, pH 6.8, 20% glycerol, 0.04% (w/v) bromophenol blue, and 1% (v/v) β-mercaptoethanol.
12. 6X agarose gel loading buffer: 0.25% (w/v) bromophenol blue, 0.25% (w/v) xylene cyanol, 30% glycerol.
13. 6X agarose gel loading buffer plus SDS: as above plus 12% SDS (w/v).
14. TE buffer: 10 mM Tris-HCl, pH 7.5, 1 mM EDTA.
15. 5M NaCl or $MgCl_2$ (or other concentrated salt solution for dissociation of the nucleoprotein complex, e.g., NaSCN).

3. Methods

The method given here covers the first three objectives outlined in Section 1.1. Experimental details for the determination of the amino acid sequence of the fragment can be found in any standard text on protein chemistry. The following protocol was used for the generation of an 11 kDa DNA binding domain from the Pf1 gene 5 protein. This protein binds cooperatively to ssDNA to produce a nucleoprotein complex of several million daltons. Different nucleoprotein complexes will require different conditions of digestion and purification, but the basic principles remain the same.

3.1. Limited Proteolysis

1. Dialyze the nucleoprotein complex into the appropriate digestion buffer (*see* manufacturer's recommendations for the buffer, temperature of reaction, and inhibitor). We routinely digest the nucleoprotein complex at approx 1 mg/mL but the concentration is not too critical.
2. Prepare 40 tubes containing 5 µL of 2X SDS loading buffer plus 1 µL of the appropriate protease inhibitor. Leave on ice.
3. Pipet 55 µL of the nucleoprotein complex (55 µg) into each of four tubes labeled 1:100, 1:200, 1:500, and 1:1000. Place on ice until needed.
4. Dissolve the protease in digestion buffer to a concentration that will give an enzyme/substrate ratio of 1:100 (w/w) when 1 µL of the protease is added to 50 µL of nucleoprotein complex (i.e., 0.5 mg/mL).
5. Prepare three dilutions of the protease. In this case the protease is diluted 1:2, 1:5, and 1:10 with digestion buffer that will result in final enzyme/substrate ratios of 1:200, 1:500, and 1:1000 (w/w).
6. Remove 5 µL of the nucleoprotein complex from each of the four tubes and add to one of the 40 tubes containing 2X loading buffer (plus inhibitor) and place on ice. This is the time = 0 tube and should be labeled accordingly.
7. Add 1 µL of the protease to the appropriate nucleoprotein solution (e.g., protease diluted 1:5 to the nucleoprotein solution marked 1:500) and incubate at the specified temperature.
8. Remove 5 µL samples every 15 min and add to the 2X loading buffer, then place on ice.
9. At the end of the experiment, boil the samples and run on an SDS polyacrylamide gel. The presence of a degraded fragment(s), resistant to further proteolysis, is evidence for a discrete domain (*see* Note 1).
10. Adjustment of the enzyme/substrate ratios, time course, and choice of enzymes is often necessary. The optimum conditions must be found by trial and error.

3.2. Purification of the DNA Binding Domain

1. Digest a large quantity (several mg) of the nucleoprotein complex under the optimized conditions determined above to produce the DNA binding domain (*see* Note 2). Add the appropriate inhibitor and run a sample on SDS PAGE to check the digestion.
2. For very large nucleoprotein complexes the proteolyzed complex can be purified away from the protease and small proteolytic fragments by ultracentrifugation. Spin the nucleoprotein complex at 229,000g (Beckman 70.1 Ti rotor) for 3 h at 4°C (*see* Note 3). Carefully wash the centrifuge tube with 4 mL of TE buffer, discard the washings, and resus-

pend the nucleoprotein complex in 2 mL of TE buffer on ice. Another ultracentifugation step can be performed to remove all traces of the protease. For smaller nucleoprotein complexes the DNA can be immobilized on a large resin (e.g., DNA cellulose) prior to interaction with the DNA binding protein. Low speed centrifugation can then be used to purify the DNA-associated domain. Sometimes limited proteolysis generates several fragments that bind DNA. These may arise from the same region of the protein and if so, this can sometimes be overcome by allowing the proteolysis to proceed further or by increasing the amount of protease.
3. Dissociate the proteolyzed nucleoprotein complex by the addition of salt to the appropriate concentration (*see* Note 4). The DNA can then be removed by ultracentrifugation (if sufficiently large) or nuclease digestion. If the DNA was originally bound on a solid support then it can be removed by low speed centrifugation (*see* Note 5).
4. Remove the high salt buffer by desalting or dialysis. If the sample contains several different domains or residual undigested protein, it will be necessary to purify the domains to homogeneity. Various chromatographic techniques are available to further purify the domains including chromatofocusing, ion exchange, and gel filtration chromatography. These techniques permit recovery of the domain in a native state for further biochemical and biophysical analysis. Alternatively, if the fragment is only to be used for sequence analysis, the mixture can be applied to a C3 reverse-phase HPLC column and separated in an acetonitrile gradient.
5. If the sequence of the native protein is known, then the sequence of the DNA binding domain can be established by N-terminal sequencing and amino acid analysis. Additionally, the mass of the fragment (determined by mass spectroscopy) should help locate the sequence of the DNA binding domain.

4. Notes

1. Often during the experiment, a protease resistant fragment is only transiently formed during complete digestion of the protein. If this occurs, vary some of the parameters (enzyme dilution, temperature) to try and prolong the lifetime of the fragment.
2. Scaling up of the digestion is not generally a problem and we routinely digest several mg (>10 mg) of nucleoprotein complex if necessary.
3. The speed and duration of centrifugation will vary depending on the size of the nucleoprotein complex. For smaller complexes, ultracentrifugation may not be appropriate.

4. In many cases a NaCl concentration between 1–2M is sufficient to dissociate the nucleoprotein complex. However, some nucleoprotein complexes remain associated above 2M NaCl and require 1M MgCl$_2$ or 1M NaSCN for dissociation *(8)*. The appropriate salt concentration can be determined by SDS PAGE analysis of the pellet and supernatant after ultracentrifugation at different ionic strengths.
5. The DNA can also be removed by DNase digestion followed by gel filtration (i.e., a desalting column) or by extensive dialysis against TE buffer.

References

1. Vita, C., Dalzoppo, D., and Fontana, A. (1987) Limited proteolysis of globular proteins: molecular aspects deduced from studies on thermolysin, in *Macromolecular Biorecognition* (Chaiken, I., Chaiancone, E., Fontana, A., and Veri, P., eds.), Humana, Clifton, NJ.
2. Merrill, B., Stone, K., Cobianchi, F., Wilson, S., and Williams, K. (1988) Phenylalanines that are conserved among several RNA-binding proteins form part of a nucleic acid-binding pocket in the heterogeneous nuclear ribonucleoprotein. *J. Biol. Chem.* **263,** 3307–3313.
3. Bandziulis, R., Swanson, M., and Dreyfuss, G. (1989) RNA binding proteins as developmental regulators. *Genes Dev.* **3,** 431–437.
4. Plyte, S. E. (1990) Ph.D. Thesis (CNAA). Portsmouth Polytechnic.
5. Reece, R. J. and Maxwell, A. (1989) Tryptic fragments of the *Escherichia coli* DNA gyrase A protein. *J. Biol. Chem.* **264,** 19,648–19,653.
6. Reece, R. J. and Maxwell, A. (1991) The C-terminal domain of the *Escherichia coli* DNA gyrase A subunit is a DNA binding protein. *Nucleic Acids Res.* **19(7),** 1399–1405.
7. Wigley, D. B., Davies, G. J., Dodson, E. J., Maxwell, A., and Dodson, G. (1991) Crystal structure of an N-terminal fragment of the DNA gyrase B protein. *Nature* **351,** 624–629.
8. Kneale, G. G. (1983) Dissociation of the Pf1 nucleoprotein assembly complex and characterisation of the DNA binding protein. *Biochim. Biophys. Acta* **739,** 216–224.

CHAPTER 14

Cloning and Expression of DNA Binding Domains Using PCR

Daniel G. Fox and G. Geoff Kneale

1. Introduction

Many DNA binding proteins are known to consist of a number of domains—discrete compact regions of a protein that often have distinct functional properties. Structural domains within a protein are generally indicated by limited proteolysis (*see* Chapter 13) and the existence of functional domains can often be tested by genetic experiments (e.g., "domain swapping"). In favorable cases, sequence homologies may also provide circumstantial evidence for domains. Once the existence of domains has been established, one may wish to express them in large quantities for structural analysis, for example, using X-ray crystallography or NMR. The domains may be more soluble than the intact protein, or have other physical properties that are advantageous. Because of their smaller size, structure determination is also likely to be more feasible. Limited proteolysis is an excellent analytical technique for the investigation of domain structure. However, it is inefficient for the large scale production of protein fragments. Cleavage is rarely at unique sites; often a number of proteolytic fragments are produced that may require extensive purification. For such studies, it is clearly an advantage to over-express the truncated proteins directly.

If the gene encoding the intact protein has been subcloned and its DNA sequence is known, then PCR can be used to amplify the DNA sequence that codes for the protein fragment. By introducing restriction sites into the PCR primers on either side of the required coding sequence, one can ligate the amplified product into an appropriate expression vector. The PCR primers must also encode translational start and stop signals, as appropriate, to define the domains.

For efficient expression, the vector used must have a strong and tightly regulated promoter, and a good Shine-Dalgarno sequence (ribosome binding site) appropriately positioned with respect to the start codon. A strategy that is now becoming standard is the use of fusion vectors, in which the coding sequence is fused to a well characterized bacterial gene. This may also help in cases where there are problems of insolubility associated with high expression, since the fusion protein is often more soluble than the isolated protein. The start codon of the cloned gene fragment is not required (assuming it is fused at its N-terminal end), since translation is initiated at the start of the gene to which it is fused. Instead, the coding sequence for a proteolytic site is placed immediately upstream of the inserted fragment and can be used to split the fusion protein into its two components after purification.

A number of fusion vectors have been developed for this purpose and generally take advantage of the properties of the parental protein to allow single step purification by affinity chromatography. The system described here utilizes the maltose binding protein (MBP) of *E. coli*, a periplasmic binding protein encoded by the *malE* gene *(1)*. Purification of the expressed protein is easily accomplished by simple chromatographic separation on an amylose binding column and elution with maltose. The gene is under control of the strong *tac* promoter, using the translational initiation signals upstream of *malE* for high expression *(2)*. The vector encodes a factor Xa cleavage site to allow subsequent release of the fused protein from the maltose binding protein *(3)*. The advantage of factor Xa cleavage is that this enzyme cleaves immediately C-terminal to its recognition sequence (Ile-Glu-Gly-Arg) and thus no additional amino acids are left at the N-terminus of the expressed protein. It is not, however, necessary to use the factor Xa site of the vector as long as such a site is incorporated into the N-terminal primer at the correct position (immediately upstream of the required coding sequence).

A number of variations on the basic vector are commercially available with multiple restriction sites in the polylinker. Fusion vectors employing the *malE* gene can direct expression into the periplasm (the normal location of MBP) or the cytoplasm (*see* Fig. 1). The lat-

PCR Cloning of Domains

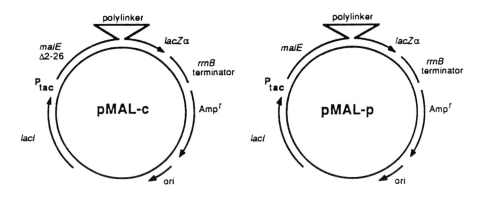

pMAL-c, pMAL-p Polylinker:

```
               ┌Sac I┐ ┌Kpn I┐ ┌Eag I┐ ┌BamHI┐                    ┌Stu I┐  Factor Xa
    malE...    TCG AGC TCG GTA CCC GGC CGG GGA TCC ATC GAG GGT AGG CCT   cleavage
                                                 Ile Glu Gly Arg Pro     site

┌EcoRI┐                              ┌BamHI┐┌Xba I┐ ┌Sal I┐ ┌Pst I┐     ┌Hind III┐
GAA TTC AGT AAA CCT ACC CTC GAT GGA TCC TCT AGA GTC GAC CTG CAG GCA AGC TTG....lacZα
```

Fig. 1. A schematic representation of the p-MAL™ expression plasmids. The two plasmids are identical, except for deletion of the N-terminal sequence of MBP in pMAL-c, to allow cytoplasmic expression of the fusion protein.

ter is accomplished by deletion of the leader sequence from the *malE* gene. The latest generation of vectors also contain an M13 origin, in addition to the original PBR322 origin. This is convenient for the production of single-stranded DNA for sequencing or mutagenesis. If problems are found in cleavage of the fusion protein or affinity purification, they may arise from the close proximity of the fragment of interest and the maltose binding protein. In such cases, it is recommended to use a spacer of up to 10 amino acids to separate the two domains.

The protocol described below has been used in our laboratory to express the Pf1 gene 5 protein *(4)*, and fragments deriving from this protein that retain the putative N-terminal DNA binding domain. Although the Pf1 gene 5 protein is known to be prone to aggregation at high concentrations, no such problems were encountered using the fusion vector.

1.1. Strategy

When designing the PCR primers for this technique four criteria must be fulfilled:

1. The N-terminal primer must be homologous to the correct region of the 5' end (on the coding strand) of the DNA sequence of interest; 10–15 bases is usually ample. The C-terminal primer is complementary to the correct region of the 3' end of the DNA sequence of interest (i.e., homologous to the 5' end on the noncoding strand). Preferably the primer should contain over 50% G and C to ensure adequate annealing.
2. The restriction sites that are incorporated into the primers to facilitate the cloning of the DNA sequence must be at the 5' end of each primer. Care must also be taken to ensure that the sequence of DNA when cloned stays in the correct reading frame in relation to that of the maltose binding protein.
3. It is essential to have sufficient bases flanking the restriction sites, otherwise the restriction enzymes will not cut the DNA. The actual number of bases required depends on the particular restriction enzyme. Most enzymes seem to need at least four bases flanking the restriction site; certain enzymes will manage with only two whereas others need as many as six bases flanking their sites. If unsure, start with four and after successful amplification ascertain that the enzymes are cutting the DNA.
4. Stable secondary structure in the template can greatly reduce the efficiency of primer annealing. If areas of secondary structure are known then they should be avoided as primer annealing sites. If, however, a specific site must be used for the primer to anneal that is known to form secondary structure, then the primer must be designed to have a higher T_d, that is, it will need to be longer and contain a higher proportion of G and C.

To illustrate the criteria that need to be taken into account, examples are given for the design of primers used to amplify fragments of the gene 5 of the Pf1 phage for cloning into a fusion vector (*see* Figs. 2 and 3). The N-terminal primer (FN-1) has a 14 base region at its 3' end homologous to the 5' end of the Pf1 gene 5. A 5' tail was added onto the primer that contained the coding region for the factor Xa protease site and a *Bgl*II restriction site. Four extra bases were included to ensure successful digestion by the restriction enzymes. A *Bgl*II site was designed into the primer rather than a *Bam*HI site; both restriction sites share the same overhang, but when they are ligated together an *Xho*II site is formed. Thus, if successful ligation takes place the recombinant plasmid now has an extra *Xho*II site, altering its restriction map when digested with this enzyme. When the clones were screened by restriction mapping a change in band patterns would be observed, indicating successful insertion of the PCR fragment.

PCR Cloning of Domains

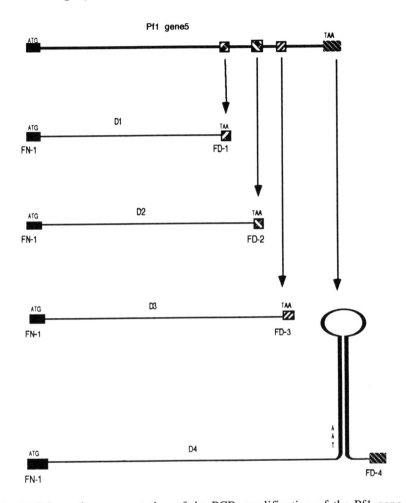

Fig. 2. Schematic representation of the PCR amplification of the Pf1 gene 5 fragments. The whole gene 5 is shown with the different binding sites of the primers (FD-1–FD-4) represented by the shaded boxes. The resulting gene fragments are indicated (D1–D4). Stop codons (TAA) are introduced at the relevant sites using the PCR primers except when amplifying the whole gene. Because of the secondary structure at the end of the gene, the primer used for this was designed to anneal beyond the hairpin, since translation is halted at the natural stop codon present in the gene.

To produce the small fragment D1, a C-terminal primer (FD-1) was synthesized. It had 15 bases that were complementary to the sequence of the gene 5. An artificial stop codon was included into the primer to terminate transcription prematurely. Finally an *Xba*I site was also introduced to facilitate cloning into the expression plasmid. Similar

FN-1 5'G GCT **AGA TCT** *ATC GAG GGT AGG* <u>ATG</u> AAC ATG TTT GC 3'

       Bgl II   Factor Xa   Start Codon   Pf1 gene 5

FD-1   5'GCC TGC **TCT AGA** <u>TTA</u> GGT GAT CCG ACC GAA 3'

                    Xba I   Stop Codon   Sequence Complementary
                                                    to Pf1 gene 5

Fig. 3. Schematic representation of two of the primers used in the amplification of different regions of the Pf1 gene 5. They demonstrate the regions of homology and complementarity depending on which strand they are to anneal to. FN-1 shows the start codon, the coding sequence for the factor Xa site, the *Bgl*II restriction site, and the extra four bases required for successful digestion by the restriction enzyme. FD-1 shows the stop codon, the restriction site, and the additional six bases required for digestion. Note that the 14 bases at the 3' end of FN-1 are identical to the N-terminus of the coding sequence, whereas the 15 bases at the 3' end of FD-1 are identical to the C-terminal region of the *reverse complement* of the coding strand.

considerations were used to design primers for the production of fragments D2, D3, and the intact gene D4. The cloning of the whole gene 5 was complicated by a secondary structure "hairpin" in the template DNA coinciding with the 3' end of the gene. This made annealing of any primer to that region during PCR very difficult. Since the gene's original stop codon is being utilized, the C-terminal primer was designed to anneal beyond the hairpin. It contained the same restriction site as the other primers.

2. Materials
2.1. Polymerase Chain Reaction

1. Oligonucleotide primers can be synthesized on any commercial DNA synthesizer (for the experiments described, a Cruachem PS 250 was used). *See* Section 1.1. for design of primers.
2. The template containing the DNA sequence of interest. Plaques, bacterial colonies, or unpurified plasmid or phage DNA all generally work well as PCR templates. However, if in doubt use purified DNA.
3. Deoxyribonucleotide triphosphate stock solution: 2 mM of each dNTP, (such as Pharmacia [Piscataway, NJ] Ultra Pure dNTPs).
4. *Taq* polymerase (e.g., Promega, Madison, WI) including supplied enzyme buffer (10X).
5. Mineral oil.
6. Double-distilled ultra pure water.

PCR Cloning of Domains

7. Thermal cycler for PCR.
8. Standard apparatus for the electrophoresis of agarose gels and a DC power supply capable of 150 V.
9. Ultra pure agarose.
10. 10X TBE buffer. Add together Tris base (109 g), boric acid (55.6 g), EDTA (50 mL of a 500 mM solution) and make up to 1 L. (**N.B.** When making up the EDTA 500 mM solution use concentrated NaOH to increase the pH to 8.3.)
11. Stock solution of ethidium bromide, 50 mg/mL.
12. Long wave UV transilluminator for visualization of agarose gels.
13. 1% Agarose gel: Add together agarose (1 g), 10X TBE (10 mL), ethidium bromide (5 µL), dissolve the agarose by heating to 100°C, and make up to 100 mL.
14. Agarose gel running buffer: Add 50 µL ethidium bromide to 100 mL of 10X TBE and make up to 1 L.

2.2. Purification and Digestion of the PCR Product

1. High pressure liquid chromatography (HPLC) apparatus with a DEAE NPR ion exchange column fitted.
2. HPLC buffers: During the run a salt gradient is produced using two different buffers. Buffer A: 20 mM Tris-HCl, pH 9.0, Buffer B: 20 mM Tris-HCl, 1M NaCl, pH 9.0.
3. Plasmid DNA at a concentration of 100 ng/µL. The expression vector used in this protocol is pMal-c™ (New England Biolabs, Beverly, MA).
4. Appropriate restriction enzymes (such as *Bam*HI, *Bgl*II, *Xho*II).
5. The relevant buffers (10X) are usually supplied with the restriction enzymes. (e.g., from New England Biolabs).

2.3. DNA Ligation and Transformation

1. PCR product and plasmid DNA of known concentration after digestion with the relevant enzymes.
2. T4 DNA ligase.
3. Ligase buffer (10X): 500 mM Tris-HCl, pH 7.4, 100 mM MgCl$_2$, 10 mM ATP, 100 mM DTT.
4. 50 mM CaCl$_2$ solution; keep refrigerated.
5. Competent *E. coli* cells (e.g., TB 1).
6. Stock solution of ampicillin (50 mg/mL); filter sterilize before use.
7. Nutrient agar (ampicillin) plates made from pouring 30 mL of L-broth/ agar into sterile Petri dishes. (10 g tryptone, 5 g yeast extract, 5 g NaCl to 15 g of agar/L). Add the agar to the bottles first before adding the L-broth

and autoclaving. Add the antibiotic to a concentration of 150 µM (3 mL of a 50 mg/mL stock solution) after the L-broth has cooled well below 60°C.
8. Two waterbaths, one at 37°C and the other at 42°C.

2.4. Expression, Purification, and Cleavage of the Fusion Protein

1. Standard SDS PAGE equipment with a DC power supply capable of 150 V.
2. SDS poly acrylamide gel stock solutions:

    Solution A; 152 g acrylamide, 4 g *bis*-acrylamide. Make up to 500 mL.
    Solution B; 2 g SDS, 91 g Tris base, pH 8.8. Make up to 500 mL.
    Solution C; 2 g SDS, 30 g Tris base, pH 6.8. Make up to 500 mL.

    When making up these three solutions they should all be degassed and filtered using a Buchner filter. They should be stored in lightproof bottles and will keep for many months.
3. 10% ammonium persulfate (APS): Dissolve 0.1 mg in 1 mL of dH_2O (*see* Note 1).
4. 15% SDS polyacrylamide gel: Mix together 8.0 mL of solution A, 4.0 mL of solution B, and 3.9 mL of dH_2O. Add 150 µL of 10% APS and 20 µL of TEMED. Mix well and then pour between plates. Immediately place a layer of dH_2O on top of the gel to create a smooth interface with the stacking gel. When the resolving gel has set, pour off the water and prepare the stacking gel. This is done by adding 750 µL of solution A and 1.25 mL of solution C to 3.0 mL of dH_2O. Finally, add 40 µL of APS and 10 µL of TEMED, pour on the stacking gel, and insert the comb. Remove the comb as soon as the gel has set to avoid the gel sticking to the comb.
5. 3X SDS loading buffer: 30% glycerol, 9% SDS, 180 mM Tris-HCl, pH 6.8. When loading the gels make up samples as follows: 1 vol loading buffer, 1 vol 150 mM DTT (make up fresh each time), and 1 vol protein sample.
6. 10X SDS running buffer: 10 g SDS, 33.4 g Tris base, 144 g glycine, made up to 1 L.
7. High methanol protein stain: 500 mL technical grade methanol, 100 mL glacial acetic acid, 0.3 g PAGE 83 stain (Coomassie blue), made up to 1 L.
8. Destain solution: 100 mL methanol, 100 mL glacial acetic acid, made up to 1 L.
9. Sonicator (e.g., M.S.E. sonic disruptor) for lysis of the cells.
10. PEI cellulose (Sigma, St. Louis, MO) to remove excess free DNA. Supplied as a powder.
11. 100 mL of amylose resin (New England Biolabs).

12. Lysis buffer: 10 mM phosphate, pH 7.0, 30 mM NaCl, 0.25% Tween 20, 1 mM DTT, 10 mM EDTA, 10 mM EGTA.
13. Column buffer: pH 7.0, 10 mM phosphate, 500 mM NaCl, 1 mM sodium azide, 1 mM DTT, 1 mM EGTA.
14. Factor Xa protease (New England Biolabs): Stock concentration 1 mg/mL.
15. Factor Xa reaction buffer: 20 mM Tris-HCl, pH 8.0, 100 mM NaCl, 2 mM CaCl$_2$, 1 mM sodium azide.
16. Standard column chromatography and centrifugation equipment are also required.

3. Methods

3.1. Polymerase Chain Reaction

1. In a 0.5-mL Eppendorf tube make up the reaction cocktail of template DNA (100 ng), primers (60 ng each), dNTPs (final concentration 0.2 mM), and 5 µL of 10X reaction buffer. Make up the final volume to 50 µL and then heat to 95°C for 5 min (*see* Note 2).
2. Cool the tubes instantly on ice to prevent nonspecific annealing of the primers to the template. Add 2.5 U (0.5 µL) of *Taq* polymerase and overlay 75 µL of mineral oil.
3. Cycle through a series of heat steps to initiate specific primer annealing, DNA extension, and further denaturation using a thermal cycler. Use 25 cycles of the following steps: Denature at 94°C for 1 min, anneal at 55°C for 30 s, extend at 71°C for 30 s.
4. Draw the aqueous phase from beneath the mineral oil. Load 5 µL of the reaction onto a 1% agarose gel and run at 100 V for 30 min. Remove the gel from the apparatus and observe under UV illumination (*see* Fig. 4).

3.2. Purification and Digestion of the PCR Product

1. Load the aqueous phase from the PCR reaction onto the HPLC column to separate the PCR products from any residual dNTPs and template (*see* Note 3). A DEAE-NPR ion exchange column can be used with a salt gradient ranging from 0 to 1M NaCl in 20 mM Tris-HCl, pH 9.0. The PCR product is eluted at approx 0.4M salt and collected in 0.5-mL fractions (*see* Fig. 5).
2. Precipitate the DNA by the addition of 3 vol of ethanol and spin in a microcentrifuge for 15 min at 4°C.
3. Resuspend the pellet in 50 µL and assess the concentration of the purified PCR product using a dot plate using DNA standards of known concentration for comparison.
4. In separate Eppendorf tubes, digest all the PCR product and 1 µg of the plasmid DNA at a concentration of approx 100 ng/mL (*see* Note 4).

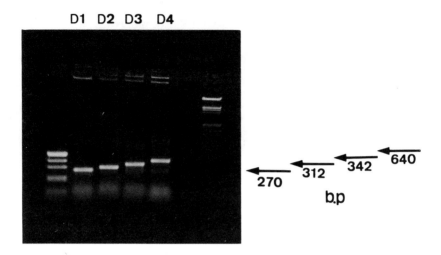

Fig. 4. A 1% agarose gel showing the different sized products of the PCR reaction producing the gene 5 fragments using the oligonucleotides FN-1 and FD-1 to FD-4. The DNA corresponds to the expected size and is produced in large quantities (1 mg/reaction).

3.3. DNA Ligation and Transformation

1. Mix 100 ng vector DNA (cut with the appropriate enzymes) with a threefold molar excess of the digested PCR product. Add 1 µL of 10X ligase buffer and make up the volume to 10 µL. Set up control ligations:
   a. Uncut plasmid and no PCR product, to verify competence of cells.
   b. Cut plasmid and no PCR product, to assess extent of digestion of the plasmid.
   c. No DNA at all, to monitor sterility of equipment and techniques used.
2. Add 1–6 U of T4 ligase and incubate at 16°C for 4–16 h (*see* Note 5). Stop the ligation reaction by heat inactivation at 75°C for 15 min.
3. Use the ligation mix directly for transformation of competent *E. coli* cells.
4. Prepare the competent cells by inoculating a 10-mL aliquot of L-broth with a single colony of *E. coli* TB1 bacteria.
5. Incubate the culture at 37°C with vigorous shaking for approx 16 h. Dilute the saturated culture 100 times with L-broth and grow until it reaches an absorbance at 600 nm of approx 0.6 OD units; this corresponds to the mid-logarithmic growth phase.
6. Pellet the cells at 3000 rpm in a Sorvall RC-3B centrifuge at 4°C for 10 min and then resuspend the cell pellet in 5 mL (i.e., half of the original volume) of 50 mM CaCl$_2$.
7. Incubate on ice for 15 min, repellet by centrifugation and resuspend in 1 mL (i.e., one tenth of the original volume) of 50 mM CaCl$_2$, 20% glyc-

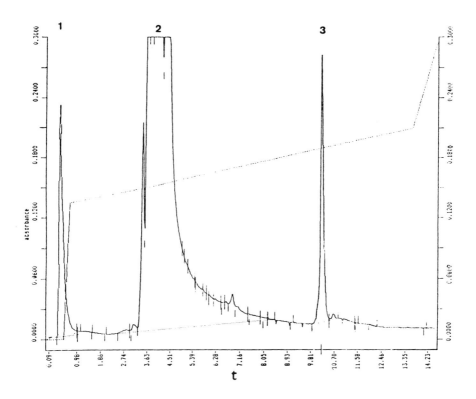

Fig. 5. HPLC trace of the purification of the PCR products. Peak 1 corresponds to the injection peak. Peak 2 corresponds to the dNTPs and excess primers. Peak 3 corresponds to the PCR product. The retention time of the PCR product will vary for different lengths of DNA.

erol (v/v). The cells are now competent for transformation with the recombinant DNA and can be used immediately or stored at –80°C until required.

8. Combine a 100-µL aliquot of competent *E. coli* cells with the ligation reaction and the controls in a sterile microcentrifuge tube. Following gentle mixing, incubate the tube on ice for 30 min.
9. Add 1 mL of L-broth and then heat shock the cells at 42°C for 2 min. Incubate the cells at 37°C for 60 min and then spin down for 10 s in a microcentrifuge.
10. Discard the supernatant and resuspend the cells in approx 150 µL of L-broth, 20 µL of a 2% solution of X-gal, and 20 µL of 100 mM IPTG and plate on an L-broth agar/ampicillin plate. The plates are then inverted and incubated at 37°C for a minimum of 16 h.
11. White colonies should be screened to check for the successful transformation of the recombinant plasmid DNA (*see* Note 6).

3.4. Expression, Purification, and Cleavage of the Fusion Protein

1. A single bacterial colony containing the recombinant plasmid is used to inoculate 100 mL of L-broth supplemented with the appropriate antibiotic and allowed to grow for 16 h. Inoculate 5 L of broth in a fermenter with 50 mL of this culture if large quantities of protein are required.
2. When the culture has reached mid-log growth phase induce with 1 mM IPTG and continue growth for a further 3–6 h to produce the desired protein. Harvest by the cells by centrifugation. Wash the cell pellet with 10 mM Tris-HCl, pH 8.2, and 100 mM NaCl, and then resuspend in lysis buffer.
3. To effect cell lysis, sonicate on ice. It is usually sufficient to use 16-µ peak to peak energy with 20 × 10 s bursts and 30 s rest intervals to allow the samples to cool.
4. After sonication centrifuge the lysate at 10,000 rpm for 15 min at 4°C in a Sorvall RC-5B centrifuge to pellet the cell debris and insoluble proteins. Assess the soluble and insoluble proteins at this stage by SDS PAGE. Take 10-µL aliquots of the samples, add equal volumes of 3X loading buffer and 150 mM DTT, and load 5 µL on the gel (*see* Fig. 6).
5. Make up the cleared lysate to 200 mM NaCl. Add 1 g of PEI for every 10 mL of lysate and gently stir. After a further 30 min of stirring at 4°C pellet the nucleic acid at 2000 rpm in the centrifuge. The PEI binds to the DNA in the solution and makes it insoluble.
6. Centrifuge the remaining protein solution at 50,000 rpm (100,000g) for 2 h at 4°C in a Beckmann L8 ultracentrifuge using a 70.1 Ti rotor to clarify the sample.
7. Add 3 vol of amylose column loading buffer supplemented with 0.1% Tween 20 and immediately load onto the amylose affinity column at a flow rate of 1 mL/min. Wash the column with 3 vol of the same buffer and then collect the run-through. Analyze the run-through by SDS PAGE to confirm the absence of the fusion protein. Step elute the fusion protein off the amylose column with 3 column vol of column elution buffer (column buffer supplemented with 10 mM maltose). Collect 2-mL fractions and analyze them by SDS PAGE. The peak fractions should contain the pure fusion protein (*see* Fig. 7). Estimate the concentration spectroscopically at 280 nm.
8. Depending on the particular fusion the concentration of fusion used in the cleavage can range from 100 µg/mL to 10 mg/mL so concentration of the eluted protein may be necessary. Use the factor Xa at an enzyme:substrate ratio of 0.1% (w/w) (*see* Note 7). Alter the buffer to the factor Xa cleavage buffer and then incubate the reaction mixture at room tem-

PCR Cloning of Domains

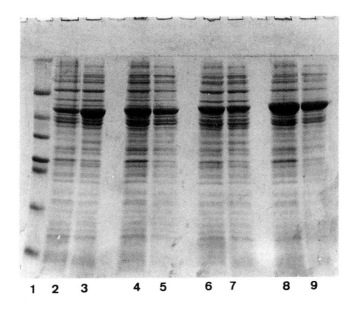

Fig. 6. 15% SDS PAGE gel illustrating the solubility of the expressed MBP-gene 5 fragment fusion protein. The proteins from the whole sample are compared with soluble-only proteins. Fusions with D1 (lanes 2, 3), D2 (lanes 4, 5), D3 (lanes 6, 7), and D4 (lanes 8, 9). Whole cell extracts were run in lanes 2, 4, 6, 8; soluble fractions (after lysis) in lanes 3, 5, 7, 9.

perature for 3 h (see Note 7). Again, analyze the products using SDS PAGE to check for the appearance of the MBP protein and the protein of interest (see Fig. 8).

9. Remove any maltose by desalting or dialysis and then pass down the agarose column to remove the MBP protein. This time the protein of interest will appear in the run-through and the MBP sticks to the column. Desalt and concentrate as appropriate.

4. Notes

1. The APS solution should be made up fresh every time and the solid APS should be kept dry over a desiccant, such as silica gel. When dissolving up the APS in water a crackling noise can be heard indicating that the APS is still viable. If it appears lumpy and slightly hydrated discard it.
2. Heating of the reaction cocktail ensures denaturation of the template if it is double-stranded and the melting of any stable secondary structures. The number of cycles will depend on how much DNA template is present at the start, and might need to be determined empirically. The

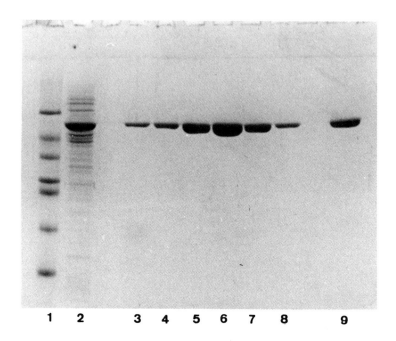

Fig. 7. 15% SDS PAGE gel illustrating the purification of the MBP-gene 5 fusion. Total soluble proteins are passed down an MBP binding column and eluted with a buffer containing maltose to yield highly purified protein. Lane 1: size marker; lane 2: total soluble proteins; lanes 3–8: fractions off the MBP binding column; lane 9: MBP-gene 5 fragment fusion marker.

    quantity of final PCR product obtained will be determined by the concentration of NTPs and primers (since these are introduced into the final product) assuming a sufficient number of amplification cycles.

3. There are a range of quick purification methods on the market that can be used as an alternative to HPLC.
4. Restriction digests of the plasmid DNA and the PCR product are carried out according to the instructions of the particular manufacturer using the recommended buffer conditions. However, in general, 1 U of restriction enzyme is sufficient to digest 1 µg of DNA in 1 h at 37°C.
5. The procedure given is for sticky-end ligation. For "blunt-end" ligation use a similar procedure, except that the number of units of T4 ligase used is increased 10-fold and 5% polyethylene glycol-8000 is included in the 1X ligation buffer.
6. Recombinant bacterial colonies can be screened in a number of ways to assess that the expression plasmid has the correct DNA fragment inserted

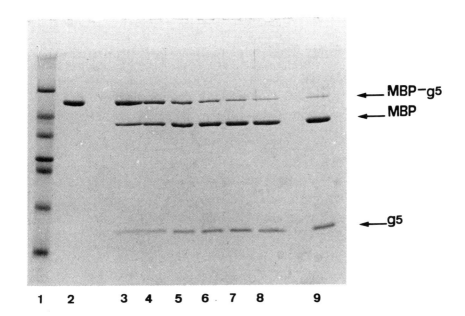

Fig. 8. 15% SDS PAGE gel illustrating the cleavage of the MBP-gene 5 fusion protein using factor Xa at various times. Lanes 2–8 correspond to digestion times of 0, 0.5, 1.0, 1.5, 2.0, 2.5, and 3 h, respectively. Lane 9: Soluble fraction after 3 h digestion, following centrifugation to remove the insoluble protein, showing that the cleaved protein retains solubility.

in it. This may be done by the traditional DNA miniprep method (the DNA is digested and the restriction maps analyzed) or PCR may be carried out using the colonies as the template and the original primers used for the amplification experiment. At some stage the DNA sequence of the entire insert should be checked by standard sequencing procedures, since PCR can occasionally introduce a mutation.
7. Depending on the particular fusion, cleavage may be carried out at higher concentrations or lower concentrations of factor Xa. Because of the high cost of factor Xa it is worth assessing the minimal concentration needed for cleavage of each fusion. Digestion periods from 3 h to 3 d have proved successful at 4°C or at room temperature.

References

1. Guan, C., Li, P., Riggs, P. D., and Inouye, H. (1987) Vectors that facilitate the expression and purification of foreign peptides in *E. coli* by fusion to maltose binding protein. *Gene* **67,** 21–30.

2. Amann, E. and Brosius, J. (1985). ATG vectors for regulated high-level expression of cloned genes in *E. coli. Gene* **40,** 183–190.
3. Nagai, K. L. and Thogerson, H. C. (1987) Synthesis and sequence specific proteolysis of hybrid proteins produced in *E. coli. Meth. Enzymol.* **153**, 461–481.
4. Kneale, G. G. (1983) Dissociation of the Pf1 nucleoprotein assembly complex and characterisation of the DNA binding protein. *Biochim. Biophys. Acta* **739,** 216–224.

CHAPTER 15

Overexpression and Purification of Eukaryotic Transcription Factors as Glutathione-S-Transferase Fusions in *E. coli*

Kevin G. Ford, Alan J. Whitmarsh, and David P. Hornby

1. Introduction

The ability to express recombinant proteins at high level in bacteria has led to dramatic increases in our understanding of protein structure and function in recent years. These techniques have provided the means to isolate the substantial quantities of recombinant protein required for both X-ray crystallographic and NMR-based structure determinations on what is becoming a nearly routine basis. The opportunity now also exists for the construction of recombinant mutants that are providing much greater insight into protein structure/function relationships. Moreover, comparisons between the activities of purified recombinant proteins and proteins purified from natural sources have often led to the discovery of previously unidentified molecules that are required for full activity of the protein under investigation. A recent example of this phenomenon has been the discovery of coactivator proteins that are internal components of the eukaryotic RNA polymerase II transcription factor TFIID *(1)*.

The cloning of eukaryotic genes into bacterial expression plasmids does not always result in high level expression of the desired protein. For example, the eukaryotic DNA may lack the specific signals necessary for interaction with the bacterial transcription and translation machinery, or may exhibit inappropriate codon usage. The most efficient bacterial expression vectors contain strong, tightly regulated promoters, efficient ribosome binding sites, and initiation codons that are optimally positioned with respect to the inserted open reading frame. Optimal insertion of any open reading frame is best obtained by PCR amplification using primers containing appropriate restriction sites. However, even such optimal expression systems do not always perform efficiently. Foreign proteins expressed in bacteria may be susceptible to degradation by host encoded proteases. Certainly, for some genes, problems also arise from additional transcription start sites occurring within the coding sequence that give rise to N-terminally truncated proteins. These often prove to be difficult to resolve from the full length protein during purification, as has been reported for overexpression and purification of the human transcription factor Sp1 *(2)*. The use of fusion vectors, in which the cloned gene is inserted upstream (N-terminal fusion) or downstream (C-terminal) of a gene encoding a stable protein that can be expressed at high levels, may often circumvent some of the problems of protein instability. A range of vectors already exist for the expression of foreign polypeptides as fusion proteins *(3)*. Fusions with *E. coli* β-galactosidase, *trp E*, and Staphylococcal protein A have been reported *(4–7)*, but subsequent purification requires denaturing conditions that may reduce both the activity and antigenicity of the fusion proteins. Alternative strategies involve tagging the 3' end of a gene with a stretch of arginine codons, thereby facilitating ion exchange chromatography of the expressed protein *(6)*, or the addition of histidine codons to either the 5' or 3' ends, which enables the expressed protein to be purified by transition metal affinity chromatography *(8)*. The addition of extra sequence information has also been used to target proteins for secretion into the growth medium or the periplasmic space of the bacterium in order to facilitate extraction and purification *(9)*.

This chapter will deal exclusively with the pGEX fusion vector system *(10)*. pGEX vectors direct the expression of foreign polypeptides as fusions with the C-terminus of the glutathione-*S*-transferase (GST) from *Schistosoma japonicum*. GST, although uncharacteristically of eukaryotic origin, can be expressed at very high levels in *E. coli* as a soluble homodimeric protein with a polypeptide mol wt of

26 kDa. The expression vector has been designed to facilitate proteolytic removal of the fusion protein after expression and purification by the insertion of either thrombin or Factor Xa cleavage sites (depending on the choice of vector) at the junction of the fusion (*see* Note 1).

Each of the original pGEX vectors (pGEX-1N, 2T, and 3X) encodes GST, followed by unique restriction sites for *Bam*HI, *Sma*I, and *Eco*RI. In addition, different translational phasing can be accommodated by the choice of appropriate derivatives. This information is given in Fig. 1. The resultant GST fusion protein can be rapidly purified, using a glutathione-agarose affinity column. The fusion protein is adsorbed onto the column (from a crude cell lysate); the column is washed extensively and the bound protein eluted with glutathione. This feature of the pGEX system has the added bonus that the protein under investigation can also be immobilized, often in a biologically active form. Thus, for example, a number of laboratories are exploring protein–protein interactions in eukaryotic gene transcription using this strategy. One other important feature of pGEX vectors is that they facilitate the isolation of proteins for which no biological or chemical assay is available.

In the following procedure, two specific examples are given of the transcription factors TFIID and Sp1. However, the method is quite general if suitable PCR primers are made for any other cloned gene.

2. Materials

2.1. Choice of Vector

The main requirement for the successful overexpression of a recombinant gene in *E. coli* is that the translational reading frame for the foreign polypeptide is in frame with that of the GST sequence. Failure to ensure this will, of course, lead to translational termination at or near the 5' end of the cloned gene (*see* Note 2). Recently, improved vectors have been reported and are usually available on request from the various authors. These include pGEX-KG, which has a much improved cloning site as well as a sequence encoding a "glycine kinker," introduced between the polylinker and the sequence encoding the thrombin cleavage site, which is said to enhance cleavage of the expressed fusion protein by thrombin *(11)*. Other authors report efficient cleavage of the expressed fusion protein by placing a sequence

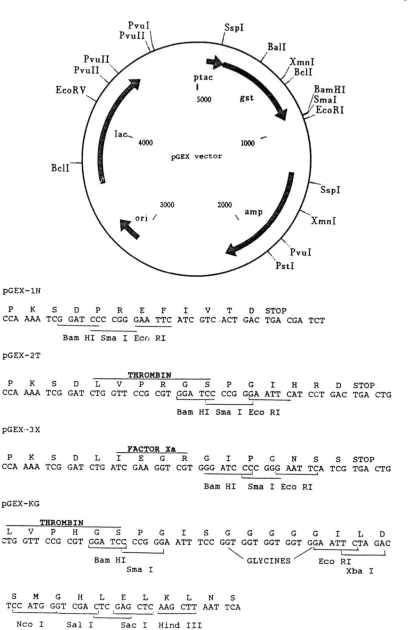

Fig. 1. Schematic of the main features of pGEX vectors. The protease, multiple cloning sites, and reading frames presented by the different vectors are also shown.

encoding a "glycine kinker" after the sequence encoding the thrombin cleavage site *(12)* (*see* Note 3).

2.2. Amplification by PCR and Subcloning of the cDNA Encoding Human TFIID

1. Plasmid DNA containing the coding sequence of interest. In this example, we used a pBS plasmid carrying the full length TFIID cDNA (kindly supplied by A. Berk, UCLA).
2. Oligonucleotide primers for TFIID amplification by the polymerase chain reaction were synthesized on an Applied Biosystems (Warrington, UK) model 381A DNA synthesizer and ethanol precipitated before use. Sequences are given below, with the indicated restriction sites underlined:

    5' CGCGAATTCATATGGATCAGAACAACAGC 3' (*Eco*RI)

    5' CGCGAAGCTTTTACGTCGTCTTCCTGAA 3' (*Hind*III)

3. 10X PCR buffer: 0.1M Tris-HCl, pH 8.8, 0.5M KCl, 0.1M NaCl, 0.015M MgCl$_2$, 0.1% Triton X-100.
4. *Taq* DNA polymerase (e.g., from Promega, Madison, WI).
5. 5 mM dATP, dTTP, dCTP, and dGTP mix (e.g., from Pharmacia, Piscataway, NJ).
6. Light mineral oil.
7. Nonionic detergent NP 40 (1% stock solution).
8. Techne (Cambridge, UK) PHC-2 thermal cycler (or similar).
9. Restriction enzymes: *Hind*III, *Eco*RI. Enyme buffers supplied by manufacturer.
10. 1X TAE buffer: 40 mM Tris-acetate, 1 mM EDTA, pH 7.6.
11. Dialysis tubing and clips.
12. Phenol/chloroform *(13)*, ether, ethanol, 70% ethanol solution, sterile H$_2$O.
13. Potassium acetate solution: a 3M solution of potassium acetate brought to pH 7.8 by the addition of glacial acetic acid.
14. pGEX-KG *(11)*.
15. T4 DNA ligase.
16. 10X ligase buffer: 0.5M Tris-HCl, pH 7.4, 0.1M MgCl$_2$, 10 mM DTT, 1 mg/mL BSA.
17. Competent *E. coli* JM109 cells (*see* Note 4).

2.3. Subcloning the cDNA Encoding Human Sp1

1. pBS plasmid carrying the full length Sp1 cDNA (kindly supplied by R. Tjian, Berkeley, CA).

2. Restriction enzymes: *Xba*I, *Sal*I, and *Spe*I.
3. Materials 10–17 as described for TFIID subcloning (*see* Section 2.2.).

2.4. Expression of Fusion Proteins

1. Luria broth: 10 g tryptone, 5 g yeast extract, and 10 g NaCl/L H_2O.
2. Isopropyl-β-D-thiogalactoside (IPTG).
3. PBS/PMSF buffer: 150 mM NaCl, 16 mM Na_2HPO_4, 4 mM NaH_2PO_4, pH 7.3, 0.1–1.0 mM phenylmethylsulfonyl fluoride (PMSF).
4. Soniprep 150 sonicator (MSE) or similar.
5. SDS-page equipment *(14)*.
6. Sarcosyl.

2.5. Purification and Cleavage of Fusion Proteins

1. Disposable 5-mL plastic chromatography columns.
2. Glutathione agarose (e.g., from Sigma, St. Louis, MO).
3. Reduced glutathione (e.g., from Sigma).
4. Column preequilibration buffer (PBS): 150 mM NaCl, 16 mM NaH_2PO_4, 4 mM Na_2HPO_4, pH 7.3.
5. Preelution buffer: 50 mM Tris-HCl, pH 8.0.
6. Elution buffer: 50 mM Tris-HCl, pH 8.0, 10 mM reduced glutathione.
7. Thrombin (e.g., from Sigma).
8. Thrombin cleavage buffer: Preelution buffer containing 150 mM NaCl, 2.5 mM $CaCl_2$.

3. Methods

3.1. Amplification by PCR and Subcloning of the cDNA Encoding Human TFIID

1. Set up the PCR reaction in 1X PCR buffer with 0.2 mM dNTPs, 10–15 ng of template DNA (pBS carrying TFIID cDNA), 0.1 µM primers, and 0.04% NP40 made up to a volume of 50 µL with sterile H_2O.
2. Heat reaction mix at 100°C for 4 min.
3. Add 1 U *Taq* DNA polymerase, mix well, overlay with 100 µL of mineral oil, and place in thermal cycler.
4. Temperature program used on thermal cycler: 1 cycle at 94°C for 5 min, 45°C for 2 min, and 72°C for 3 min; 4 cycles at 94°C for 1 min, 45°C for 2 min, and 72°C for 3 min; and then 25 cycles at 94°C for 1 min, 50°C for 2 min, and 72°C for 3 min.
5. Carry out electrophoresis of the PCR product on a 1% agarose gel (1 g agarose/100 mL 1X TAE buffer) at 100 V for 1 h.
6. Cut the PCR product band out of the gel and place the gel slice inside a piece of dialysis tubing in 0.5 mL 1X TAE buffer.

Eukaryotic Transcription Factors in E. Coli

7. Put the dialysis tubing in a horizontal gel tank with the gel slice toward the negative anode and electroelute for 30 min at 100 V. Reverse the current polarity for 1 min and then remove the DNA solution from the dialysis tubing.
8. Add an equal volume of phenol/chloroform to the DNA solution, mix by vortexing, and spin at 12,000 rpm in a microcentrifuge for 5 min. Remove and keep the top aqueous layer. Repeat two more times.
9. Add an equal volume of ether, mix by vortexing, and spin at 12,000 rpm for 1 min. Discard the top ether layer.
10. Precipitate the DNA by addition of 1 vol 10X potassium acetate solution, pH 4.8, and 2 vol of ice-cold ethanol. Leave at –20°C for 30 min, and then spin at 12,000 rpm for 10 min. Wash the pellet in 70% ethanol solution, spin again for 2 min, and dry under vacuum for 10 min. Resuspend pellet in 10–20 µL of sterile H_2O.
11. Digest 2–3 µg of purified PCR product with 25 U each of *Eco*RI and *Hin*dIII at 37°C overnight in the buffer supplied by the manufacturer.
12. Phenol/chloroform extract 3X, wash with ether, and ethanol precipitate the DNA as in steps 6–8. Resuspend DNA in 20 µL sterile H_2O.
13. Digest 0.5 µg of pGEX-KG with 10 U each of *Eco*RI and *Hin*dIII for 3 h at 37°C in the buffer supplied by the manufacturer.
14. Purify the cut DNA from an agarose gel as described in steps 3–8. Resuspend the DNA pellet in 20 µL sterile H_2O.
15. Set up ligation reaction containing 1.5 µL of 10X ligase buffer, 10 µL of PCR product, 1 µL of cleaved vector, 1 U of T4 DNA ligase, and 1.5 µL of sterile water. Incubate at room temperature for 2 h.
16. Competent *E. coli* JM109 cells were transformed with half the ligation mixture. Appropriate clones were initially identified by restriction analysis.

3.2. Subcloning the cDNA Encoding Human Sp1

1. Digest 1 µg of pBS plasmid containing Sp1 cDNA with 10 U each of *Spe*I and *Sal*I at 3°C for 3 h.
2. Digest 0.5 µg of pGEX-KG with 10 U each of *Xba*I and *Sal*I at 37°C for 3 h (*Spe*I and *Xba*I are complementary sites).
3. Purify the DNA from an agarose gel and carry out the ligation as described for the TFIID DNA (*see* Section 3.1.).

3.3. Expression of Fusion Proteins

The choice of growth and induction conditions will vary from protein to protein (*see* Note 5). Our own experience has shown, for example, that it is not always best to grow and induce the samples at 37°C. We generally find that growth at 25–30°C results in a much higher percentage of expressed protein being soluble. Many proteins, however, are readily soluble, and so the strain can be grown and induced at 37°C. Conditions need to be determined empirically for each protein. The following protocol should therefore only be considered as a general guide.

1. Grow up a 5-mL culture of recombinant bacterial strain overnight.
2. Inoculate 500 mL of Luria broth with the 5-mL culture, and grow cells at 30°C until an absorbance at 540 nm of 0.5 is obtained.
3. Add IPTG (or often lactose will work just as effectively and is considerably less expensive than IPTG) to a final concentration of 0.1 mM.
4. Grow the strain at 30°C for a further 2 h. In some instances, this time will vary in order to optimize solubility and level of expression.
5. Harvest cells by centrifugation (5000 rpm in a Sorvall GS-3 rotor for 10 min).
6. Resuspend cells in a small volume of ice-cold PBS/PMSF (0.1–1.0 mM), and transfer to a small metal sonication bucket. Sonicate on ice, in short bursts (10–15 s) using a medium-sized probe (5–10 mm diameter) set at 14 µm amplitude for 2–3 min. The metal bucket should ensure rapid heat transfer from the sample. The temperature should not rise above 10°C.
7. Centrifuge the extract at 18,000 rpm in a Sorvall SS34 rotor for 60 min.
8. Assay both pellet and supernatant by SDS-PAGE for the presence of induced protein.

    If at this stage the protein is found in the insoluble pellet, the sonication can be repeated with a fresh culture as listed in step 6, but with the addition of detergents, such as NP40, Tween, Triton X-100, or Sarcosyl, to the sonication buffer. Once again, whichever detergent is most suitable for any given protein needs to be determined empirically. Alternatively, one can add deterent to the pellet and resuspend as detailed in step 9. We have found that although the insoluble pellet contains less protein overall and therefore is potentially a better starting point for purification of the fusion protein, higher detergent levels are required to obtain the comparative levels of solubility found on detergent addition to a presonicated culture. The following paragraph deals with extraction of insoluble protein from the pellet.
9. To the insoluble pellet, add ice-cold PBS/PMSF and 0.1–1.0% Sarcosyl. Resuspend the pellet by hand pipeting and gently agitate at intervals over a period of 1–2 h, keeping the sample on ice. The choice and level of detergent addition is once again a matter of trial and error. Low levels of most detergents do not interfere with subsequent purification procedures, although higher levels do reduce the percentage of fusion protein

that binds the affinity matrix (*see* Note 6). The most commonly used detergent in this laboratory is Sarcosyl, which we have used to a level of 0.5% without loss of fusion protein binding ability to the affinity matrix. Some authors report that Sarcosyl addition may affect protein activity. Therefore, as a general precaution it is advisable to wash the column or beads thoroughly prior to elution, thereby removing all traces of detergent in the eluent. Our own experience shows that transcription factors, such as TFIID, retain their DNA binding ability after purification using a Sarcosyl based approach.

10. Clarify the soluble phase by centrifugation at 18,000 rpm for 60 min (Sorvall SS34 rotor). Assess the quality of the preparation by SDS-PAGE.

3.4. Purification and Cleavage of the Fusion Proteins

Purification of glutathione-*S*-transferase fusion proteins can be achieved in a simple one-step procedure from crude bacterial lysates. Figure 2, for example, shows the results of a rapid purification and cleavage experiment on TFIID-GST fusion protein. Trial purifications are best performed in microcentrifuge tubes using 100 µL of resin/1.5 mL of bacterial culture. Conditions such as buffer pH, buffer ionic strength, and detergent levels can be rapidly explored, leading to a final optimization of parameters. Given below is a large-scale purification protocol that works well for both TFIID and Sp1 fusion proteins. The method involves the use of 1-mL columns, but can be performed equally as well in 10-mL Sterilin tubes.

1. Pour a 2-mL (50%) slurry of preswollen glutathione agarose into a 3-mL disposable column and leave to settle. Equilibrate with 4 × 10 mL washes of ice-cold PBS.
2. Stop the flow of the column and apply the clarified protein solution. Resuspend the resin with a Pasteur pipet and allow to resettle on ice for 20 min. Restart the column and reapply the flowthrough (*see* Note 7).
3. Wash the column in 4 × 10 mL of ice-cold PBS.
4. Wash the column in 2 × 10 mL of 50 m*M* Tris-HCl, pH 8.0.
5. Stop the column flow and add 1 mL of elution buffer. Agitate resin by pipeting the slurry up and down. Leave the resin to settle for 10 min and collect the eluent fraction. Add another 1 mL elution buffer and repeat this procedure. Assay the fractions on SDS-PAGE and/or by gel retardation assay (*see* Note 8).
6. Dialyze column eluent against 1000 vol of thrombin cleavage buffer for 4–5 h, 4°C.

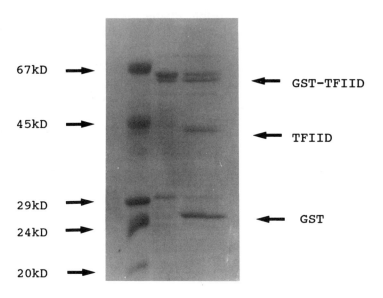

Fig. 2. SDS-PAGE gel showing glutathione agarose affinity purification of GST-TFIID fusion protein and results of subsequent thrombin cleavage. Purification was performed in one round in 10-mL sterilin plastic tubes as described in the text. Lane 1: Dalton markers VII-L (Sigma). Lane 2: Affinity purified GST-TFIID. Lane 3: Thrombin cleaved GST-TFIID protein.

7. Add 4 µg of thrombin (2 µg thrombin/mg of fusion protein; see Note 9). Leave at room temperature for 0.5–2.0 h. Once again, optimum conditions for proteolysis need to be determined for each fusion protein examined (see Note 10).
8. Pass the cleavage reaction through a fresh 1.0-mL glutathione–agarose column and collect the flowthrough (see Note 11). Alternatively, glutathione–agarose resin can be added directly to the tube containing the cleavage reaction. After brief incubation and centrifugation, cleaved GST is retained by the beads and the recombinant protein is liberated into the supernatant (see Note 12).

4. Notes

1. We have found that generally thrombin gives greater reproducibility and cleavage efficiency than Factor Xa. The latter has a limited shelf life, and can vary in quality from batch to batch and from supplier to supplier.

2. We have occasionally observed that some constructs do not express the desired protein owing to an introduction of a frame shift mutation. This can often be remedied by simply filling in the *Bam*HI site and retransforming. If this fails, then it is simple enough to repeat the PCR cloning step. In order to determine whether such mutations have arisen, it is sensible to sequence across the junction of the recombinant plasmid using a primer based on the 3' sequence of the GST gene.
3. The glycine kinker is thought to facilitate proteolytic cleavage by "pushing" the fusion protein away from the compact globular structure of GST. A recently described pGEX vector has the glycine kinker 5' to the sequence encoding the protease cleavage site and the expressed protein is reportedly more readily cleaved as a result. We ourselves have found no detectable differences between the rate of cleavage at glycine kinker containing sites to normal protease cleavage sites. A recent observation that is most critical to efficient protease cleavage is the identity of the amino acids immediately distal to the thrombin cleavage site *(12)*.
4. An important point to note when using pGEX vectors for expression of potentially lethal fusion proteins is that expression under control of the *tac* promoter is extremely tightly repressed in the absence of IPTG. Maintenance of the *lacI^q* allele of the *lac* repressor on the pGEX vector means that most strains of *E. coli* are suitable as an expression host.
5. The insolubility of GST fusion proteins may not always be a function of the physical properties of the recombinant protein fused to the GST moiety. Often growth rate, level of aeration, choice of host strain, and volume of cell extraction can influence the solubility of the protein.
6. The effect of detergent on the efficiency of binding of fusion protein to the affinity matrix has been evaluated for a number of detergents. Protein binding is unaffected, for example, by 1% Triton X-100, 1% Tween-20, 1% CTAB, and up to 0.5% Sarcosyl.
7. The capacity of the glutathione affinity resin is substoichiometric with the amount of GST fusion protein loaded under most circumstances, and therefore it is best to load concentrated material and to recycle the effluent from the first pass through the column until most of the protein is recovered.
8. Estimation of yield can be achieved by directly loading a small aliqout of the agarose beads onto an SDS-gel. This should be sufficiently accurate to calculate the required thrombin level if cleavage is to be performed while the fusion protein is still bound to the column. Purified protein yields can vary from well over 10 mg/L in the best instance to a few µg/L in the worst. However, whatever the level of expression, homogeneous protein can be obtained by passing the crude cell extract through the glutathione affinity column.

9. Levels of protease required for complete cleavage will vary from protein to protein. However, as a rough guide the following values should be used in initial trials. Thrombin: 0.2–1% w/w fusion protein; Factor Xa: 1% wt/wt fusion protein.
10. If full sequence data are available, prior to subcloning the DNA into a particular pGEX vector, the sequence should be checked for the presence of internal protease cleavage sites. The likelihood of such sites being accessible to the proteases will obviously increase with the number of sites found. However, even when a number of sites are present within the target protein, the rate of cleavage at the designed linker site is likely to be much greater unless the other sites are exposed to solvent in the three-dimensional structure of the protein.
11. The affinity resin can be regenerated by thorough washing in high salt buffers and used several times. There is a gradual loss of performance, however, presumably because of incomplete removal of glutathione. It is perhaps prudent to use fresh resin at the point at which cleaved protein is to be purified.
12. A variation on the general method outlined above requires that the protein should be left attached to the column during proteolytic cleavage. Indeed, some authors report this method as being more efficient than cleavage off the column. In this method, the resin is removed from the column to a 10-mL sterilin tube and resuspended in 2 mL of thrombin cleavage buffer. Thrombin is added (2 µg/mg of fusion protein) and the tube shaken gently for 2 h at room temperature. The suspension is briefly centrifuged and then cleaved. Protein is released into the supernatant.

References

1. Dynlacht, B. D., Hoey, T., and Tjian, R. (1991) Isolation of coactivators associated with the TATA-binding protein that mediate transcriptional activation. *Cell* **66**, 563–576.
2. Kadonaga, J., Courey, A., Ladika, J., and Tjian, R. (1988) Distinct regions of Sp1 modulate DNA binding and transcriptional activation. *Science* **242**, 1566–1570.
3. Uhlen, M. and Moks, T. (1990) Gene fusions for purpose of expression: an introduction. *Meth. Enzymol.* **185**, 129–143.
4. Ruther, U. and Muller-Hill, B. (1983) Easy identification of cDNA clones. *EMBO J.* **2**, 1791–1794.
5. Stanley, K. K. and Luzio, J. (1984) Construction of a new family of high efficiency bacterial expression vectors: identification of cDNA clones coding for human liver proteins. *EMBO J.* **3**, 1429–1434.
6. Abrahmsen, L., Moks, T., Nilsson, B., and Uhlen, M. (1986) Secretion of heterologous gene products to the culture medium of *Escherichia coli*. *Nucleic Acids Res.* **14**, 7487–7500.

7. Sassenfield, H. M. and Brewer, S. J. (1984) A polypeptide fusion designed for the purification of recombinant proteins. *Biotechnology* **2,** 76–81.
8. Van Dyke, M. V., Sirito, M., and Sawadago, M. (1992) Single-step purification of bacterially expressed polypeptides containing an oligo-histidine domain. *Gene* **111,** 99–104.
9. Stader, J. A. and Silhavy, T. (1990) Engineering *Escherichia coli* to secrete heterologous gene products. *Meth. Enzymol.* **185,** 166–187.
10. Smith, D. and Johnson, K. S. (1988) Single-step purification of polypeptides expressed in *Escherichia coli* as fusions with glutathione-*S*-transferase. *Gene* **67,** 31–40.
11. Guan, K. and Dixon, J. (1991) Eukaryotic proteins expressed in *Escherichia coli*: an improved thrombin cleavage and purification procedure of fusion proteins with glutathione-*S*-transferase. *Anal. Biochem.* **192,** 262–267.
12. Haun, R. S. and Moss, T. (1992) Ligation-independent cloning of glutathione-*S*-transferase fusion genes for expression in *Escherichia coli*. *Gene* **112,** 37–43.
13. Maniatis, T., Fritsch, E. F., and Sambrook, J. (1982) *Molecular Cloning: A Laboratory Approach,* Cold Spring Harbor Laboratories, Cold Spring Harbor, NY.
14. Laemmli, U. K. (1970) Cleavage of structural proteins during assembly of the head of bacteriophage T4. *Nature* **227,** 680.

CHAPTER 16

Site-Directed Mutagenesis by the Cassette Method

Andrew F. Worrall

1. Introduction

Cassette mutagenesis is a technique for altering a protein sequence at the DNA level by replacing a section of genetic information with an alternative sequence, normally provided by a synthetic DNA duplex. First, the gene contained in a suitable vector is cleaved with two restriction enzymes. This releases a small section of DNA from the gene. The prerequisite for this method to be practical is that these restriction enzymes cut at sites that flank the area of DNA to be changed and are unique in the gene/vector system. In this way only the desired cutting occurs. A synthetic duplex is then ligated in place of the released cassette and the resultant construct is sequenced through the cassette and the reformed restriction sites to check that the mutated gene now has the intended sequence. This technique can be used to make single or multiple amino acid changes to the protein sequence and to insert sequences, or indeed, delete them from the protein structure. The changes are only limited by the available size of the synthetic DNA cassette. With current DNA synthesis technology and the expertise of the average DNA synthesis service, duplexes of up to 100 bp are readily available. This methodology has been used extensively in this laboratory to perform site-directed mutagenesis experiments on a number of synthetic genes including bovine pancreatic DNase 1 *(1)*. Synthetic genes are normally designed with a number of unique restriction sites in the sequence, making cassette mutagenesis particularly applicable in these cases. In addition to performing amino acid changes to the protein sequence, the technique can be utilized to alter the noncoding portion of genes to increase expression levels *(2)*. When designing synthetic genes, it is useful to flank the Shine-Dalgarno sequence and the first few codons at the beginning of the gene by restriction sites. This enables the Shine-Dalgarno sequence, its spacing to the ATG start codon, and the initial coding nucleotides to be easily altered. This often has a dramatic effect on expression levels *(3)*. We routinely mutate genes cloned in M13mp18/19 before subcloning into an expression vector for the following reasons:

1. Mutation can be performed and verified without the complication of possible gene expression; dealing with recombinant proteins that are often toxic to the host expression system is thus made simpler.
2. Single-stranded sequencing template is readily available to confirm the mutation experiment.
3. A full restriction map of M13mp18/19 is available (*see* Note 1) that makes the choice of unique restriction sites possible. There are some 60 restriction enzymes that do not cut M13mp18/19, and it is therefore often possible to identify the restriction sites needed.

However, if gene product toxicity does not present a problem and a restriction map of the expression vector is available, the use of plasmid sequencing techniques *(4)* allows mutation of the gene to be performed in the expression vector, eliminating the need for a subsequent subcloning step. Finally, for the method to reach high efficiency of mutation, it is important that the synthetic DNA cassette is rigorously purified to eliminate the truncated sequences arising from the oligonucleotide synthesis. Failure to do this can result in the production of unplanned and unwanted mutations.

2. Materials

The majority of equipment needed for this technique will be found in any well equipped biochemistry laboratory. However, the list below gives the more specialized apparatus needed for this technique. All chemicals should be of the best available quality. In those cases in which only one source of a product is known, this is stated.

2.1. Equipment

1. Savant (Hicksville, NY) SpeedVac concentrator, or equivalent.
2. For the purification of oligonucleotides <40 bases in length (Section

3.1.1.): HPLC capable of running gradient elution profiles, fitted with an octadecylsilica reverse-phase column.
3. For the purification of oligonucleotides >40 bases in length (Section 3.1.2.): Hand-held UV lamp capable of emitting radiation at 254 nm.
4. Electroelution apparatus for isolating DNA from agarose gels.

2.2. Reagents Needed Exclusively for the Purification of Oligonucleotides <40 Nucleotides Long

1. Buffer A: 50 mL 2M triethylammonium acetate, 50 mL HPLC grade acetonitrile, made up to 1 L with distilled water.
2. Buffer B: 50 mL 2M triethylammonium acetate, 650 mL HPLC grade acetonitrile, made up to 1 L with distilled water.
3. Triethylamine: 1% solution in distilled water.
4. Trifluoroacetic acid: 2% solution in distilled water.

2.3. Reagents Needed Exclusively for the Purification of Oligonucleotides >40 Nucleotides Long

1. 0.1M Triethylammonium acetate obtained by diluting a 2M stock solution.
2. Deionized formamide (*see* Note 2).
3. 0.05% Bromophenol blue in deionized formamide.
4. Plastic wrap.
5. TLC Plate containing a fluorophore that fluoresces at 254 nm (e.g., Merck [Rahway, NJ] product no. 5554).
6. Gel elution buffer: 100 mM Tris-HCl, 0.5M NaCl, 5 mM EDTA, pH 8.0.
7. Acrylamide solution containing 30% w/v acrylamide and 0.8% w/v *bis*-acrylamide in water.
8. Ammonium persulfate solution: 10% w/v in distilled water.
9. N,N,N',N'-Tetramethylethylenediamine (TEMED).

2.4. Other Reagents Needed

1. 2M Triethylammonium acetate (Applied Biosystems [Foster City, CA] cat. no. 400613).
2. HPLC grade acetonitrile.
3. Oligonucleotide purification cartridges (OPCs, Applied Biosystems cat. no. 400771).
4. Restriction enzymes, T4 DNA ligase, and appropriate buffers (most manufacturers supply these with the enzymes).
5. TE buffer: 10 mM Tris-HCl, 1 mM EDTA, pH 7.5.

6. TBE buffer: 54 g Tris base, 27.5 g boric acid, 20 mL 0.5M EDTA, pH 8.0, dissolved in 1 L of distilled water gives a 5X concentrated stock of this buffer.
7. H-plates and H-top agar: Dissolve 10 g Bacto tryptone and 8 g NaCl in distilled water, make up to 1 L, and autoclave. For pouring plates, add 12 g agar before autoclaving (for H-top agar add 8 g agar).

3. Methods

Two oligonucleotides are required to form the synthetic cassette, one for the coding 5'–3' strand and the other for the anticoding 3'–5' strand. They should be designed to form the correct sticky (or blunt) ends when annealed together so that they can be ligated into the cut gene. It should be remembered that attempting to perform cassette mutagenesis with a cassette carrying two blunt ends means that only a maximum of 50% of the clones obtained will carry the cassette in the correct orientation. Because of this, it is normal to design the experiments to use a cassette that has two different sticky ends or, at worst, one sticky and one blunt end. As stated above, high efficiency mutation is dependent on purification of the synthetic oligonucleotides, and the protocols for this are now presented.

3.1. Oligonucleotide Purification

3.1.1. Purification of Oligonucleotides <40 Nucleotides Long

The method of choice for producing really pure oligonucleotides in this size range is reverse-phase HPLC. The column used contains an octadecylsilica (ODS or C-18) support and the crude oligonucleotide is separated using a gradient of increasing concentration of acetonitrile in a triethylammonium acetate buffer. This protocol assumes that the oligonucleotide is provided to the reader as a crude solution in concentrated ammonia solution and is the product of a 0.2 μmole scale, trityl-on synthesis using standard phosphoramidite chemistry (*see* Note 3).

1. Evaporate the oligonucleotide solution in a microcentrifuge tube to dryness using a Savant SpeedVac or equivalent.
2. Redissolve the product in 500 μL of 1% aqueous triethylamine. Not all the solid will dissolve. Whirly-mix the suspension for 2 min and then centrifuge in a benchtop microcentrifuge for 5 min and carefully pipet the supernatant into a fresh microcentrifuge tube. This solution now

contains the crude oligonucleotide. Only a portion needs to be purified and the rest can be stored frozen at −20°C indefinitely.
3. Place buffers A and B into 2-L Buchner flasks, seal the tops of the flasks with large rubber bungs, and attach the sidearms to water aspirated pumps. Stirring of the buffers with magnetic followers results in dissolved air being removed from solution. Continue the degassing process until effervescence stops (about 20–30 min) (*see* Note 4).
4. Set up the HPLC to deliver a gradient of 25–75% buffer B over 20 min at a flow-rate of 1 mL/min. Set the UV detector on the HPLC to detect at 260 nm at a sensitivity of 0.2 AUFS.
5. Equilibrate the column to the buffer conditions by running the gradient through the system and then reset to the start conditions of 25% buffer B. Leave running for 20 min to complete the equilibration process.
6. Inject 5 µL of the solution of crude oligonucleotide obtained in step 2 and start the gradient as a test run. The chromatogram obtained may be simple or quite complex depending on the quality of the oligonucleotide synthesis. However, the only important peak will be the major product that elutes with a retention time of 10–14 min, depending on length, sequence, and exact conditions. This first test run can be discarded (*see* Note 5).
7. After the gradient has finished, reset the HPLC to the initial 25% buffer B and continue pumping at 1 mL/min for 20 min to reequilibrate the column. Change the sensitivity on the UV detector to 1 or 2 AUFS.
8. Inject 100 µL of the crude oligonucleotide and this time collect the peak corresponding to the purified oligonucleotide.
9. The oligonucleotide thus purified still has the hydrophobic dimethoxytrityl group attached to the 5'-hydroxyl. This must now be removed using trifluoroacetic acid. The most convenient method for doing this is using an oligonucleotide purification cartridge (OPC). The OPC is conditioned by passing 5 mL of HPLC grade acetonitrile followed by 5 mL of 2M triethylammonium acetate through the cartridge. The solution from step 8 is slowly passed twice through the OPC using a syringe (*see* Note 6). This loads between 1 and 5 A_{260} U of oligonucleotide onto the cartridge. Again, the unused oligonucleotide at this step can be stored at −20°C indefinitely.
10. Wash the cartridge slowly by passing 10 mL of distilled water followed by 1 mL of 2% aqueous trifluoroacetic acid. Leave this solution in the OPC for 5 min, then follow it with a slow wash with a further 4 mL of the acid solution. After another slow wash with 10 mL of distilled water, the purified, detritylated, desalted oligonucleotide can be eluted from the cartridge with 1 mL of 20% aqueous HPLC grade acetonitrile.

11. Evaporate the oligonucleotide to dryness in a SpeedVac and redissolve in 100 µL of distilled water. Dilute 10 µL of this solution to 1 mL with distilled water and measure the absorbance in a UV spectrometer at 260 nm. Calculate the concentration of the purified oligonucleotide using the fact that 1 µmol of an oligonucleotide n-bases long, when in 1 mL of solution in a 1-cm path length cell, will have an absorbance at 260 nm of approx $10 \times n$.
12. Dilute a portion of the purified oligonucleotide with TE buffer to a concentration of 10 pmol/µL. All oligonucleotide solutions can be stored at –20°C.

3.1.2. Purification of Oligonucleotides >40 Nucleotides Long

The yields of these oligonucleotides from HPLC columns is often low and HPLC does not have the resolution required to separate the desired full-length oligonucleotide from the failure sequences. It is necessary to use preparative gel electrophoresis on denaturing polyacrylamide gels to achieve this result. For this method, oligonucleotides must be synthesized without the dimethoxytrityl group (trityl-off). As before, it is assumed that the oligonucleotide is provided crude in concentrated ammonia solution.

1. Evaporate the crude oligonucleotide solution to dryness in a Savant SpeedVac or equivalent.
2. Redissolve the solid in 500 µL of distilled water. Not all the solid will dissolve. Whirly-mix the solution for 2 min and then spin in a microcentrifuge for 5 min. Decant the solution carefully into a fresh microcentrifuge tube.
3. Dilute 10 µL of the supernatant to 1 mL with distilled water and measure the absorbance at 260 nm in a UV spectrometer.
4. Before purification by electrophoresis, the sample must be desalted to ensure good resolution between the desired full length oligomer and the failure sequences. Again, many products exist that will desalt oligonucleotides, but OPCs are routinely used in this laboratory. An aliquot of the crude oligonucleotide containing 10 A_{260} U is diluted to 500 µL and this solution is slowly passed twice through an OPC, previously conditioned with HPLC grade acetonitrile and $2M$ triethylammonium acetate as described above (Section 3.1.1., step 9). This will load the majority of the oligonucleotide onto the cartridge.
5. Wash the cartridge slowly with 15 mL of $0.1M$ triethylammonium acetate and then elute the desalted oligonucleotide with 1 mL of 50% aqueous HPLC grade acetonitrile.

6. Evaporate the oligonucleotide to dryness in a SpeedVac and redissolve the product (which may by now be invisible) in 20 µL of deionized formamide.
7. Cast a denaturing polyacrylamide gel. The composition of the polyacrylamide gel required for the purification depends on the length of the oligonucleotide. Those of <80 nucleotides can be purified on 12% polyacrylamide gels, whereas those >80 nucleotides require 8% gels. All the gels should be 1.5-mm thick, 40-cm long, and contain $7M$ urea as a denaturant, and a standard TBE buffer system. The gels are cast with 1 × 1.5-cm wells to accommodate the oligonucleotide solution in formamide. Load half of the solution from step 6 (about 5 A_{260} U of crude oligonucleotide) into one of the wells in the gel. Do not include any marker dyes with the oligonucleotide. Instead, load a separate well with 10 µL of 0.05% bromophenol blue in deionized formamide as a marker for the electrophoresis. This ensures that any impurities in the dye are not transferred to the oligonucleotide. Run the gel at 40 W constant power with appropriate cooling until the bromophenol blue dye has reached the bottom of the plates (*see* Note 7).
8. Carefully separate the electrophoresis plates to leave the gel attatched to one of them. Gels of this thickness are fairly robust so the following procedures are quite straightforward. Cover the gel with plastic cling film (e.g., Saran Wrap®), invert the glass plate with the gel, and remove the plate to leave the gel attatched to the plastic film.
9. Place a 20 × 20-cm TLC plate coated with a fluorophore beneath the plastic film to form a gel-film-TLC plate stack. In a darkened room, shine UV light at 254 nm from a hand-held lamp vertically through the gel and plastic film onto the TLC plate. The products of oligonucleotide synthesis are clearly visible as a ladder of bands corresponding to products differing in one nucleotide in length. The product of highest molecular weight (lowest mobility) should be the most intense band on the gel (*see* Note 8). This is the full-length product required. Cut the gel with a clean razor blade to remove this band, making sure that the incisions occur within the boundaries of the band to avoid any contamination of the product with other shorter oligonucleotides. Also make sure that the lamp is held vertically over the band being excised—the image of the band on the TLC plate is a shadow and therefore its precise observed position is subject to parallax effects. This technique for band visualization is known as UV shadowing.
10. Place the small piece of gel into a microcentrifuge tube and cover it with 400 µL of gel elution buffer, and incubate at 37°C for 16 h or overnight. Placing the tube on a rotating wheel ensures good mixing of

the buffer and gel slice. Carefully pipet off the solution now containing the purified oligonucleotide and wash the gel slice with a further 100 µL of the elution buffer. The combined oligonucleotide solutions can now be desalted using an OPC cartridge precisely as described in steps 4 and 5. Evaporate the desalted oligonucleotide solution to dryness in a SpeedVac and then redissolve the oligonucleotide in 100 µL TE buffer. Dilute a 10-µL aliquot of this solution to 1 mL with distilled water and measure the absorbance at 260 nm using a UV spectrometer. Calculate the concentration of the desalted oligonucleotide using the relationship given above (Section 3.1.1., step 12) and dilute the oligonucleotide to a final concentration of 10 pmol/µL. Store frozen at –20°C.

3.2. Mutagenesis of the Target Gene

It is assumed that the gene is cloned into M13 and a sample of double-stranded RF DNA is available. This must be of "cloning quality," that is, from a cesium chloride gradient or a good quality minipreparation. Five micrograms of DNA is sufficient for the experiment and less is often used. By its very nature, this protocol involves the use of many standard molecular biological techniques. Only those that are particularly unique to cassette mutagenesis are described in detail. Other instructions may be obtained from books in the Further Reading section of this chapter.

1. Set up an incubation containing 5 µg of the double-stranded M13 DNA containing the gene of interest in a total volume of 50 µL of an appropriate buffer and 10 U of each of the restriction enzymes (*see* Note 9). Incubate at 37°C for 2 h.
2. The "double cut" M13 DNA can now be mixed with loading buffer and loaded into a slot cast in a 0.5% agarose gel and the digestion products run out. After staining with ethidium bromide, visualization of the gel on a UV transilluminator normally shows only a band corresponding to the linearized M13 DNA. Cut the band corresponding to this DNA from the gel with a clean razor blade and isolate the DNA from the agarose. Again, there are many commercial products that enable this to be achieved including low melting agarose. In this laboratory, we normally use an electroelution device called "Bio-Trap" manufactured by Schleicher and Schuell (Dassell, Germany). In this apparatus the DNA is eluted under the influence of an electric field into a small volume of TBE buffer. The DNA can then be recovered by phenol extraction and ethanol precipitation (*see* Note 10). Whichever method is available, the recovered DNA should end up in 100 µL of TE buffer.

3. It is normal at this point to check the recovery process by running a 5-µL aliquot of DNA on a 0.5% agarose gel. Mix the 5 µL of DNA solution with 1 µL of a loading buffer and apply to a small agarose gel. After electrophoresis and staining, the DNA should be visible as a tight, single band (*see* Note 11). If desired, the molecular weight may be estimated by running DNA molecular weight standards in an adjacent well on the gel.
4. Mix 5 µL of each of the purified oligonucleotides prepared above (Section 3.1.), heat to 80°C for 2 min, and cool slowly over about 30–60 min to below 30°C. This is conveniently done by floating the microcentrifuge tube containing the oligonucleotides on 200 mL of water in a beaker that is initially at 80°C. This heating/slow cooling cycle ensures that the two oligonucleotides anneal properly together to form the double-stranded mutagenic cassette.
5. A number of ligation experiments must now be set up to join the synthetic DNA cassette to the cleaved M13/gene construct. The insert is included at a range of concentrations that ensures that the correct ligation occurs in at least one of the ligation trials. Control ligations are also included to test:
    a. Whether any uncut M13 remains from steps 1 and 2 (tube 7);
    b. Whether any M13 that has only been cut with one enzyme rather than two remains from these steps (tube 6);
    c. The competence of the *E. coli* cells used in the transformation (tube 8); and
    d. The lack of phage contamination in any of the buffers used (tube 9).
   Make up the nine ligation tubes according to the matrix given below (Table 1). Incubate these tubes at 4°C for sticky end ligations or at 16°C for blunt end ligations for 16 h (or overnight) (*see* Note 12).
6. The total amount of these ligation mixes is used to transform a suitable host (*E. coli* TG1 is used in this laboratory for M13) normally made competent by the calcium chloride method. Plate these cells in H-top agar onto H-plates and incubate overnight at 37°C. Plates from ligation tubes 6, 7, and 9 should have no plaques on them. Tube 8 should give a large number (>200) of plaques and tubes 1–5 should have varying numbers of plaques. The highest concentration of synthetic cassette rarely gives the largest number of plaques. It is normally the mid-range of concentrations that lead to the most ligation events.
7. Pick phage clones from the plate that has the largest number of plaques on it and prepare single-stranded template for dideoxy sequencing. This normally shows that about 70% of the clones contain the gene carrying the desired mutation.

Table 1
Ligation Experiments
to Join Synthetic DNA Cassette to Cleaved M13/Gene Construct

Tube	1	2	3	4	5	6	7	8	9
Cut M13 (μL)	10	10	10	10	10	10	10	0	0
Cassette[a] (μL)	10	2	0.2	0.02	0.002	0	0	0	2
10X Ligation buffer (μL)	3	3	3	3	3	3	0	0	3
Ligase [1 U/μL (μL)]	1	1	1	1	1	1	0	0	0
Uncut M13[b] (μL)	0	0	0	0	0	0	0	30	0
Water (μL)	6	14	14	14	14	15	20	0	25

[a]For tubes 3–5, use 2 μL of an appropriate serial dilution of the 5 pmol/μL double-stranded stock from step 4.

[b]This M13 is only used to test for competence of the *E. coli*, therefore any double-stranded circular M13RF DNA at approx 1 ng/μL will suffice for tube 8.

4. Notes

1. Good sources of restriction maps of various cloning vectors are the catalogs distributed by the various molecular biology supplies companies. In particular, New England Biolabs (Boston, MA) provides a complete map of M13mp18/19.
2. Formamide is easily deionized by storing a freshly opened sample over a mixed bed ion exchange resin (1 g of resin for 50 mL of formamide).
3. "Trityl-on" refers to the presence of a large hydrophobic protecting group (dimethoxytrityl) at the 5'-end of the oligonucleotide. All the incomplete sequences in the mixture of products provided by the DNA synthesizer do not have this group in place. Therefore the hydrophobicity of the full-length product results in it being retained on the HPLC column longer than the other more hydrophilic sequences. The trityl group is removed after HPLC purification.
4. Prolonged degassing of buffer B will result in slow evaporation of the acetonitrile. It is therefore important not to leave this buffer under vacuum for longer than 30 min.
5. If the oligonucleotide contains any secondary structure (hairpins, and so forth), this can result in the splitting of the product peak on the HPLC into a cluster of peaks. If this seems to be the case, place the HPLC column in a waterbath at 65°C while the gradient is running. This will melt out the secondary structure and the cluster of peaks should coalesce. If the HPLC column is heated, it is important not to stop the flow of eluant through the column without cooling it first. This is to prevent

evaporation of the eluent in the column and the formation of bubbles in the packing material.
6. "Passing slowly" in this protocol means 1–2 drops/s.
7. Normally six oligonucleotides can be purified at the same time on the one gel.
8. If the highest molecular weight band is not the most intense band on the gel, or if there are many equally intense bands on the gel, there has probably been some failure in the DNA synthesis, although nowadays this happens extremely rarely. Consult your DNA synthesis service for advice if you are worried about the quality of the oligonucleotide product.
9. It is normally possible to perform digestions with two different restriction enzymes at the same time using a compromise buffer suggested by the enzyme suppliers. However, if this proves not to be the case, the incubation with the enzyme requiring the lower ionic strength should be performed first, the solution heated to 65°C for 10 min to kill this enzyme, the ionic strength adjusted for the second enzyme, and the second incubation carried out. This will ensure that each enzyme is cutting under near optimal conditions.
10. Experience has shown that many batches of agarose, even those described as molecular biology grade, contain impurities that copurify with the DNA during the recovery process. These impurities then inhibit the T4 DNA ligase used in the subsequent cloning step. Extraction of the DNA solution three times with phenol and two rounds of ethanol precipitation is normally sufficient to stop these problems.
11. If the DNA band has smearing to lower molecular weight, suspect a nuclease contamination in one of your buffers or enzymes. If no DNA is present, suspect low recovery from the agarose gel. Try again or try an alternative method of DNA recovery.
12. If the mutagenic cassette has one blunt end and one sticky, incubate the ligation reaction at 16°C for 5 h and then 4°C for 16 h (or overnight).

Further Reading

Berger, S. L. and Kimmel, A. R., eds. (1987) Guide to molecular cloning techniques, in *Methods in Enzymology,* vol. 152, Academic, London.

Sambrook, J., Fritsch, E. F., and Maniatis, T. (1989) *Molecular Cloning: A Laboratory Manual,* Cold Spring Harbor Laboratory, Cold Spring Harbor, NY.

References

1. Worrall, A. F. and Connolly, B. A. (1990) The chemical synthesis of a gene coding for bovine pancreatic DNase1 and its cloning and expression in *Escherichia coli. J. Biol. Chem.* **265,** 21,889–21,895.

2. Worrall, A. F., Evans, C., and Wilton, D. C. (1991) Synthesis of a gene for rat liver fatty-acid-binding protein and its expression in *Escherichia coli. Biochem. J.* **278,** 365–368.
3. Barnes, H. J., Arlotto, M. P., and Waterman, M. R. (1991) Expression and enzymatic activity of recombinant cytochrome P450 17α-hydroxylase in *Escherichia coli. Proc. Natl. Acad. Sci. USA* **88,** 5597–5601.
4. Mierendorf, R. C. and Pfeffer, C. (1987) Direct sequencing of denatured plasmid DNA. *Meth. Enzymol.* **152,** 556–562.

CHAPTER 17

Site-Directed and Site-Saturation Mutagenesis Using Oligonucleotide Primers

Michael J. O'Donohue and G. Geoff Kneale

1. Introduction

A very powerful method of probing DNA–protein interactions is the technique of site-directed mutagenesis. Using this approach one can introduce specific amino acid changes at any given position in the amino acid sequence of a DNA-binding protein, and test the functional consequences of these mutations in vitro. Furthermore, it is also possible to investigate the resulting mutations in vivo if a suitable phenotypic change can be measured. If so, strategies that introduce random base changes to a given codon (or indeed to a larger region of the protein) are particularly valuable.

Since the use of primer mutagenesis was first reported a little more than a decade ago *(1)* the technique has rapidly developed. The more recent alterations to the basic technique have largely been concerned with achieving greater mutagenic efficiency. The principle of primer directed mutagenesis is relatively simple. A short oligonucleotide primer is hybridized to a single-stranded circular DNA template containing the gene of interest. The oligonucleotide is completely complementary to a region of the template except for a mismatch that directs the mutation. The oligonucleotide primer is extended by a DNA polymerase in the presence of nucleotide triphosphates. Closed, circular, double-stranded DNA molecules are formed by ligation with DNA ligase. On transformation of *E. coli* cells with the in vitro synthesized DNA, a population of mutant and wild-type molecules are produced. Mutant molecules are then distinguished from wild-type molecules using a suitable screening procedure (Fig. 1).

In order to increase the efficiency of the mutagenesis procedure many methods for selecting against the wild-type DNA strand have been devised. Among the most widely used of these techniques are those of Kunkel *(2)* and Taylor et al. *(3)*. Both of these techniques select against the wild-type DNA, the former by incorporating uracil into the wild-type DNA strand in place of thymine (thus rendering this strand liable to excision in vivo), and the latter by incorporating thionucleotides into the mutant DNA strand (which offers protection against cleavage by certain restriction enzymes in vitro). Both methods depend on the gene being cloned into a vector that produces single-stranded DNA (e.g., M13, or one of a number of "phagemid" expression vectors that are now available). The practical protocol that will be described here is based on the method of Taylor et al. *(3)* (*see* Note 1). A schematic diagram showing the individual stages in this mutagenesis method are shown in Fig. 2.

An alternative method relies on the excision of a small region of DNA surrounding the amino acid of interest using restriction enzymes. The excised DNA is then replaced with a double-stranded DNA fragment containing a base substitution in the codon specifying the amino acid one wants to mutate. This method is known as cassette mutagenesis and is described elsewhere (*see* Chapter 16). The disadvantage of this method for general use, however, is that the gene of interest must be cloned with suitable unique restriction enzyme sites surrounding the region that one wants to mutate. The method is most suitable when a synthetic or wild-type gene has been cloned, and in which appropriate restriction sites have been introduced, making use of the degeneracy of the genetic code.

As well as being used to generate single base substitutions, any of these mutagenesis techniques can be extended to carry out more complex mutagenic studies, such as site-saturation mutagenesis. In this strategy, one attempts to substitute a codon defining a particular amino acid in a protein with all other 19 amino acids. We will describe a method based on primer mutagenesis *(4,5)*. In this technique a synthetic oligonucleotide primer is employed in which the three bases that specify an amino acid codon of interest are degenerate. Synthesis of

Primer Mutagenesis

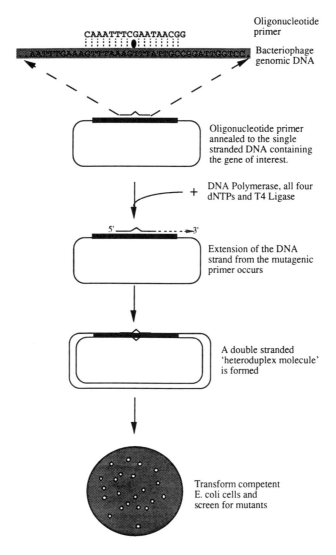

Fig. 1. Site-directed mutagenesis. An oligonucleotide primer is used to direct the mismatching of bases when annealed to the single-stranded DNA template. Extension of the primer using DNA polymerase I (or Klenow fragment) converts the molecule into a double-stranded heteroduplex that can be used to directly transform a suitable *E. coli* host.

the oligonucleotide is straightforward because either the codon that one desires to be degenerate can be entered into the sequence as "NNN" (where N is G, A, T, or C) on the DNA synthesizer, or, if this option is unavailable, an equimolar mixture of all four phosphoroamidites can be

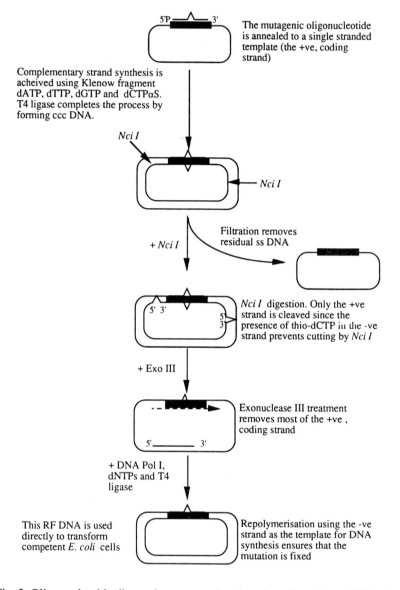

Fig. 2. Oligonucleotide directed mutagenesis using phosphorothioate DNA. Polymerization of the complementary strand is carried out in the presence of dCTPaS. Following the removal of residual single-stranded template molecules, the double-stranded heteroduplex is incubated with *NciI*. Since the restriction enzyme cannot cleave the phosphorothioate DNA, the heteroduplex is nicked in the template strand. The template strand is removed by incubation with exonuclease III and subsequent repolymerization fixes the mutation.

used on the "X port," which is normally present on commercial DNA synthesizers. This method can also be made more efficient with the use of strand selection procedures, and will be discussed below as a variant of the standard site-directed mutagenesis experiment since, with the exception of the oligonucleotide design, the steps are identical.

2. Materials

1. Agarose gel: 0.8% electrophoresis grade agarose (w/v) in distilled water. Dissolve the agarose in 100 mL of TBE buffer by boiling the solution for 3 min. Add the ethidium bromide to a final concentration of 0.5 mg/L using a concentrated stock solution, and allow the agarose solution to cool to 60°C before pouring it into a gel casting tray. The comb is put in place and the gel is allowed to set.
2. Buffer A: 700 mM NaCl, 700 mM Tris-HCl, pH 8.0.
3. Buffer B: 100 mM DTT, 60 mM MgCl$_2$, 300 mM NaCl.
4. Buffer C: 100 mM DTT, 60 mM MgCl$_2$, 800 mM NaCl, 400 mM Tris-HCl, pH 8.0.
5. Centrex™ filter unit-centrifugal filter (Schleicher and Schuell, Dassel, Germany) with 0.45-µm pore size nitrocellulose filters.
6. Centrifugal evaporator: Savant SpeedVac, or similar.
7. Competent *E. coli* cells: *E. coli* TG1 can be made competent for transformation by DNA using any of the recommended techniques *(6)*. A protocol is described in Section 3.5. Other *E. coli* strains displaying an F$^+$ genotype, such as JM109 and MV1193, may be used in place of TG1 if this strain is not available.
8. 50 mM CaCl$_2$: Prepare using ultrapure H$_2$O. Prior to use, the solution is sterilized by autoclaving or filter sterilization.
9. 50 mM CaCl$_2$/Glycerol (1:1 v/v): Prepare using ultrapure H$_2$O. As with 50 mM CaCl$_2$, this solution should also be sterilized by autoclaving or filter sterilization before use.
10. Double-stranded (RF) DNA marker: This marker is prepared using the same method as that employed to prepare the single-stranded DNA marker. In this case, a 100 µg/µL RF DNA solution is used instead of the single-stranded DNA solution.
11. Enzymes: Polynucleotide kinase, Klenow fragment, DNA polymerase I and exonuclease III are routinely purchased from Pharmacia (Piscataway, NJ). The preferred source of the restriction enzyme *NciI* is New England Biolabs (Beverly, MA).
12. Ethanol: Analytical reagent grade absolute ethanol is used and is stored, in small quantities, at –20°C.

13. 80% Ethanol solution: Prepare using distilled H$_2$O and absolute ethanol (analytical reagent grade) and store at –20°C.
14. Ethidium bromide solution: Prepare a concentrated stock solution of ethidium bromide (10 mg/mL solution) using ultrapure water.
15. H$_2$O: Routinely we use ultrapure water produced by an Elga water purification system.
16. 10X Kinase buffer: 1M Tris-HCl, pH 8.0, 100 mM MgCl$_2$. Prepare the buffer using H$_2$O and sterilize by autoclaving. When cool, add dithiothreitol and ATP to final concentrations of 70 mM and 10 mM, respectively.
17. 5X Loading buffer: 50% glycerol (v/v), 100 mM EDTA, pH 8.0, 0.125% bromophenol blue (w/v), 0.125% xylene cyanol (w/v).
18. Microcentrifuge: A centrifuge capable of reaching a maximum speed of approx 11,000g will be required that can accommodate 1.5-mL microcentrifuge tubes.
19. 100 mM MgCl$_2$: Prepare using ultrapure H$_2$O. Prior to use, the solution is sterilized by autoclaving.
20. 5M NaCl: Prepare using ultrapure H$_2$O. Before use, the solution is sterilized by autoclaving.
21. 500 mM NaCl: Dilute the 5M sodium chloride solution ten times with ultrapure water.
22. NEN-sorb oligonucleotide purification column (optional): These columns can be purchased from NEN-Dupont (Wilmington, DE). Comprehensive instructions for their use are supplied by the manufacturer.
23. Nucleotide mix A: 3.1 mM each of dATP, dGTP, dTTP, and dCTPαS, 12.5 mM rATP.
24. Nucleotide mix B: 200 mM each of dATP, dGTP, dTTP, and dCTPαS, 8.3 mM rATP.
25. rATP solution: We routinely use fresh rATP of molecular biology grade (or equivalent).
26. Single-stranded template DNA: A 1 µg/µL solution of single-stranded template DNA containing the gene of interest is prepared using a standard protocol for the preparation of bacteriophage or phagemid DNA (6). The single-stranded DNA is dissolved in H$_2$O before use and the concentration of the solution is determined by measuring the absorbance at 260 nm. One can assume that 1 A$_{260nm}$ unit represents a concentration of approx 33 µg/µL.
27. Single-stranded DNA marker: An aliquot of a single-stranded template DNA solution (100 µg/µL) is combined with an equal aliquot of 5X loading buffer. The volume of the mixture is then adjusted to give a final concentration of 1X loading buffer. A 10-µL aliquot of this solution can be loaded onto an agarose gel to act as a marker.

28. Sodium acetate solution: $3M$ CH_3COONa, pH 6.5.
29. TBE buffer: 90 mM Tris base, 90 mM boric acid, 2.5 mM EDTA, pH 8.3.
30. The mutagenic oligonucleotide: For automated synthesis of oligonucleotides we routinely use a Cruachem P250A automated DNA synthesizer that is capable of synthesizing randomized oligonucleotides, if required. *See* Notes 2–5 for the design of a suitable oligonucleotide sequence.
31. Tris buffer: 500 mM Tris-HCl, pH 8.0.
32. Waterbaths: At least three waterbaths will be required. One should be set at 16°C in the cold room and the other two should be set at 37°C and 70°C, respectively.

3. Method
3.1. Oligonucleotide Synthesis

An oligonucleotide is synthesized using an automated DNA synthesizer. If the oligonucleotide is to be used to generate a single base mutation then the oligonucleotide will be fully complementary to a region in the gene of interest, apart from a single base mismatch that will define the mutation. Oligonucleotides used for this purpose should have a minimum length of 17 bases but may be longer depending on circumstances (*see* Note 2). For site-saturation mutagenesis, the oligonucleotide will be complementary to a region in the gene of interest but will contain a degenerate sequence of three bases in the center of the oligonucleotide (if the coding sequence is on the plus strand, then these three bases correspond to the anticodon of the amino acid that one wants to mutate; *see* Note 5). Whether the trityl group is left on the oligonucleotide or not depends on whether purification is to be attempted (*see* Section 3.2.1.).

3.2. Oligonucleotide Purification and Phosphorylation

1. For primer directed mutagenesis we find that postsynthesis purification of the mutagenic primer is unnecessary. However, if purification is desirable we have found that NEN-Sorb purification columns (NEN-Dupont) produce good results. In order to employ these columns, however, the last trityl group to be added during the oligonucleotide synthesis must be left on.
2. For site-saturation mutagensis, purification may be disadvantageous. Since the mutagenic oligonucleotide is degenerate in three positions the synthesis product will represent a mixture of oligonucleotide species that may be differentially purified.

3. To phoshorylate the oligonucleotide, combine 0.125 A_{260nm} units of the mutagenic oligonucleotide with 3 µL of 10X kinase buffer in a microcentrifuge tube. Add 25 µL of ultrapure water and 2 U of T4 polynucleotide kinase. Mix by repeated pipeting and incubate at 37°C for 15 min (see Note 6).
4. Terminate the reaction by incubating the reaction mixture at 70°C for 10 min. Store the sample at –20°C until required.

3.3. Heteroduplex Formation

1. Combine 5 µg of single-stranded template DNA (M13 or phagemid DNA for example) with 4 pmol of phosphorylated mutagenic oligonucleotide and 3.4 µL of buffer A in a screw-top microcentrifuge tube. Adjust the total volume of the mixture to 17 µL with H_2O.
2. Place the tube in a 100°C waterbath. Remove the waterbath from the heat source and cool by incubation at 4°C, until the temperature of the water has fallen below 35°C. Finally, place the tube on ice. This procedure ensures optimal annealing of the oligonucleotide to the template.
3. To the mixture add a 6.25-µL aliquot of 100 mM $MgCl_2$, 5 µL of 500 mM Tris buffer, and 5 µL of nucleotide mix A. Add 6 U each of Klenow fragment and T4 DNA ligase and adjust the total volume to 50 µL with H_2O. Mix briefly by gentle repeated pipeting and incubate the reaction at 16°C overnight (at least 16 h). This is sufficient to ensure that the primer has been extended all the way around and ligated to form covalently closed circular double-stranded DNA.
4. Combine 1 µL of the reaction mixture with 2 µL of 5X loading buffer and adjust the sample volume to 10 µL with water. Load the sample into a well on a 0.8% agarose gel and electrophorese, along with the double- and single-stranded DNA markers, for 30 min at a constant voltage of 100 V, using TBE buffer containing ethidium bromide (0.5 mg/L) as the electrophoresis buffer. Observe the gel under UV illumination. If one finds that the conversion of the single-stranded template DNA to heteroduplex DNA has been successful the mutagenesis procedure can continue (see Note 6).
5. Add 170 µL of water and 30 µL of 5M NaCl to the heteroduplex DNA mixture. Apply this solution to the top of a Centrex™ filter unit. With the lower half of the filter unit in place, centrifuge at 1500 rpm for 10 min at room temperature in a Sorvall RC-5B employing an HB-4 swing out rotor. Wash the filter by adding 100 µL of 500 mM NaCl to the upper part of the filter unit. Centrifuge again using the same conditions as before. Remove the lower half of the filter unit containing the sample. This procedure removes excess single-stranded template DNA (see Note 7).

6. Add 700 µL of ice-cold ethanol and 28 µL of sodium acetate solution and incubate at −20°C for 30 min. Recover the precipitated DNA by centrifugation in a microcentrifuge at 13,000 rpm for 15 min at 4°C (*see* Note 8). Wash the DNA pellet with 1 mL of ice-cold 80% ethanol and centrifuge as before. Carefully discard the supernatant and dry the pellet in a centrifugal evaporator.

3.4. Strand Selection

1. Resuspend the DNA in 25 µL of 90 m*M* Tris-HCl, pH 8.0. At this stage 15 µL of the DNA solution is stored at −20°C and the mutagenesis is continued using only a 10-µL aliquot of the DNA sample.
2. Add a 7.5-µL aliquot of buffer B to the 10-µL sample and adjust the total volume to 75 µL with H_2O. Add 5 U of *NciI* and incubate at 37°C for 90 min. After incubation, remove a 10-µL sample of this reaction mixture for subsequent analysis by agarose gel electrophoresis.
3. Combine the remainder of the solution (65 µL) with 9 µL of buffer C and adjust the total volume to 88 µL with H_2O. Add 50 U of exonuclease III (25 U/µL). After mixing by repeated pipetings, incubate the reaction mixture at 37°C. The time of this incubation is determined by the position of the *NciI* site(s) with respect to the mutagenic mismatch. If M13 DNA is used, a digestion time of 30 min should suffice. However, exonuclease III will digest about 100 bases/min so it is possible to estimate a suitable digestion time (*see* Note 9). Following digestion, it is necessary to heat inactivate the exonuclease III by incubation at 70°C for 15 min. Following a 30 s centrifugation at 13,000 rpm at room temperature in a microcentrifuge, remove a 15-µL sample for subsequent analysis by agarose gel electrophoresis.
4. Add 9.5 µL of nucleotide mix B, 6 µL of 100 m*M* $MgCl_2$ solution, 3 U of DNA polymerase I, and 2 U of T4 DNA ligase to the reaction mixture. Adjust the total volume to 95 µL with H_2O and incubate the sample at 16°C for at least 3 h. Remove a final 15-µL sample for analysis by agarose gel electrophoresis.
5. At this stage, to each of the samples taken from steps 2–4 should be added 5X loading buffer (giving a final concentration of 1X loading buffer). Electrophorese the samples on a 0.8% agarose gel using the same conditions as those described in step 4 of Section 3.3. Again, the double- and single-stranded DNA markers should be also loaded onto the gel (*see* Note 10 and Fig. 3).
6. Use aliquots of the remainder of the reaction mixture (e.g., 1 and 10 µL) to transform competent *E. coli* cells in separate experiments using any suitable transformation procedure.

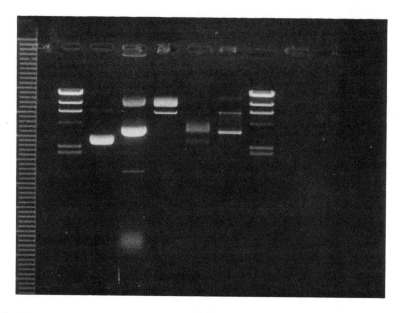

Fig. 3. Analysis of the samples of agarose gel electrophoresis. The samples from the mutagenesis procedure were electrophoresed on a 1% agarose gel containing ethidium bromide, at a constant voltage of 100 V. The DNA bands were visualized by illuminating the gel on a UV transilluminator (254 nm). Lanes (1,7) λ *Hind*III marker DNA; (2) single-stranded template DNA marker DNA (20 ng/µL); (3) double-stranded (replicative form) template DNA marker (20 ng/µL); (4) DNA sample after filtering and nicking with *NciI*; (5) DNA sample after digestion with exonuclease III; (6) DNA sample after repolymerization and ligation.

7. In order to screen for mutants it is prudent to sequence ten isolates by Sanger dideoxy sequencing. Of course, if a phenotypic selection procedure is available then the task of finding mutant derivatives may be simpler (*see* Note 11).

3.5. Transformation with the Heteroduplex

The following protocol may be used to make competent *E. coli* for transformation:

1. Prepare a saturated bacterial culture by inoculating a 10-mL aliquot of L-broth with a single colony of *E. coli* TG1 bacteria. Incubate the culture at 37°C with shaking for approx 16 h.
2. This saturated culture is then diluted 100 times with L-broth and is grown until an absorbance at 600 nm of approx 0.6 OD units is reached. This corresponds to the mid-logarithmic growth phase.

3. Pellet the cells at 3000 rpm in a Sorvall RC-3B centrifuge (or similar apparatus) at 4°C for 10 min. Discard the supernatant and resuspend the cell pellet in 5 mL (i.e., half of the original volume) of 50 mM CaCl$_2$. Incubate the cells on ice for 15 min and then repellet them by centrifugation.
4. Once again, discard the supernatant and resuspend the cells in 1 mL (i.e., one tenth of the original volume) of 50 mM CaCl$_2$, 20% glycerol (v/v). The cells are now competent for transformation with DNA and can be either used immediately or stored at –80°C until required.

4. Notes

1. The protocol described here has also been adopted in a site-directed mutagenesis kit that is marketed by Amersham Plc (Arlington Heights, IL). Some of the buffers and both of the nucleotide mixes as well as all of the enzymes required to perform the mutagenesis reaction are supplied in this kit.
2. For a single base substitution, an oligonucleotide composed of as few as 17 bases can be used. For a more complex procedure, such as site-saturation mutagenesis, a larger oligonucleotide of at least 27 bases should be used. When designing any mutagenic oligonucleotide the bases that will direct the mutation should be placed at the center of the oligonucleotide. The primer sequence should be checked to ensure it does not contain an *NciI* site. An oligonucleotide primer containing an *NciI* site will allow the heteroduplex to be cut since the primer does not contain phosphorothioate nucleotides. If an *NciI* site is found in the primer sequence one should either redesign the primer (if this is possible) or employ an alternative restriction enzyme in the place of *NciI*. Eckstein and coworkers *(7–9)* have investigated several restriction enzymes that display an inability to cleave DNA containing phosphorothioate nucleotides. However, if an alternative restriction enzyme is employed several alterations to the basic protocol may have to be made.
3. Once the mutagenic oligonucleotide has been designed it is advisable, but not essential, to examine the sequence of the template DNA in order to check for any other sites that may compete for the mutagenic oligonucleotide. Generally, if any contiguous sequence, other than the sequence that is complementary to the mutagenic oligonucleotide, is found that has more than 70% of bases in common with the mutagenic oligonucleotide then the sequence of the mutagenic oligonucleotide should be altered. This can be achieved either by increasing the length of the oligonucleotide or by shifting the sequence of the oligonucleotide with respect to the target site.

Competing sites on the template for the mutagenic oligonucleotide can also be detected by Sanger dideoxy sequencing. The template is employed in a sequencing reaction in which the mutagenic oligonucleotide is used as the sequencing primer. If there is more than one target site for the oligonucleotide, multiple DNA sequences will be observed on the resulting autoradiograph. Normally, one can expect a sequence as good as that which could be obtained using a normal, fully complementary, primer. However, as long as the correct sequence is predominant, with other sequences being only faintly represented, the mutagenic oligonucleotide may still be useful.

4. When designing the mutagenic oligonucleotide one should ensure that the A-T content of the oligonucleotide is not too high since this may lead to instability of the oligonucleotide when it anneals to the template. As a rule of thumb, the melting temperature (T_m) of each "arm" of the mutagenic oligonucleotide (the sequences either side of the mismatch) should not be lower than 37°C. The T_m can be estimated for short oligonucleotides (16–40-mers) using the Wallace rule: $T_m = 4 (G-C) + 2(A-T)$.

5. When designing an oligonucleotide for site-saturation mutagenesis, careful consideration should be given to the following points:
    a. If the oligonucleotide employs the degenerate sequence NNN (where N is G, A, T, or C) to specify the codon that is to be mutated, on mutagenesis all 64 codons will be made. Therefore, theoretically 1 in 64 mutant isolates will display a wild-type genotype. Futhermore, if the mutagenesis is not 100% efficient (which is probable) some isolates that display the wild-type genotype will represent unmutated templates, which will be indistinguishable from the mutated wild-type isolates in their DNA sequence. If it is necessary to calculate the efficiency of mutagenesis, this may represent a problem (but *see below*).
    b. It may be desirable to limit the number of codons that can be created if one wants to observe the relative frequency of occurrence of any particular amino acid substitution. This may be especially important in experiments where one is employing a positive screening strategy (i.e., screening for protein activity) in order to identify those amino acid substitutions that do not abolish the functional viability of the protein. In this case, restricting the number of possible mutant codons, usually at the third base position in the codon (i.e., the first base of the "anticodon" in the mutagenic primer in its usual orientation) will facilitate normalization of the results, so that the wild-type codon is not represented in the primer population. Any wild-type sequences that are found, therefore, must represent the background of unmutated template DNA. An alternative strategy is to introduce a silent "marker" base change into the oligonucleotide primer outside of the codon of interest *(10)*.

On many automated DNA synthesizers it is possible to specify nucleotide mixtures other than those specified by N (25% of each of the four bases). For example, the random sequence NNS (where S is G or C) would specify only 32 codons but would still direct the substitution of an amino acid by all 20 amino acids in a mutagenesis experiment. To take a concrete example, if the wild-type codon were TAT (tyrosine), then this particular codon would be eliminated by allowing only G or C (A could also be included if required) at the third base (i.e., C or G at the first base of the "anticodon"). All 20 amino acids can be coded by this combination although their frequencies will be different. Tyrosine would also be allowed, but from the codon TAC; any transformants sequenced as TAT must therefore be a result of incomplete destruction of the template, and these can be eliminated from any statistical analysis.

6. Successful 5' phosphorylation of the oligonucleotide is essential if one is to obtain closed circular double-stranded DNA after the initial primer extension reaction. Therefore, on analyzing sample 1 by agarose gel electrophoresis, one should take particular note of the amount of nicked circle DNA that is present. The nicked circle DNA will not be suitable for the mutagenesis reaction. The efficiency of the kinasing reaction can be maximized by using fresh, good quality enzyme and fresh rATP. If only 40–50% of the double-stranded DNA from the primer extension reaction is in a closed circular form then one may still proceed with the mutagenesis and expect to obtain reasonable mutant yields.

7. We have found the centrex filter unit does not fit perfectly into the centrifuge bucket. However, it appears that as long as the unit can remain in an approximately vertical position in the bucket when the centrifuge is at rest no other action is necessary. Routinely, we use a HB 4 (Dupont) swing out rotor with the buckets modified using plastic inserts that would normally accommodate a 14-mL polypropylene tube.

   WARNING: Take care to ensure that the filter unit cannot fall out of the buckets during the centrifugation process.

8. At this stage a DNA pellet may be visible. However, it is good practice to orientate the microcentrifuge tube in the same way whenever a centrifugation step is required. This ensures that even when the pellet is invisible the approximate position of the DNA is known, therefore reducing the possibility of an accidental loss of the pellet.

9. If necessary, a series of exonuclease III digestions can be performed using small samples of the nicked heteroduplex DNA that was stored (*see* step 2 of Section 3.4.). The digestions can be terminated after different incubation times and the results can be analyzed on a 1% agarose gel in order to determine the optimum digestion time (*see* Note 10).

The exonuclease III digestion is unlikely to completely destroy the nonmutant strand and therefore small "primers" for the repolymerization reaction should always remain. However, if the subsequent repolymerization reaction fails as a result of the complete destruction of the nonmutant strand by exonuclease action, then one can reprime the single-stranded template. This can be achieved using the M13 universal primer or any other suitable primer that may be available.

10. The analysis of each stage of the mutagenesis procedure by agarose gel electrophoresis is extremely important and should therefore be carried out routinely. Figure 3 shows an example of the results one should obtain if the mutagenesis procedure is successful. After filtration and nicking there should be only one DNA species corresponding to the nicked circle DNA that is present in the double-stranded marker. If single-stranded template DNA or double-stranded closed circular DNA remains at this stage one can expect that the efficiency of mutagenesis will be considerably decreased. In this case, it may be necessary to screen more isolates in order to identify mutants if a phenotypic screen is unavailable. In the example shown there is also a small amount of linear DNA present; this does not affect the efficiency of mutagenesis.

Likewise, the incidence of wild-type isolates will be increased if the exonuclease III treatment is inefficient. Following incubation with exonuclease III, the nicked circle DNA band seen in the previous sample should disappear, being replaced by a mixed population of smaller DNA molecules similar to those in lane 5 of the example. These degraded double-stranded DNA molecules will migrate further on the agarose gel than the double-stranded nicked circle DNA.

Finally, the repolymerization and ligation step should regenerate double-stranded closed circular DNA. This reaction does not always produce only double-stranded closed circular DNA. Often after this step, a significant proportion of the DNA may be in a nicked circle form. This is not important, however, since the DNA will still be suitable for transformation of competent *E. coli* cells. Equally, even if there is only a small amount of DNA visible one can still obtain a large quantity of isolates since 1 ng of double-stranded closed circular DNA should produce more than 100 transformants.

11. If it is necessary to screen large quantities of isolates by Sanger dideoxy sequencing it may be better initially to employ a single track protocol. Using this strategy, only one dideoxynucleotide is used and mutants are identified by the loss or gain of one or more bands on the autoradiograph. The advantages of using this protocol are that more isolates can be screened using one quarter of the sequencing reagents that one would

normally employ. This strategy is applicable whenever one is sure of the nature of the mutation that is being created (i.e, normal primer directed mutagenesis). In the case of site-saturation, whether one uses a single track strategy for screening will depend on the design of the mutagenic oligonucleotide. Clearly, if the sequence directing the mutation is 5' NNN 3' then the mutagenesis could result in the loss or gain of a band in either of the four tracks and therefore only a four track sequencing protocol will suffice for the screening procedure.

References

1. Hutchison, C. A., Phillips, S., Edgell, M., Gillam, S., Jahnke, P., and Smith, M. (1978) Mutagenesis at a specific position in a DNA sequence. *J. Biol. Chem.* **18**, 6551–6560.
2. Kunkel, T. A. (1985) Rapid and efficient site-specific mutagenesis without phenotypic selection. *Proc. Natl. Acad. Sci. USA* **82**, 488–492.
3. Taylor, J. W., Ott, J., and Eckstein, F. (1985) The rapid generation of oligonucleotide-directed mutations at high frequency using phosphorothioate-modified DNA. *Nucleic Acids Res.* **13**, 8765–8785.
4. Alber, T., Bell, J. A., Dao-Pin, S., Nicholson, H., Wozniak, J. A., Cook, S., and Matthews, B. W. (1988) Replacements of Pro86 in phage T4 lysozyme extend an α-helix but do not alter the protein stability. *Science* **239**, 631–635.
5. Zabin, H. B. and Terwilliger, T. C. (1991) Isolation and in vitro characterisation of temperature sensitive mutants of the bacteriophage f1 gene V protein. *J. Mol. Biol.* **279**, 257–275.
6. Sambrook, J., Fritsch, E. F., and Maniatis, T. (1989) *Molecular Cloning: A Laboratory Manual*, Cold Spring Harbor Laboratory, Cold Spring Harbor, NY.
7. Taylor, J. W., Schmidt, W., Costick, R., Okrusek, A., and Eckstein, F. (1985) The use of phosphorothioate-modified DNA in restriction enzyme reactions to prepare nicked DNA. *Nucleic Acids Res.* **13**, 8749–8764.
8. Nakayame, K. L. and Eckstein, F. (1986) Inhibition of restriction endonuclease Nci I cleavage by phosphorothioate groups and its application to oligonucleotide directed mutagenesis. *Nucleic Acids Res.* **14**, 9679–9698.
9. Sayers, J. R., Schmidt, W., and Eckstein, F. (1988) 5'-3' exonucleases in phosphorothioate-based oligonucleotide-directed mutagensis. *Nucleic Acids Res.* **16**, 791–802.
10. Hermes, J. D., Parekh, S. M., Blacklow, S. C., Koster, H., and Knowles, J. R. (1989) A reliable method for random mutagenesis: the generation of mutant libraries using spiked oligonucleotide primers. *Gene* **84**, 143–151.

CHAPTER 18

UV Laser-Induced Protein–DNA Crosslinking

Stefan I. Dimitrov and Tom Moss

1. Introduction

1.1. The Method

Photochemical crosslinking is a powerful method for studying all types of protein–nucleic acids interactions. In particular, UV-induced crosslinking has been successfully applied to the study of protein–DNA interactions *(1)*. Ultraviolet (UV) light is a zero-length crosslinking agent. It is therefore not subject to the steric problems that can be associated with chemical crosslinking agents and provides strong evidence for close protein–DNA interactions. However, to achieve an acceptable degree of crosslinking with conventional UV light sources, exposure times ranging from minutes to several hours have had to be used *(1–3)*. Such prolonged irradiation allows for redistribution of proteins and the artifactual crosslinking of UV-damaged molecules, and it also precludes kinetic studies. The use of UV lasers overcomes these difficulties, since crosslinking is achieved after only nano- or picosecond exposures *(4,5)*.

The UV-crosslinking reaction consists of two distinct steps. The first is the light absorption by and excitation of the bases of the DNA. The second step is the chemical crosslinking reaction, which proceeds via radicals of the bases. The time during which excitation takes place is simply defined by the time of UV-irradiation. The crosslinking reaction is completed in <1 µs *(6)*. Since the microconformational transitions of macromolecules take more than 100 µs *(7)*, nano- or

From: *Methods in Molecular Biology, Vol. 30: DNA–Protein Interactions: Principles and Protocols*
Edited by: G. G. Kneale Copyright ©1994 Humana Press Inc., Totowa, NJ

picosecond UV laser-irradiation can essentially freeze protein–DNA interactions and hence even allow us to take "snap-shots" during the assembly of protein–DNA complexes. Besides the advantage of the extremely short irradiation times, laser-induced reactions, as opposed to those generated by conventional UV sources, proceed by way of the higher (S_n and T_n) excited states of nucleotide bases that are induced by the rapid sequential absorption of two photons. This leads to a high quantum yield of cationic radicals and hence a high efficiency of crosslinking. It is also probable that UV laser-irradiation permits mechanisms of crosslinking that do not occur when conventional sources are used, although the photochemistry of these mechanisms is not yet understood *(6,8)*. Together, these factors lead to highly efficient crosslinking when UV lasers are used *(6,8–10)*. According to the data so far reported, 5–15% of UV laser-irradiated protein–DNA complexes are crosslinked *(6,8)*. This exceeds, by nearly two orders of magnitude, the yield of crosslinking obtained using conventional UV sources *(4,5,8,9,11)*. In fact, every protein that interacts specifically with DNA can be crosslinked to it (only nonspecific protein–DNA interactions involving association constants weaker than about $10^3 M^{-1}$ will not be detected by the crosslinking procedures *[4,6]*). Laser-irradiation of multiprotein–DNA complexes induces protein–DNA crosslinks only; no protein–protein crosslinks are formed *(4,6,9,10)*.

1.2. Practical Applications

UV laser crosslinking has been used to investigate several different aspects of protein-nucleic acids interactions, such as binding constant measurements *(6)*, determination of the size of a protein–nucleic acid binding site *(6)*, interactions of macromolecular complexes of several proteins with DNA *(5,6)* or RNA *(10)*, and the determination of protein–DNA contact points *(12)*. It can provide information that is not accessible through other techniques. For example, the DNA–ATPase complex in the T4 DNA replication system is greatly weakened in the presence of ATP, such that DNase I footprinting no longer works. UV laser crosslinking has allowed a transient DNA–ATPase complex to be frozen and hence its protein–DNA contacts studied *(6)*.

A procedure for crosslinking chromosomal proteins to DNA by laser-irradiation of nuclei and whole cells has been developed *(9)*.

Protein–DNA Crosslinking

The essentially instantaneous crosslinking produced by the UV laser avoids all possibility of redistribution of the DNA bound proteins. Several experiments have studied the in vivo crosslinking and the distribution of chromosomal proteins. It was found possible to crosslink the histones and the high mobility group 1 proteins to DNA and to detect the presence of the histones on various regions of the *Xenopus* ribosomal DNA *(9,13,14)*. It was further shown that only the N-terminal domains of the histones were crosslinked to the DNA and that hyperacetylation of these domains did not disturb their interaction with the DNA *(13–15)*.

The ability of the laser to "freeze" protein–DNA complexes generated in very rapid reactions, the high yield of crosslinks and the applicability of the method even for in vivo studies, means that a large number of problems can be addressed by this technique. An excellent guide to the practical application of in vitro laser protein–DNA crosslinking will be found in the recent methodological review by von Hippel and coauthors *(6)*.

1.3. The Basic Approach

Below we describe a procedure used to induce histone–DNA crosslinking in nuclei and the subsequent determination of DNA sequences of interest crosslinked to the histones. The procedures used are quite general and could well be applied to study the interactions of many types of proteins with DNA both in vivo and in vitro. The approach consists of the following steps:

1. UV laser-irradiation of nuclei.
2. Separation and isolation of the crosslinked protein–DNA complexes.
3. Detection of specific proteins crosslinked to bulk DNA using immunochemical techniques.
4. Immunoprecipitation of the crosslinked protein–DNA complex.
5. Identification and quantitation of the DNA sequences covalently attached to a given protein using hybridization with specific DNA probes.

Since the amount of any given DNA sequence recovered by antibody precipitation depends directly on the amount of antigen crosslinked to it, the procedures allow the quantitation of specific DNA–protein interactions present in vivo.

2. Materials

2.1. Lasers

As a source of powerful UV radiation we usually use a passively mode-locked picosecond neodymium-yttrium-aluminum-garnet (Nd:YAG) laser *(9)* (*see* Note 1). The Nd:YAG lasers produce pulse radiation at 1064 nm that can be converted to 266 nm (the wavelength at which the samples are irradiated) by quadrupling the main frequency. The parameters of the laser radiation at 266 nm were as follows: pulse duration 30 ps (in a Gaussian pulse shape assumption), pulse energy 4 mJ, diameter of the beam 0.5 cm, repetition rate 0.5 Hz. The intensity of irradiation was controlled by focusing and defocusing fused silica lenses. The energy of radiation was measured with pyroelectrical detectors calibrated with Model Rj7200 energy meter (Laser Precision Analytical, Irvine, CA). The electric signal from the pyrodetectors was transmitted to an Apple II microcomputer for further processing and handling.

For irradiation of the samples a nanosecond Nd:YAG laser (Model DCR-3J, Spectra-Physics Inc. (Mountain View, CA) or YAGMaster YM1000/YM1200, Lumonics (Kanata, Ontario, Canada) could also be used *(6)*. Pulses in the UV are about 5 ns in duration, and the energy per pulse at 266 nm is typically 80 mJ.

2.2. Reagents and Solutions

2.2.1. Irradiation Techniques

1. 8M urea.
2. 1% SDS.

2.2.2. Separation of Covalently Crosslinked Histone–DNA Complexes

1. 10 mM Tris-HCl, pH 7.5, 1 mM CaCl$_2$.
2. 0.48M Sodium phosphate buffer, pH 6.8.
3. 0.12M Sodium phosphate buffer, pH 6.8.
4. 0.12M Sodium phosphate buffer, pH 6.8, 2M NaCl, 5M urea.
5. Four CsCl solutions of ρ = 1.76 g/mL, 1.57 g/mL, 1.54 g/mL, and 1.32 g/mL.
6. Microccocal nuclease.
7. Hydroxyapatite.

2.2.3. Dot Immunoassay

1. 1X Phosphate buffered saline, 150 mM sodium chloride, 10 mM sodium phosphate, pH 7.3 (PBS).
2. PBS containing 0.05% Triton X-100 (PBS-T).
3. 1% Bovine serum albumin in PBS-T.
4. PBS containing 0.4% Triton X-100.
5. 0.3% 4-chloro-1-naphtol, 0.03% H_2O_2 in 50 mM Tris-HCl, pH 7.5, 150 mM NaCl.
6. Nitrocellulose filter (Schleicher and Schuell, Dassel, Germany).
7. Peroxidase-conjugated goat antirabbit IgG (Sigma, St. Louis, MO).

2.2.4. Immunoprecipitation of Crosslinked Protein–DNA Complexes

1. IgGsorb (The Enzyme Center, Malden, MA).
2. Antibody buffer: 50 mM HEPES, pH 7.5, 2M NaCl, 0.1% SDS, 1% Triton X-100, 1% sodium deoxycholate, 5 mM EDTA, 0.1% bovine serum albumin.
3. Rinse buffer: 50 mM HEPES, pH 7.5, 0.15M NaCl, 5 mM EDTA.
4. 3.5M potassium thiocyanate, 20 mM Tris-HCl, pH 7.5.
5. RNase (1 mg/mL).
6. Pronase (1 mg/mL).
7. Ethanol.
8. TE buffer: 10 mM Tris-HCl, pH 7.5, 0.25 mM EDTA.

2.2.5. Hybridization Analysis

1. Zeta-Probe blotting membranes (Bio-Rad, Richmond, CA).
2. 100X Denhardt's solution: 10 g Ficoll 400, 10 g polyvinylpyrrolidone, 10 g bovine serum albumin, final volume 500 mL in H_2O. Filter and store at $-20°C$ in 25 mL aliquots.
3. 6X SSC: 0.9 g NaCl, 0.09M sodium citrate, pH 7.0.
4. Prehybridization buffer: 6X SSC, 10X Denhardt's solution, 0.1 mg/mL denatured *E. coli* DNA, 1% SDS, 0.2% sodium pyrophosphate, 50% formamide.
5. 0.5X SSC, 0.5% SDS.
6. 0.1% SSC.

3. Methods

3.1. Irradiation Techniques

For irradiation 1–2 mL of the nuclei/sample suspension should be placed in a standard rectangular fused silica cuvet, thermostatted at

4°C, and the sample constantly stirred. The optical density of the solution should be kept in the range of $2 < A_{260} < 5$ (i.e., optically thick samples; the optical density of the nuclei can be determined in $8M$ urea or 1% SDS) (*see* Note 2).

In the case of a picosecond laser the conditions of irradiation have to be chosen in such a way as to achieve 10–20 absorbed photons per nucleotide (about 500 mJ of incident light per A_{260} of optically thick sample) at a constant laser intensity 0.7 GW/cm^2. In the case of a nanosecond UV laser use 250 mJ of incident light per A_{260} U of optically thick sample.

3.2. Separation of Covalently Crosslinked Histone–DNA Complexes

When working with the laser crosslinked protein–DNA complexes never use acids, since the crosslinked adducts are unstable under acidic conditions.

1. To reduce the DNA size, digest the irradiated nuclei with microccocal nuclease (5 enzyme units/A_{260} U for 15 min at 37°C) in the presence of 10 mM Tris-HCl, pH 7.5, 1 mM CaCl$_2$ (*see* Note 3).
2. Add 0.12M phosphate buffer, pH 6.8, 2M NaCl, 5M urea to stop the reaction.
3. Load the irradiated sample on a hydroxyapatite column (1 g hydroxyapatite per mg of DNA) equilibrated with 0.12M phosphate buffer, pH 6.8, 2M NaCl, 5M urea (*see* Note 4).
4. Wash the column with 5 vol of the same buffer, then with 0.12M phosphate buffer only.
5. Elute the free DNA and the crosslinked complex with 0.48M phosphate buffer, pH 6.8).
6. Load the material on a preformed CsCl gradient (four layers, 2.5 mL each, ρ= 1.76, 1.57, 1.54, and 1.32 g/mL), and run in SW 41 Beckman rotor (or equivalent) at 15°C for 35–40 h at 35,000 rpm (150,000g).
7. Collect 6–8 drop fractions and measure the optical densities of the fractions at 260 nm.
8. Draw the sedimentation profile of the gradient—a clear shoulder should be observed on the light side of the peak. The material in this shoulder is a highly enriched fraction of protein–DNA crosslinked complexes (*see* Note 5).
9. Collect the material from the peak (or better from the shoulder only), and dialyze extensively against 10 mM Tris-HCl, pH 7.5, 0.25 mM EDTA (*see* Note 5).

3.3. Dot-Immunoassay to Identify Abundant Proteins Crosslinked to the Bulk DNA

1. Dot the crosslinked material (about 0.5 mg DNA) on nitrocellulose filters.
2. Wash twice, 5 min each time, with gentle shaking in PBS-T buffer to remove unbound antigen.
3. Repeat step 2, but with PBS buffer only.
4. Incubate the filters successively with BSA (1% in PBS-T) for 1 h at 37°C and then overnight at 4°C with a suitable dilution of specific antibody in PBS-T, 1% BSA.
5. Wash the filters extensively (3 times with PBS-T then 3 times with PBS only) under gentle shaking and incubate for 4 h at 37°C with peroxidase-conjugated goat antirabbit IgG (Sigma, 1:1000 dilution) in PBS containing 1% BSA.
6. After extensive washings (as in step 5), develop the dots using 0.3% 4-chloro-1–naphtol, 0.03% H_2O_2 in 50 mM Tris-HCl, pH 7.5, 150 mM NaCl.

3.4. Isolation of Specific Crosslinked Protein–DNA Complexes by Immunoprecipitation

1. Suspend 0.05 mL IgGsorb in 0.5 mL of 1% BSA in PBS and shake for 30 min at room temperature to block the sites of nonspecific absorption.
2. After 30 s microcentrifugation, suspend the pellet in 0.5 mL of a mixture of the specific antibody and the crosslinked DNA–protein complexes (w:w ratio 1:2.5) in antibody buffer (usually 20–50 mg of crosslinked material was used) (*see* Note 6).
3. Following 2 h shaking at room temperature, wash the suspension five times with antibody buffer alone, and three times with 0.5 mL of rinse buffer, centrifuging for 30 s between each wash.
4. Elute the bound material with 0.1 mL 3.5M potassium thiocyanate, 20 mM Tris-HCl, pH 8.2.
5. Treat the eluate with RNase (0.015 mg per sample) for 30 min at room temperature and digest with pronase (1 mg/mL) for at least 4 h again at room temperature.
6. Finally, precipitate with ethanol (3 vol per sample volume, overnight at –20°C).
7. After 10 min centrifugation in a microcentrifuge, resuspend the pellet in 10 mM Tris-HCl, pH 7.5, 0.25 mM EDTA.

3.5. Identification of Crosslinked DNA Sequences by DNA Hybridization

1. The DNA samples from Section 3.4. are alkali denatured by the addition of 1 vol of 1M NaOH and heating at 37°C for 10 min. They are then chilled on ice and dotted onto Zeta-Probe membranes (Bio-Rad) by repeatedly applying 1 to 2-µL aliquots and allowing the membrane to dry between applications (*see* Note 7).
2. Prehybridize the membranes in prehybridization buffer for 3–5 h at 42°C.
3. Hybridize at 42°C for 16–20 h with 50–100 ng of DNA probe, ^{32}P-labeled by random priming *(16,17)*.
4. Wash the filters extensively with 0.5X SSC, 0.5% SDS at 42°C, and finally with 0.1X SSC at 65°C and autoradiograph at –70°C using Cronex Lightning-plus intensifying screens (Dupont, Wilmington, DE).

4. Notes

1. We have used different lasers in our work and have found that the most suitable for protein–DNA crosslinking are the Nd-YAG lasers. For molecular biologists the description of the lasers and the irradiation techniques may sound obscure. However, most laser spectroscopy laboratories have the equipment necessary for UV laser-induced crosslinking. Thus, molecular biologists interested in using the approach are best advised to locate the nearest laser laboratory.
2. Do not use 2-mercaptoethanol or dithiothreitol in solutions to be irradiated, since at 5 mM these reagents reduce the efficiency of crosslinking by about 50–60%.
3. Micrococcal nuclease can be used to reduce the molecular weight of DNA in irradiated samples. Alternatively, sonication could be used. Recover the irradiated nuclei by centrifugation and resuspend them in 300–400 µL of 1% sarkosyl in an Eppendorf tube (A_{260} = 10–12). Sonicate the sample with a Model W-35 sonicator (Heat Systems Inc., Farmingdale, NY) or equivalent, using a microtip at a power setting of 5 for ten 30-s bursts in an ice bath. Under these conditions the size of DNA is reduced to about 150–200 bp.
4. Separation of the covalently crosslinked protein–DNA complexes. Hydroxyapatite is used to remove excess protein, and the like, from the crude nuclear lysate. To avoid large losses, the molecular weight of the DNA must be reduced to 200–300 bp before it is applied to the hydroxyapatite column. Even when purifying low molecular material on this column we sometimes lose 20–25% of the loaded material. This step is not essential and hence is best omitted if only small quantities of the irradiated samples are available.

5. The shoulder in the CsCl sedimentation profile of the irradiated material contains an enriched fraction of the protein–DNA complexes. However, some crosslinked material is also present in other regions of the DNA peak. Thus, if it is essential to recover all crosslinked material, the complete DNA peak must be pooled. Free DNA does not, however, significantly interfere with the immunoprecipitation procedure.
6. Immunoprecipitation of the crosslinked protein–DNA complexes. When washing the IgGsorb, microcentrifuge for 30 s only. Longer centrifugation causes the pellet to become very compact and difficult to resuspend. Usually 20–50 µg of crosslinked material (from the whole CsCl peak) were taken for the immunoprecipitations. The results with smaller samples were found to be irreproducible.
7. In our studies different membranes were used for the hybridization. Best results were obtained with the Zeta-Probe membrane (Bio-Rad). If a quantitative estimation of the protein(s) present on a specific DNA sequence(s) is required, as for example, was made for the histones on the *Xenopus* rDNA *(15)*, the following procedure is recommended. Dot aliquots from the antibody-precipitated DNA preparations on a Zeta-Probe filter. Apply increasing amounts of genomic DNA (in the range 50–1000 ng) to the same filter to produce a calibration curve and repeat exactly the same set of dots on a second filter. Hybridize one filter with the specific DNA probe and the second with labeled total genomic DNA. Exposure of the autoradiogram must be made within the linear range of the film. Scan the hybridization signals and estimate (in ng bulk DNA) the amount of DNA precipitated by the antibody, using the respective calibration curve (genomic hybrization signal vs ng of total DNA or sequence-specific signal vs ng total DNA). If the crosslinked protein does not exhibit sequence-specific DNA binding, the signals obtained with the two hybridization probes, although differing in magnitude, should indicate the same amount of bulk DNA. If, however, the protein is a sequence-specific one, the two results will differ.

References

1. Welsh, J. and Cantor, C. R. (1984) Protein-DNA crosslinking. *TIBS* **9**, 505–507.
2. Markovitz, A. (1972) Ultraviolet light-induced stable complexes of DNA and DNA polymerase. *Biochim. Biophys. Acta* **281**, 522–534.
3. Labbé, S., Prévost, J., Remondelli, P., Leone, A., and Séguin, C. (1991) A nuclear factor binds to the metal regulatory elements of the mouse gene encoding metallothionein-I. *Nucleic Acids Res.* **19**, 4225–4231.
4. Hockensmith, J. W., Kubasek, W. L., Vorachek, W. R., and von Hippel, P. H. (1986) Laser crosslinking of nucleic acids to proteins. Methodology and first applications to the phage T4 DNA replication system. *J. Biol. Chem.* **261**, 3512–3518.

5. Harrison, C. A., Turner, D. H., and Hinkle, D. C. (1982) Laser crosslinking of E. coli RNA polymerase and T7 DNA. *Nucleic Acids Res.* **10,** 2399–2414.
6. Hockensmith, J. W., Kubasek, W. L., Vorachek, W. R., Evertsz, E. M., and von Hippel, P. H. (1991) *Meth. Enzymol.* **208,** 211–236.
7. Careri, G., Fasella, P., and Gratton, E. (1975) Statistical time events in enzymes: a physical assessment. *CRC Crit. Rev. Biochem.* **3,** 141–164.
8. Pashev, I. G., Dimitrov, S. I., and Angelov, D. (1991) Crosslinking proteins to nucleic acids by ultraviolet laser-irradiation. *Trends Biochem. Sci.* **16,** 323–326.
9. Angelov, D., Stefanovsky, V. Yu., Dimitrov, S. I., Russanova, V. R., Keskinova, E., and Pashev, I. G. (1988) Protein-DNA crosslinking in reconstituted nucleohistone, nuclei and whole cells by picosecond UV laser-irradiation. *Nucleic Acids Res.* **16,** 4525–4538.
10. Budowsky, E. I., Axentyeva, M. S., Abdurashidova, G. G., Simukova, N. A., and Rubin, L. B. (1986) Induction of polynucleotide-protein crosslinkages by ultraviolet irradiation. Peculiarities of the high-intensity laser pulse irradiation. *Eur. J. Biochem.* **159,** 95–101.
11. Dobrov, E. N., Arbieva, Z. K., Timofeeva, E. K., Esenaliev, R. O., Oraevsky, A. A., and Nikogosyan, D. N. (1989) UV laser induced RNA-protein crosslinks and RNA chain breaks in tobacco mosaic virus RNA *in situ*. *Photochem. Photobiol.* **49,** 595–598.
12. Buckle, M., Geiselmann, J., Kolb, A., and Buc, H. (1991) Protein-DNA crosslinking at the lac promoter. *Nucleic Acids Res.* **19,** 833–840.
13. Stefanovsky, V. Y., Dimitrov, S. I., Russanova, V. R., Angelov, D., and Pashev, I. G. (1989) Laser-induced crosslinking of histones to DNA in chromatin and core particles: implications in studying histone-DNA interactions. *Nucleic Acids Res.* **17,** 10,069–10,081.
14. Stefanovsky, V. Yu., Dimitrov, S. I., Angelov, D., and Pashev, I. G. (1989) Interactions of acetylated histones with DNA as revealed by UV laser induced histone-DNA crosslinking. *Biochem. Biophys. Res. Commun.* **164,** 304–310.
15. Dimitrov, S. I., Stefanovsky, V. Yu., Karagyozov, L., Angelov, D., and Pashev, I. G. (1990) The enhancers and promoters of the *Xenopus laevis* ribosomal spacer are associated with histones upon active transcription of the ribosomal genes. *Nucleic Acids Res.* **18,** 6393–6397.
16. Feinberg, A. P. and Vogelstein, B. (1983) A technique for radiolabeling DNA restriction endonuclease fragments to high specific activity. *Anal. Biochem.* **132,** 6–13.
17. Feinberg, A. P. and Vogelstein, B. (1984) A technique for radiolabeling DNA restriction endonuclease fragments to high specific activity. Addendum. *Anal. Biochem.* **137,** 266, 267.

CHAPTER 19

Ultraviolet Crosslinking of DNA–Protein Complexes via 8-Azidoadenine

Rainer Meffert, Klaus Dose, Gabriele Rathgeber, and Hans-Jochen Schäfer

1. Introduction

In biological systems photoreactive derivatives have been widely applied to study specific interactions of receptor molecules with their ligands by photoaffinity labeling *(1–3)*.

Receptor + Ligand ⇌ [Receptor · Ligand]

Receptors are generally proteins like enzymes, immunoglobulins, or hormone receptors, for example. The ligands, however, differ widely in their molecular structure (e.g., sugars, amino acids, nucleotides, or oligomers of these compounds).

The advantage of photoaffinity labeling compared with affinity labeling, or chemical modification with group-specific reagents is that photoactivatable nonreactive precursors can be activated at will by irradiation (Fig. 1). These reagents do not bind covalently to the protein unless activated. On irradiation of the precursors, highly reactive intermediates are formed that react indiscriminately with all surrounding groups. Therefore, after activation a photoaffinity label—interacting at the specific binding site—can label all the different amino acid residues of the binding area. Today aromatic azido com-

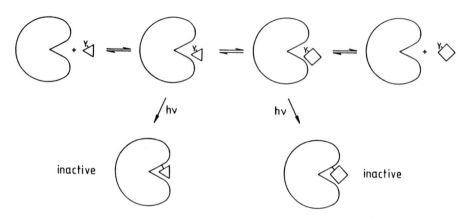

Fig. 1. Photoaffinity labeling of receptor proteins (e.g., enzymes) by photoactivatable ligand analogs (e.g., v Y substrate analog, h Y product analog). In the dark (upper line) the biological interactions of the protein with the ligand analog can be studied. On irradiation (lower line) the protein (enzyme) is labeled and inactivated.

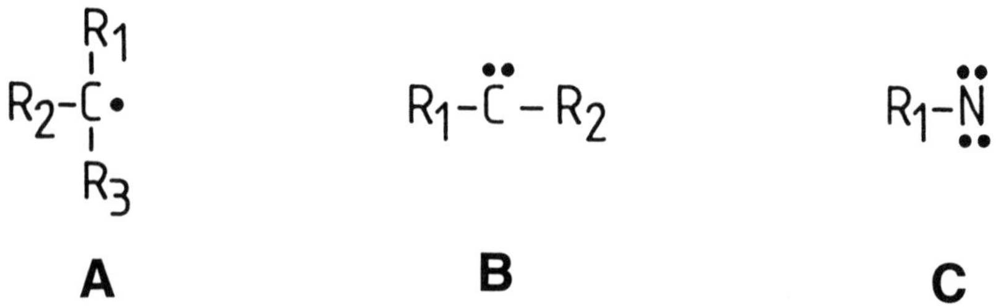

Fig. 2. Highly reactive photogenerated intermediates: radical (**A**), carbene (**B**), nitrene (**C**).

pounds are mostly used as photoactivatable ligand analogs. They form highly reactive nitrenes on irradiation because of the electron sextet in the outer electron shell of these intermediates (Fig. 2).

In addition to the azido derivatives photoreactive precursors forming radicals or carbenes on irradiation can be used as photoaffinity labels as well. All of these intermediates (nitrenes, for example) vehemently try to complete an electron octet (Fig. 3).

To produce covalent crosslinks between proteins and DNA, various methods have been applied *(4–11):* UV irradiation, γ-irradiation,

DNA–Protein Photocrosslinking

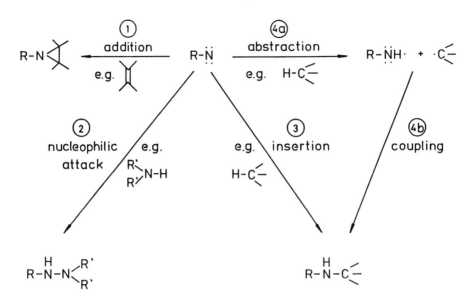

Fig. 3. Reactions of nitrenes. Cycloaddition to multiple bonds forming three membered cyclic imines (1), addition to nucleophiles (2), direct insertion into C–H bonds yielding secondary amines (3), and hydrogen atom abstraction followed by coupling of the formed radicals to a secondary amine (4a,4b).

chemical methods, and even vacuum or extreme dryness. Besides these methods, photoaffinity labeling should be a helpful tool for the study of specific interactions between proteins and deoxyribonucleic acids. Nevertheless, to date only a few successful attempts have been made to photocrosslink proteins to nucleic acids using photoactivatable deoxynucleotides. 5-Bromo- and 5-azido-2'-deoxyuridine-5'-monophosphate have been incorporated into deoxyribonucleic acids to bind DNA covalently to adjacent proteins *(12,13)*.

Here we describe the synthesis of 8-azido-dATP (8-N_3dATP), its incorporation into DNA by nick translation, and the procedure to photocrosslink azido-modified DNA to proteins *(14,15)*.

2. Materials
2.1. Synthesis of 8-N_3dATP
1. dATP (disodium salt, Boehringer Mannheim, Mannheim, Germany).
2. 1M Potassium acetate buffer: pH 3.9.
3. Bromine.

4. Sodium disulfite ($Na_2S_2O_5$).
5. Ethanol.
6. DEAE-Sephadex A-25.
7. 0.7M Triethylammonium bicarbonate buffer, pH 7.3.
8. Dimethylformamide.
9. 1M Hydrazoic acid in benzene.
10. Triethylamine.

2.2. Characterization of 8-N_3dATP

1. Silica gel plates F254 (Merck, Darmstadt, Germany).
2. Cellulose plates F (Merck).
3. Isobutyric acid/water/ammonia (66:33:1 v/v).
4. n-Butanol/water/acetic acid (5:3:2 v/v).

2.3. Preparation of Azido-Modified DNA

1. DNA (e.g., pBR322 or pWH106).
2. Deoxyribonucleotides (dATP, dGTP, dCTP, dTTP, α-[^{32}P]-dCTP).
3. DNase I (*E. coli,* 2000 U/mg, Boehringer Mannheim) in 0.15M NaCl, 50% glycerol.
4. 50 mM Tris-HCl: pH 7.2.
5. Magnesium sulfate.
6. Bovine serum albumin.
7. DNA polymerase I (*E. coli,* Boehringer Mannheim, No. 104493, which is purchased containing definite amounts of DNase I).
8. Ethylenediaminetetraacetic acid, disodium salt (EDTA).
9. Sephadex A-25 column.

2.4. Photocrosslinking

UV lamp (e.g., Mineralight handlamp UVSL 25 at position "long wave") emitting UV light at wavelengths of 300 nm and longer.

3. Methods

3.1. Synthesis of 8-N_3dATP

The synthesis of 8-N_3dATP (Fig. 4) is performed principally by analogy to the synthesis of 8-N_3ATP *(16)* (*see* Note 1). In the first step bromine exchanges the hydrogen at position 8 of the adenine ring. Then the bromine is substituted by the azido group.

1. Dissolve 0.2 mmol (117.8 mg) of dATP in 1.6 mL of potassium acetate buffer (1M, pH 3.9) and add 0.29 mmol (15 µL) of bromine. Keep the

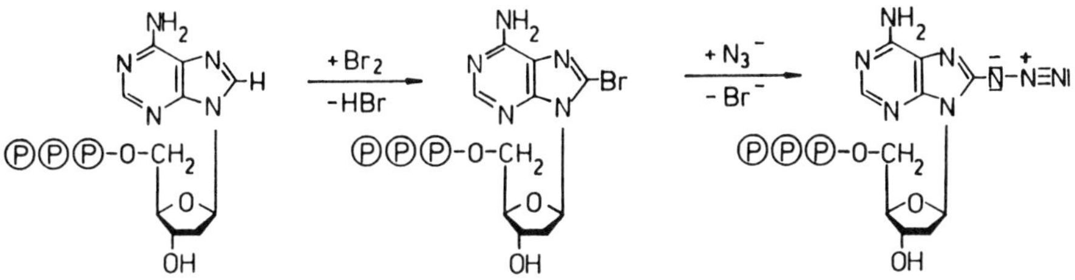

Fig. 4. Synthesis of 8-N$_3$dATP.

  reaction mixture in the dark at room temperature for 6 h (the absorption maximum shifts from 256 nm to 262 nm; see Note 2).
2. Reduce excessive bromine by addition of traces (≈5 mg) of Na$_2$S$_2$O$_5$ until the reaction mixture looks colorless or pale yellow. Pour the reaction mixture into 20 mL of cold ethanol (–20°C) and allow to stand for at least 30 min at –20°C in the dark.
3. Collect the precipitated deoxynucleotide by centrifugation and redissolve the residue in 0.5 mL of double-distilled water. Further purification is achieved by ion exchange chromatography over DEAE-Sephadex A-25 column (50 × 2 cm) with a linear gradient of 1000 mL each of water and triethylammonium bicarbonate (0.7M, pH 7.3).
4. Combine the fractions containing 8-bromo-dATP (8-BrdATP) (main peak of the elution profile) and dry the solution by lyophilization. 8-BrdATP is obtained as the triethylammonium salt. The expected yield should be 65% (spectroscopically).
5. Dissolve 0.1 mmol (87.3 mg) of dried 8-BrdATP (triethylammonium salt) in 3 mL of freshly distilled dimethylformamide (see Notes 3 and 3). Add a dried solution of 0.8 mmol (34.4 mg) of hydrazoic acid (HN$_3$) in 800 µL of benzene and 0.8 mmol (111.3 µL) of freshly distilled triethylamine. Keep the reaction mixture in the dark at 75°C for 7 h (the absorption maximum shifts from 262 to 280 nm).
6. Evaporate the solvents under vacuum and redissolve the residue in 1 mL of water. Further purification is achieved by ion exchange chromatography over DEAE-Sephadex A-25 as described in Section 3.1.3. Figure 5 shows the elution profile of the chromatography (see Notes 4 and 5).
7. Combine the fractions containing 8-N$_3$dATP and dry the solution by lyophilization. 8-N$_3$dATP is obtained as the triethylammonium salt. Yield: 30% (spectroscopically). 8-N$_3$dATP can be stored at –20°C in the dark (see Notes 6–8), freeze-dried, or frozen in aqueous solution, pH 7.0.

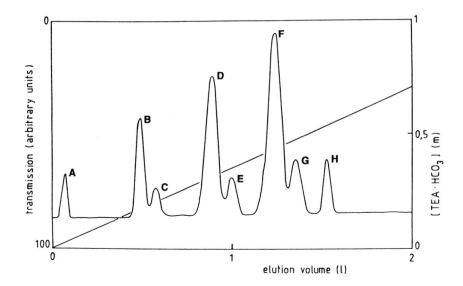

Fig. 5. Elution profile (anion exchange chromatography on DEAE-Sephadex A 25; elution buffer: linear gradient of 1000 mL each of water and 0.7M triethylammonium bicarbonate (pH 7.3), of the reaction products of 8-N_3dATP synthesis: front (**A**), 8-N_3dAMP (**B**), 8-BrdAMP (**C**), 8-N_3dADP (**D**), 8-BrdADP (**E**), 8-N_3dATP (**F**), 8-BrdATP (**G**), and probably a higher phosphorylated 8-azidodeoxyadenosine derivative (**H**).

3.2. Characterization of 8-N_3dATP

1. Thin layer chromatography. TLC is carried out on silica gel plates F_{254} or cellulose plates F. The development is performed in either isobutyric acid/water/ammonia (66:33:1 v/v) or *n*-butanol/water/acetic acid (5:3:2 v/v).
2. UV absorbance. Record the UV absorbance spectrum of 8-N_3dATP. It shows a maximum at 280 nm. The UV absorbance of 8-N_3dATP is pH dependent (*see* Note 9).
3. Photoreactivity. The photoreactivity of 8-N_3dATP is tested twice by different methods. It can either be demonstrated by the spectroscopic observation of the photolysis (Fig. 6; *see* Note 10) or by the ability of the photolabel to bind irreversibly to cellulose on thin layer plates on UV irradiation (Mineralight handlamp UVSL 25) prior to the development of the chromatogram. After development, most of the irradiated label is detected at the origin of the chromatogram in contrast to the nonirradiated control, which has completely migrated.

DNA–Protein Photocrosslinking

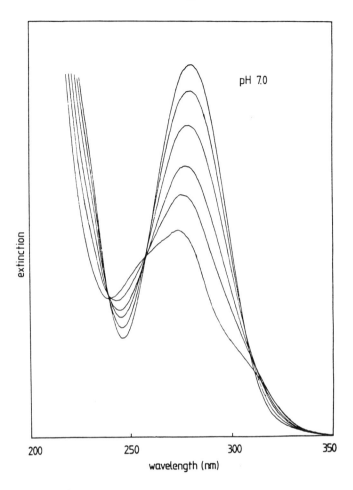

Fig. 6. Change of the optical absorption spectrum of 8-N_3dATP on UV irradiation in Tris-HCl buffer (0.01M, pH 7.0, 20°C). The irradiation time between two subsequent absorption spectra was 2 min initially. It was increased up to 10 min toward the end of photolysis. The final spectrum was taken after 30 min of irradiation. During the photolysis the absorbance at 280 nm decreased; two new absorbance maxima at 248 and 305 nm are formed.

3.3. Preparation of Azido-Modified DNA

Azido-modified and [^{32}P]-labeled DNA are prepared by nick translation. For this purpose the detailed and exact composition of the reaction medium depends strongly on the size as well as on the amount of the DNA to be modified. The optimal ratio of DNA, DNase I, and

DNA polymerase I (Kornberg enzyme) has to be tested out in preliminary experiments (see Notes 11 and 12).

Here we describe the well-tested reaction conditions for the modification of plasmid pBR322 (4363 bp). The preparation of azido-modified pWH106 (4970 bp) can be performed analogously.

1. Add 17.3 pmol of pBR322 to a mixture of 50 nmol of dGTP, 50 nmol of dTTP, 50 nmol of 8-N_3dATP, and 500 pmol of dCTP; prepare on ice.
2. Add 370 kBq of α-[^{32}P]-dCTP (110 TBq/mmol) and 20 pg of DNase I (freshly prepared out of a stock solution of 1 mg of DNase I in 1 mL of 0.15M NaCl/50% glycerol).
3. Adjust the reaction medium to an end concentration of 50 mM Tris-HCl, pH 7.2, 10 mM MgSO$_4$, and 50 mg/mL of bovine serum albumin (standard reaction volume: 100 µL).
4. Start the nick translation reaction by adding 100 U of DNA polymerase I from *E. coli* (see Section 2.3.7.).
5. Incubate for 1 h at 15°C in the dark.
6. Stop the reaction by adding EDTA (final concentration: 20 mM).
7. Separate the excess deoxyribonucleotides not incorporated during the nick translation from photoreactive [^{32}P]-labeled pBR322 by gel filtration over a Sephadex A-25 column using a 1 mL syringe.
8. Precipitate photoreactive pBR322 by adding twice the volume of cold ethanol and redissolve the residue in double-distilled water. Store the aqueous solution at –20°C in the dark.

Nonphotoreactive DNA (control) can be prepared analogously replacing the 8-N_3dATP by 50 nmol of dATP.

3.4. Photocrosslinking

1. Prepare 20–30 µL aqueous solutions containing the photoreactive DNA (0.5 pmol) and the protein (1–25 pmol) planned to be crosslinked (see Notes 13 and 14).
2. Incubate the reaction mixture for 10 min at 37°C in the dark.
3. Expose the sample to UV irradiation (see Notes 15 and 16). The irradiation times can be choosen in a range from 1 s to 60 min (see Note 17).
4. Keep the solutions in the dark before and after photolysis (see Note 6).

3.5. Analysis of DNA–Protein Adducts

Analysis of DNA–protein adducts can be done, for example, by polyacrylamide gel electrophoresis of the irradiated samples followed by autoradiography. SDS polyacrylamide gel electrophoresis should be performed immediately after photocrosslinking according to Laemmli *(17)* with some variations. After the addition of 20 mg/mL of bromophenol blue the samples are loaded onto a SDS polyacrylamide gel of 5% polyacrylamide (separating gel) with an overlay of 3.5% polyacrylamide (stacking gel) containing 1% of SDS, respectively. After the electrophoretic separation the gels are silver-stained according to Adams and Sammons *(18)*, dried, and exposed to X-ray films at −70°C. The autoradiograms are then developed. A quantitative determination of the DNA–protein adducts is possible by densitometric measurement of the autoradiogram *(19)*. Figure 7 shows a typical result on photocrosslinking of *Eco*RI-digested plasmid pWH106 with a specific interacting protein (Tet repressor). Another possibility to detect the DNA–protein adducts is the application of the nitrocellulose filter binding assay according to Braun and Merrick *(20)* (*see also* Chapter 20).

4. Notes

1. Experiments to synthesize α-$[^{32}P]$ or U-$[^{14}C]$-labeled 8-N_3dATP by starting the synthesis with α-$[^{32}P]$ or U-$[^{14}C]$dATP, respectively, failed. This is most probably because of the formation of bromine radicals induced by radioactive irradiation. These radicals could react unspecifically with the deoxyribonucleotide, suppressing the very specific electrophilic substitution of the hydrogen in position 8 of the adenine ring by the bromine ion.
2. Do not stop the reaction of dATP with bromine before the end of the given 6 h even if the absorption maximum is next to 262 nm after 1 or 2 h (otherwise a significant reduction of the yield of 8-BrdATP may occur; *see* Section 3.1.1.).
3. 8-BrdATP is obtained as triethylammonium salt, which is soluble in dimethylformamide in contrast to the alkali salts of this nucleotide. This is advantageous for the following substitution of bromine by the azido group yielding 8-N_3dATP (*see* Section 3.1.5.).
4. The exchange reaction of bromine by the azido group requires absolute dryness. However, the formation of 8-N_3dAMP and 8-N_3dADP is usually observed, owing to a limited hydrolytic cleavage of 8-N_3dATP (*see* Fig. 5; *see also* Sections 3.1.5. and 3.1.6.).
5. Besides the three azidoadenine deoxyribonucleotides, minor amounts of unreacted 8-bromoadenine deoxyribonucleotides are eluted as well (*see* Fig. 5; *see also* Section 3.1.6.).
6. Because of the photoreactivity of azido compounds, samples containing 8-N_3dATP should always be kept in the dark if possible. However,

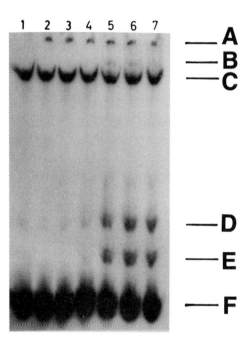

Fig. 7. Photocrosslinking of proteins to DNA (pWH106). Autoradiogram of a denaturing 5% SDS-polyacrylamide gel electrophoresis showing photocrosslinking of Tet repressor to azido-activated 187 bp and 3848 bp fragments of pWH106 (radioactive labeled by ^{32}P). Each 187 bp-fragment contains two *tet* operator sequences, the 3848 bp-fragment none. UV irradiation of azidomodified 187 bp fragment in the presence of Tet repressor results in a reduced migration of the 187 bp-fragment because of covalent crosslinking of the DNA to one or two Tet repressor dimers.

In each of lanes 1–7 0.06 pmol pWH106 (cleaved by *Eco*RI) and 20 pmol Tet repressor were applied. Lane 1: photoactive fragments of pWH106 without protein (30' UV); lane 2: nonphotoactive fragments of pWH106 with Tet repressor (30' UV); lanes 3–7: photoactive fragments of pWH106 with Tet repressor (0', 4', 10', 30' UV).

Fractions: origin of sample loading (**A**); traces of 3848 bp fragment covalently crosslinked (unspecifically) to Tet repressor (**B**); 3848 bp fragment (no Tet repressor bound) (**C**); 187 bp fragment covalently crosslinked to two Tet repressor dimers (**D**); 187 bp fragment covalently crosslinked to one Tet repressor dimer (**E**); 187 bp fragment (no Tet repressor bound) (**F**).

short exposure of azido compounds to normal daylight in our laboratory never falsified the results obtained (*see* Section 3.1.7.).
7. 8-N$_3$dATP can be stored frozen at –20°C in aqueous solution, pH 7.0, in the dark for at least 2 yr without significant loss of photoreactivity as demonstrated by subsequent photocrosslinking experiments (*see* Section 3.1.7.).

DNA–Protein Photocrosslinking

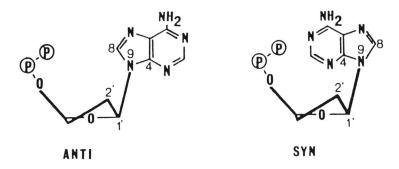

Fig. 8. Conformation of adenine nucleotides *(26)*.

8. Exclude dithiothreitol from any buffers or other solutions that contain 8-N_3dATP *(see* Section 3.1.7.). It is well known that dithiothreitol reduces azido groups to the corresponding amines *(22)*. In addition, the UV absorbance of dithiothreitol resembles that of 8-N_3dATP because of the formation of disulfide bonds by oxidation on storage in aqueous solution. This results in a less efficient photocrosslinking rate by the UV irradiation.
9. The UV absorption of 8-N_3dATP shows a maximum at 280 nm. The absorbance at 280 nm increases with decreasing pH value *(see* Section 3.2.2.). A second absorption maximum at 219 nm shifts to 204 nm in acidic solution. Both effects are a result of the protonation at N^1 of the purine ring *(21)*. The UV absorption spectrum of 8-N_3dATP resembles that of 8-N_3ATP *(16)*.
10. Take into account that the photolysis of 8-N_3dATP (to test its photoreactivity) is pH dependent *(see* Section 3.2.3.). Exhaustive irradiation in neutral solution yields new absorption maxima at 248 and 305 nm whereas in acidic or basic solution the destruction of the purine ring is observed as indicated by the disappearance of the absorbance between 240 and 310 nm (data not shown).
11. 8-N_3dATP should prefer the *syn* conformation (Fig. 8) by analogy with 8-azidoadenine nucleotides because of the bulky substituent in position 8 of the purine ring *(23)*. This seems to be contradicted by our results indicating that DNA polymerase I *(E. coli)* accepts 8-N_3dATP for a successful nick translation. It has been suggested that the enzyme only interacts with 2-deoxynucleoside triphosphates in the *anti* conformation *(24)*. This discrepancy may be explained in two ways: First, the steric requirements for the binding of 8-N_3dATP by DNA polymerase I is less restrictive than assumed *(25)*, or second, 8-N_3dATP interacts in the anticonformation with the binding site of the enzyme. This could be demonstrated for the interaction of 8-N_3ATP with the F_1ATPase from mitochondria *(26)*.

12. The preparation of azido-modified and [^{32}P]-labeled DNA by nick translation is critically dependent on the ratio of DNA, DNase I, and DNA polymerase I in the reaction medium (*see* Section 3.3.). High concentrations of DNase I on the one hand result in a very efficient incorporation rate of the azido-modified and radioactive labeled deoxynucleotides, but on the other hand, the degradation of the DNA probes by DNase I has to be evaluated. Application of too small amounts of DNase I results in inefficient incorporation of the photoactivatable deoxynucleotides and in an insufficient photocrosslinking to the interacting proteins during subsequent irradiation of the azido-modified DNA.
13. Tris-HCl buffer, 50 mM, pH 7.2, can be used instead of double-distilled water without any significant effect on the photocrosslinking efficiency (*see* Section 3.4.1.).
14. The amount of protein planned to be photocrosslinked to photoreactive DNA can be varied over a wide range (*see* Section 3.4.1.). Too high an excess of proteins, however, should be avoided because the absorbance maximum of proteins at 280 nm will lead to inefficient crosslinking rates.
15. One way to expose the samples to UV light is to deposit the probes (typically 30–50 µL) in plastic wells (normally used for RIA or ELISA tests). The UV lamp is positioned directly above the samples; thus more than one probe can be irradiated simultaneously (*see* Section 3.4.3.).
16. The emitted light of the UV lamp used for photocrosslinking should not contain light of shorter wavelengths than 300 nm because of the possibility of photo-damaging DNA or protein. For example, the Mineralight handlamp UVSL 25 (long wave) emits UV light of mainly 366 nm. The small portion of UV light of wavelengths between 300 and 320 nm emitted allows the photoactivation of the azido group without any significant photo-damage of DNA or protein (*see* Section 3.4.3.).
17. UV irradiation times for photocrosslinking can be chosen over a wide range (*see* Section 3.4.3.). Optimal UV fluence rates (flux per unit area) have to be tested out. In our experiments (using the Mineralight handlamp UVSL 25 fixed in a position resulting in a fluence rate of 4 J/m^2/s at the position of the sample) first slight amounts of DNA-protein adducts are detected after irradiation times of 10–30 s; irradiation periods longer than 15–20 min do not improve the yield of photocrosslink products.

Acknowledgments

The authors thanks Marianne Schüz (Universität Mainz) for editing the manuscript. This work was supported by the Bundesministerium für Forschung und Technologie (07QV8942) and by the Deutsche Forschungsgemeinschaft (Sch 344/1–3).

References

1. Bayley, H. and Knowles, J. R. (1977) Photoaffinity labeling. *Meth. Enzymol.* **46,** 69–114.
2. Bayley, H. (1983) Photogenerated reagents in biochemistry and molecular biology, in *Laboratory Techniques in Biochemistry*, vol. 12 (Work, T. S. and Burdon, R. H., eds.), Elsevier, Amsterdam.
3. Schäfer, H.-J. (1987) Photoaffinity labeling and photoaffinity crosslinking of enzymes, in *Chemical Modifications of Enzymes, Active Site Studies* (Eyzaguirre, J., ed.), Ellis Horwood Ltd, Chichester, UK, pp. 45–62.
4. Smith, K. C. (1962) Dose-dependent decrease in extractability of DNA from bacteria (by UV-light). *Biochem. Biophys. Res. Commun.* **8,** 157–163.
5. Shetlar, M. D. (1980) Crosslinking of proteins to nucleic acids by UV-light. *Photochem. Photobiol. Rev.* **5,** 105–197.
6. Welsh, J. and Cantor, C. R. (1984) Protein-DNA crosslinking. *Trends Biochem. Sci.* **9,** 505–508.
7. Ekert, B., Giocanti, N., and Sabattier, R. (1986) Study of several factors in RNA-protein crosslink formation induced by ionizing radiations within 70S ribosomes of *E. coli* MRE 600. *Int. J. Radiat. Biol.* **50,** 507–525.
8. Lesko, S. A., Drocourt, J. L., and Yang, S. U. (1982) DNA-protein- and DNA interstrand crosslinks induced in isolated chromatin by H_2O_2 and Fe-EDTA-chelates. *Biochemistry* **21,** 5010–5015.
9. Wedrychowsky, A., Ward, W. S., Schmidt, W. N., and Hnilica, L. S. (1985) Chromium-induced crosslinking of nuclear proteins and DNA. *J. Biol. Chem.* **260,** 7150–7155.
10. Summerfield, F. W. and Tappel, A. L. (1984) Crosslinking of DNA in liver and testis of rats fed 1,3-propanediol. *Chem. Biol. Interact.* **50,** 87–96.
11. Dose, K., Bieger-Dose, A., Martens, K.-D., Meffert, R., Nawroth, T., Risi, S., Steinborn, A., and Vogel, M. (1987) Survival under space vacuum—biochemical aspects. *Proc. 3rd Eur. Symp. Life Sci. Res. in Space (ESA SP-271)* 193–195.
12. Lin, S. Y. and Riggs, A. D. (1974) Photochemical attachment of lac repressor to bromodeoxyuridine-substituted *lac* operator by UV radiation. *Proc. Natl. Acad. Sci. USA* **71,** 947–951.
13. Evans, R. K., Johnson, J. D., and Haley, B. E. (1986) 5-Azido–2'-deoxyuridine-5'-triphosphate: a photoaffinity labeling reagent and tool for the enzymatic synthesis of photoactive DNA. *Proc. Natl. Acad. Sci. USA* **83,** 5382–5386.
14. Meffert, R. and Dose, K. (1988) UV-induced crosslinking of proteins to plasmid pBR322 containing 8-azidoadenine 2'-deoxyribonucleotides. *FEBS Lett.* **239,** 190–194.
15. Meffert, R., Rathgeber, G., Schäfer, H.-J., and Dose, K. (1990) UV-induced crosslinking of Tet repressor to DNA containing *tet* operator sequences and 8-azidoadenines. *Nucleic Acids Res.* **18,** 6633–6636.
16. Schäfer, H.-J., Scheurich, P., and Dose, K. (1978) Eine einfache Darstellung von 8-N_3ATP: ein Agens zur Photoaffinitätsmarkierung von ATP-bindenden Proteinen. *Liebigs Ann. Chem.* **1978,** 1749–1753.
17. Laemmli, U. K. (1970) Cleavage of structural proteins during the assembly of the head of bacteriophage T4. *Nature* **227,** 680.

18. Adams, L. D. and Sammons, D. W. (1981) A unique silver staining procedure for color characterization of polypeptides. *Electrophoresis* **2,** 155–165.
19. Westermeier, R., Schickle, H., Thesseling, G., and Walter, W. W. (1988) Densitometrie von Gelelektrophoresen. GIT Labor-Medizin **4/88,** 194–202.
20. Braun, A. and Merrick, B. (1975) Properties of UV-light-mediated binding of BSA to DNA. *Photochem. Photobiol.* **21,** 243–247.
21. Koberstein, R., Cobianchi, L., and Sund, H. (1976) Interaction of the photoaffinity label 8-azido-ADP with glutamate dehydrogenase. *FEBS Lett.* **64,** 176–180.
22. Staros, J. V., Bayley, H., Standring, D. N., and Knowles, J. R. (1978) Reduction of aryl azides by thiols: implications for the use of photoaffinity reagents. *Biochem. Biophys. Res. Commun.* **80,** 568–572.
23. Vignais, P. V. and Lunardi, J. (1985) Chemical probes of the mitochondrial ATP synthesis and translocation. *Ann. Rev. Biochem.* **54,** 977–1014.
24. Czarnecki, J. J. (1978) PhD Thesis, University of Wyoming, Laramie, WY.
25. Englund, P. T., Kelly, R. B., and Kornberg, A. (1969) Enzymatic synthesis of DNA: binding of DNA to DNA polymerase. *J. Biol. Chem.* **244,** 3045–3052.
26. Garin, J., Vignais, P. V., Gronenborn, A. M., Clore, G. M., Gao, Z., and Bäuerlein, E. (1988) ^{1}H-NMR studies on nucleotide binding to the catalytic sites of bovine mitochondrial F_1-ATPase. *FEBS Lett.* **242,** 178–182.

CHAPTER 20

Filter-Binding Assays

Peter G. Stockley

1. Introduction

Membrane filtration has a long history in the analysis of protein–nucleic acid complex formation, having first been used to examine RNA–protein interactions *(1)*, before being introduced to DNA–protein interaction studies by Jones and Berg in 1966 *(2)*. The principle of the technique is straightforward. Under a wide range of buffer conditions, protein-free nucleic acids pass freely through membrane filters, whereas proteins and their bound ligands are retained. Thus, if a particular protein binds to a specific DNA sequence, passage through the filter will result in retention of a fraction of the protein–DNA complex by virtue of the protein component of the complex. The amount of DNA retained can be determined by using radioactively labeled DNA to form the complex and then determining the amount of radioactivity retained on the filter by scintillation counting. The technique can be used to analyze both binding equilibria and kinetic behavior, and, if the DNA samples retained on the filter and in the filtrate are recovered for further processing, the details of the specific binding site can be probed by interference techniques.

The technique has a number of advantages over footprinting and gel retardation assays, although there are also some relative disadvantages, especially where multiple proteins are binding to the same DNA molecule. However, filter-binding is extremely rapid, reproducible, and in principle can be used to extract accurate equilibrium and rate constants *(3–5)*. We have used the technique to examine the interaction between the *E. coli* methionine repressor, MetJ, and various operator sites cloned into restriction fragments (*6,7; see* also Chapter 10). Results from these studies will be used to illustrate the basic technique.

Before discussing the experimental protocols it is important to understand some fundamental properties of the filter-binding assay. The molecular basis of the discrimination between nucleic acids and proteins during filtration is still not fully understood. Care should therefore be taken to characterize the assay with the system under study. Nucleic acid–protein complex retention occurs with differing efficiencies depending on the lifetime of the complex, the size of the protein component, the buffer conditions, and the extent of washing of the filter. Experiments with the *lac* repressor system have shown that prior filtration of protein followed by passage of DNA containing operator sites does not result in significant retention of the nucleic acid, presumably because filter-bound protein is inactive for further operator binding. The DNA retained on filters is therefore a direct reflection of the amount of complex present when filtration began. Furthermore, incubation of *lac* repressor with large amounts of DNA that does not contain an operator site followed by filtration also does not lead to significant retention. Since *lac* repressor (and indeed essentially all DNA-binding proteins) binds nonsequence-specifically to DNA, forming short-lived complexes, it is clear that these are not readily retained. The experiments with *lac* repressor *(3–5)* can therefore be used as a guide when designing experimental protocols. The repressor is a large protein (being a tetramer of 38 kDa subunits) but the basic features seem to apply even to short peptides with mol wt <2 kDa *(8)*.

In any particular system the percentage of the DNA-protein complex in solution retained by the filter should ideally be constant throughout the binding curve, and this is known as the retention efficiency. Experimental values range from 30 to >95%. An example of the sort of results obtained with the MetJ repressor is shown in Fig. 1.

2. Materials

2.1. Preparation of Radioactively End-Labeled DNA

1. Plasmid DNA carrying the binding site for a DNA-binding protein on a convenient restriction fragment (usually <200 bp).

Filter-Binding Assays

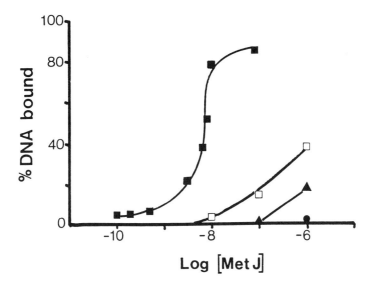

Fig. 1. Filter binding affinity curves of met repressor binding to the single met-box construct, *00045*, (triangles); the double met-box construct, *00048*, (squares); and the control pUC18 polylinker construct (circles), in the presence (solid symbols) and absence (open symbols) of 1 mM SAM. Only data points above baseline are plotted.

2. Restriction enzymes and the appropriate buffers as recommended by the suppliers.
3. Phenol: redistilled phenol equilibrated with 100 mM Tris-HCl, pH 8.0.
4. Chloroform.
5. Solutions for ethanol precipitation of DNA: 4M NaCl and ethanol (absolute and 70% v/v).
6. Calf intestinal phosphatase (CIP).
7. CIP reaction buffer (10X): 0.5M Tris-HCl, pH 9.0, 0.01M MgCl$_2$, 0.001M ZnCl$_2$.
8. TE buffer: 10 mM Tris-HCl, pH 8.0, 1 mM ethylenediaminetetraacetic acid (EDTA).
9. Sodium dodecyl sulfate (SDS) 20% w/v.
10. 0.25M EDTA, pH 8.0.
11. T4 polynucleotide kinase (T4PNK).
12. T4PNK reaction buffer, 10X: 0.5M Tris-HCl, pH 7.6, 0.1M MgCl$_2$, 0.05M dithiothreitol.
13. Radioisotope: γ-[^{32}P]-ATP.
14. 30% w/v acrylamide stock (29:1 acrylamide: *N,N'*-methylene-*bis*-acrylamide).

15. Polyacrylamide gel elution buffer: 0.3M sodium acetate, 0.2% w/v SDS, 2 mM EDTA.
16. Polymerization catalysts: Ammonium persulfate (10% w/v) and N,N,N',N'-tetramethylethylene diamine (TEMED).
17. X-ray film, autoradiography cassette, and film developer.
18. Plastic wrap and scalpel.

2.2. Filter-Binding Assays

1. Nitrocelluose filters: We use HAWP (00024) filters from Millipore (Bedford, MA) but suitable filters are available from a number of other manufacturers, such as Schleicher and Schuell (Dassel, Germany). Filters tend to be relatively expensive. Some manufacturers produce sheets of membrane that can be cut to size and are thus less expensive.
2. Filter-binding buffer (FB): 100 mM KCl, 0.2 mM EDTA, 10 mM Tris-HCl, pH 7.6.
3. Binding buffer (BB): This is FB containing 50 µg/mL bovine serum albumin (BSA, protease and nuclease free; see Note 1).
4. Filtration manifold and vacuum pump: We use a Millipore 1225 Sampling Manifold (cat. no. XX27 025 50), which has twelve sample ports.
5. Liquid scintillation counter, vials, and scintillation fluid.
6. Siliconized glass test tubes.
7. TBE buffer: 89 mM Tris, 89 mM boric acid, 10 mM EDTA, pH 8.3.
8. Formamide/dyes loading buffer: 80% v/v formamide, 0.5X TBE, 0.1% w/v xylene cyanol, 0.1% w/v bromophenol blue.
9. Sequencing gel electrophoresis solutions and materials: 19% w/v acrylamide, 1% w/v *bis*-acrylamide, 50% w/v urea in TBE.
10. Acetic acid (10% v/v).

3. Methods

3.1. Preparation of End-Labeled DNA

1. Digest the plasmid (≈20 µg in 200 µL) with the restriction enzymes used to release a suitably sized DNA fragment (usually <200 bp). Extract the digest with an equal volume of buffered phenol and add 2.5 vol of ethanol to the aqueous layer in order to precipitate the digested DNA. If preparing samples for interference assays, only one restriction digest should be carried out at this stage.
2. Add 50 µL 1X calf intestinal phosphatase (CIP) reaction buffer to the ethanol-precipitated DNA pellet (<50 µg). Add 1 U CIP and incubate at 37°C for 30 min followed by addition of a further aliquot of enzyme and incubate for a further 30 min. Terminate the reaction by adding SDS and EDTA to 0.1% (w/v) and 20 mM respectively in a final vol-

ume of 200 µL and incubate at 65°C for 15 min. Extract the digest with buffered phenol, then with 1:1 phenol:chloroform, and finally ethanol precipitate the DNA from the aqueous phase as above.
3. Redissolve the DNA pellet in 18 µL 1X T4PNK buffer. Add 20 µCi γ-[^{32}P]-ATP and 10 U T4PNK and incubate at 37°C for 30 min. Terminate the reaction by phenol extraction and ethanol precipitation (samples for interference assays should be digested with the second restriction enzyme) at this point. Redissolve the pellet in nondenaturing gel loading buffer and electrophorese on a nondenaturing polyacrylamide gel.
4. After electrophoresis, separate the gel plates, taking care to keep the gel on the larger plate. Cover the gel with plastic wrap and in the darkroom, under the safe-light, tape a piece of X-ray film to the gel covering the sample lanes. With a syringe needle puncture both the film and the gel with a series of registration holes. Locate the required DNA fragments by autoradiography of the wet gel at room temperature for several minutes (≈10 min). Excise slices of the gel containing the bands of interest using the autoradiograph as a guide. Elute the DNA into elution buffer overnight (at least) at 37°C. Ethanol precipitate the eluted DNA by adding 2.5 vol of ethanol, wash the pellet thoroughly with 70% v/v ethanol, dry briefly under vacuum, and rehydrate in a small volume (≈50 µL) of TE. Determine the radioactivity of the sample by liquid scintillation counting of a 1-µL aliquot.

3.2. Filter-Binding Assays
3.2.1. Determination of the Equilibrium Constant

1. Presoak the filters in FB at 4°C for several hours before use. Care must be taken to ensure that the filters are completely "wetted." This is best observed by laying the filters carefully onto the surface of the FB using blunt-ended tweezers.
2. Prepare a stock solution of radioactively labeled DNA fragment in an appropriate buffer, such as FB. We adjust conditions so that each sample to be filtered contains roughly 20 kcpm. Under these conditions the DNA concentration is <1 pM. Aliquot the stock DNA solution into plastic Eppendorf tubes. It is best at this stage if relatively large volumes are transferred in order to minimize errors caused by pipeting. We use 180 µL/sample. If the DNA-binding protein being studied requires a cofactor, it is best to add it to the stock solution at saturating levels so that its concentration is identical for every sample.
3. Prepare a serially diluted range of protein concentrations diluting into binding buffer (BB). A convenient range of concentrations for the initial assay is between 10^{-11} and $10^{-5}M$ protein.

4. Immediately add 20 µL of each protein concentration carefully to the sides of the appropriately labeled tubes of stock DNA solution. When the additions are complete centrifuge briefly (5 s) to mix the samples and then incubate at a temperature at which complex formation can be observed (37°C for MetJ). For each binding curve it is important to prepare two control samples. The first contains no protein in the 20 µL of BB and is filtered to determine the level of background retention. The second is identical to the first but is added to a presoaked filter in a scintillation vial (*see* step 6) and is dried directly without filtering. This gives a value for 100% input DNA.
5. After an appropriate time interval to allow equilibrium to be established, recentrifuge the tubes to return the liquid to the bottom of the tube and begin filtering.
6. The presoaked filters are placed carefully on the filtration manifold ensuring that excess FB is removed and that the filter is not damaged. Cracks and holes are easily produced by rough handling. The sample aliquot (200 µL) is then immediately applied to the filter, where it should be held stably by surface tension. Apply the vacuum. If further washes are used they should be applied as soon as the sample volume has passed through the filter. Remove the filter to a scintillation vial and continue until all the samples have been filtered.
7. The scintillation vials should be transferred to an oven at 60°C to dry the filters thoroughly ($\approx$20 min) before being allowed to cool to room temperature and 3–5 mL of scintillation fluid added. The radioactivity associated with each filter can now be determined by counting on an open channel (*see* Note 2).
8. Correct the value for each sample by subtracting the counts in the background sample (no protein). Calculate the percentage of input DNA retained at each protein concentration using the value for 100% input from the unfiltered sample. Plot a graph of percentage retained vs the logarithm of the protein concentration (e.g., Fig. 1). The binding curve should increase from left to right until a plateau is reached. This is rarely 100%. The plateau value can be assumed to represent the retention efficiency, and for quantitative measurements the data points can be adjusted accordingly. There is not enough space here to describe in detail the form of the binding curve or how best to interpret the data. (For an authoritative yet accessible account *see* ref. 9). For our purposes the protein concentration at 50% saturation can be thought of as the equilibrium dissociation constant.
9. Once an initial binding curve has been obtained, the experiment should be repeated with sample points concentrated in the appropriate region.

Filter-Binding Assays 257

Control experiments with DNAs that do not contain specific binding sites and the like should also be carried out to prove that binding is sequence-specific. Highly diluted protein solutions appear to lose activity in our hands possibly because of nonspecific absorption to the sides of tubes, among other things. We therefore produce freshly diluted samples daily. BB can be stored at 4°C for several days without deleterious effect. Ideally binding curves should be reproducible. However, there is some variability between batches of filters and we therefore recommend not switching lot numbers during the course of one set of experiments.

3.2.2. Kinetic Measurements

Kinetic analysis of the binding reaction depends on prior determination of the equilibrium binding curve, especially the concentration of DNA-binding protein required to saturate the input DNA. This information allows a reaction mixture to be set up containing a limiting amount of protein, for example, at a protein concentration that produces 75% retention. Both association and dissociation kinetics can be studied. The major technical problem arises because of the relatively rapid sampling rates that are required. However, it is almost always possible to adjust solution conditions such that sampling at 10 s intervals is all that is needed. Dissociation measurements often need to be made over periods of up to 1 hr, whereas association reactions are usually complete within several minutes.

3.2.2.1. DISSOCIATION

Repeat steps 1 and 2 of Section 3.2.1. but do not aliquot the stock DNA solution. Add to this sample the appropriate concentration (i.e., which produces ≈75% retention) of stock protein and allow to equilibrate. Add a 20-fold excess of unlabeled DNA fragment containing the binding site and begin sampling (≈200-µL aliquots) by filtration. Plots of radioactivity retained vs time can then be analyzed to derive kinetic constants. In the simplest case of a bimolecular reaction, a plot of the natural logarithm of the radioactivity retained at time t divided by the initial radioactivity vs time yields the first order dissociation constant from the slope. An important control experiment is to repeat the experiment with DNA that does not contain a specific binding site to show that dissociation is sequence-specific.

A variation of this experiment can be used in which the concentration of protein in the reaction mix is diluted across the range where most complex formation occurs. In this case it is necessary to prepare the initial complex in a small volume (≈50 µL) and then dilute 100X with BB, followed by filtering 500-µL aliquots.

3.2.2.2. ASSOCIATION

Set up a stock DNA concentration in a single test tube as above (Section 3.2.1., steps 1 and 2). Incubate both this DNA and the appropriate solution of protein at the temperature at which complexes form. Add the appropriate volume of protein (e.g., 200 µL) to the DNA stock solution (1800 µL) and immediately begin sampling (10 × 200-µL aliquots).

3.2.3. Interference Measurements

Experiments of this type can be used to gain information about the site on the DNA fragment being recognized by the protein. The principle is identical to that used in gel retardation interference assays but has the advantage that the DNA does not have to be eluted from gels after fractionation.

1. Modify the purified DNA fragment radiolabeled (≈100 kcpm) at a single site with the desired reagent; for example, hydroxyl radicals, which result in the elimination of individual nucleotide groups *(11)*, dimethyl sulfate, DMS *(10)*, which modifies principally guanines, or ethyl nitrosourea, ENU (*see* Chapter 10), which ethylates the nonesterified phosphate oxygens. These techniques are covered in detail in Chapters 4, 6, and 10, respectively.The extent of modification should be adjusted so that any one fragment has no more than one such modification. This can be assessed separately in test reactions and monitored on DNA sequencing gels.
2. Ethanol precipitate the modified DNA, wash twice with 70% (v/v) ethanol and then dry briefly under vacuum. Resuspend in 200 µL FB. Remove 20 µL as a control sample. Add 20 µL of the appropriate protein concentration to form a complex and allow equilibrium to be reached. Filter as usual but with a siliconized glass test tube positioned to collect the filtrate. (The Millipore manifold has an insert for just this purpose.) Do not overdry the filter.
3. Place the filter in an Eppendorf tube containing 250 µL FB, 250 µL H_2O, 0.5% (w/v) SDS. Transfer the filtrate into a similar tube and then add SDS and H_2O to make the final volume and concentration the same as the filter-retained sample. Add an equal volume of buffer-saturated phenol to each tube, vortex, and centrifuge to separate the phases. Remove the aqueous

top layers, reextract with chloroform:phenol (1:1), and then ethanol precipitate. A Geiger counter can be used to monitor efficient elution of radioactivity from the filter, which can be reextracted if necessary.

4. Recover all three DNA samples (control, filter-retained, and filtrate) after ethanol precipitation and, if necessary, process the modification to completion (e.g., piperidine for DMS modification, NaOH for ENU, and so on). Ethanol precipitate the DNA, dry briefly under vacuum, and then redissolve the pellets in 4 µL formamide/dyes denaturing loading buffer. At this stage it is often advisable to quantitate the radioactivity in each sample by liquid scintillation counting of 1-µL aliquots. Samples for sequencing gels should be adjusted to contain roughly equal numbers of counts in all three samples.

5. Heat the samples to 90°C for 2 min and load onto a 12% w/v polyacrylamide sequencing gel alongside Maxam-Gilbert sequencing reaction markers *(12)*. Electrophorese at a voltage that will warm the plates to around 50°C. After electrophoresis, fix gel in 1 L 10% v/v acetic acid for 15 min. Transfer gel to 3MM paper and dry under vacuum at 80°C for 60 min. Autoradiograph the gel at –70°C with an intensifying screen.

6. Compare lanes corresponding to bound, free, and control DNAs for differences in intensity of bands at each position (*see* Note 3). A dark band in the "free fraction" (and a corresponding reduction in the intensity of the band in the "bound fraction") indicates a site where prior modification interferes with complex formation. This is interpreted as meaning that this residue is contacted by the protein or a portion of the protein comes close to the DNA at this point. (*See* Chapters 4, 6, and 10 for more extensive discussions of interference experiments.)

4. Results and Discussion

Figure 1 shows a typical filter-binding curve for the *E. coli* methionine repressor binding to its idealized operator site of $(dAGACGTCT)_2$ cloned into a pUC-polylinker. In the presence of saturating amounts of cofactor (SAM) a sigmoidal binding curve is produced, whereas in the absence of SAM the binding curve does not saturate in the protein concentration range tested. Similar binding curves have been analyzed to produce Scatchard and Hill plots *(9)* in order to examine the cooperativity with respect to protein concentration *(6)*. However, such multiple binding events should also be studied by gel retardation assays that can give data about the individual complex species.

Table 1 shows the results obtained for binding to a series of variant operator sites and illustrates the apparent sensitivity of the technique.

Table 1
Relative K_ds of a Number of Variant *Met* Operator Sites

Variant	Operator Sequence	Relative K_d
00045	A G A C G T C T	>12.2
00048	A G A C G T C T A G A C G T C T	1.0
00184	A G A C G T C a t G A C G T C T	2.8
00299	g t c t A G A C G T C T a g a c	4.9

The K_d is the concentration of protein that produce 50% binding of input DNA. Values are averages of several experiments and are quoted relative to the two met-box perfect consensus sequences (00048) which, under the conditions used, had an apparent K_d of 82 ± 5 n*M* MetJ monomer.

However, in order to make such comparisons it is essential to determine the binding curves accurately and with the same batches of protein and filters to minimize minor differences between experiments.

The table lists the affinities of a number of variant *met* operator sites cloned into pUC-polylinkers determined by filter-binding in the presence of saturating levels of corepressor, SAM. The repressor binds cooperatively to tandem arrays of an 8 bp met-box sequence (dAGACGTCT) with a stoichiometry of one repressor dimer per met-box. The variant operators were designed to examine both the tandem binding and the alignment of repressor dimers with the two distinct dyads in tandem met-box sequences *(6)*.

Operator variants are:

1. 00045—A single 8 bp met-box or half-site. The binding curve does not saturate because singly bound repressor dimers dissociate very rapidly.
2. 00048—Two perfect met-boxes representing the idealized minimum operator sequence. Repressors bind cooperatively with high affinity.
3. 00184—Two met-boxes with the central T-A step reversed.

The crystal structure of the repressor–operator complex shows that the central T-A step is not contacted directly by the repressors, rather the pyrimidine-purine step promotes a sequence-dependent DNA distortion that results in protein–DNA contacts elsewhere in the operator fragment. The A-T step has less tendency to undergo this conformational change and this is reflected in its lowered affinity. 00299—A "shifted" two met-box operator used to define the alignment between the repressor twofold axis and the operator dyads. The low affinity of this construct compared to 00048 confirms that each repressor dimer is centered on the middle of a met-box.

5. Notes

1. None of the radioactivity is retained by the filter. This again can be caused by a variety of factors. Check that the preparation of DNA-binding protein is still functional (if other assays are available) or that the protein is still intact by SDS-PAGE. Check the activity/concentration of the cofactor if required. A common problem we have encountered arises because of the different grades of commercially available BSA. It is always advisable to use a preparation that explicitly claims to be nuclease and protease free.
2. All of the radioactivity is retained by the filter. This is a typical problem when first characterizing a system by filter-binding and can have many causes. Check that the filters being used "wet" completely in FB and do not dry significantly before filtration. Make sure that the DNA remains soluble in the buffer being used by simple centrifugation in a benchtop centrifuge. If the background remains high add dimethyl sulfoxide to the filtering solutions. Classically 5% (v/v) is used but higher concentrations (≈20% v/v) have been reported with little, if any, effect on the binding reaction. We have experienced excessive retention when attempting to analyze the effects of divalent metal ions on complex formation, and in general it is best to avoid such buffer conditions.
3. Poor recoveries from the filter-retained samples in interference assays, or other problems in processing such samples further, can often be alleviated by addition of 20 µg of tRNA as a carrier during the SDS/phenol extraction step.

Acknowledgments

I am grateful to Yi-Yuan He for providing the data shown in Table 1 and Fig. 1.

References

1. Nirenberg, M. and Leder, P. (1964) RNA codewords and protein synthesis. The effect of trinucleotides upon the binding of sRNA to ribosomes. *Science* **145,** 1399–1407.
2. Jones, O. W. and Berg, P. (1966) Studies on the binding of RNA polymerase to polynucleotides. *J. Mol. Biol.* **22,** 199–209.
3. Riggs, A. D., Bourgeois, S., Newby, R. F., and Cohn, M. (1968) DNA binding of the *lac* repressor. *J. Mol. Biol.* **34,** 365–368.
4. Riggs, A. D., Suzuki, H., and Bourgeois, S. (1970) *lac* repressor-operator interaction. I. Equilibrium studies. *J. Mol. Biol.* **48,** 67–83.

5. Riggs, A. D., Bourgeois, S., and Cohn, M. (1970) The *lac* repressor-operator interaction. III. Kinetic studies. *J. Mol. Biol.* **53,** 401–417.
6. Phillips, S. E. V., Manfield, I., Parsons, I., Davidson, B. E., Rafferty, J. B., Somers, W. S., Margarita, D., Cohen, G. N., Saint-Girons, I., and Stockley, P. G. (1989) Cooperative tandem binding of Met repressor from *Escherichia coli*. *Nature* **341,** 711–715.
7. Old, I. G., Phillips, S. E. V., Stockley, P. G., and Saint-Girons, I. (1991) Regulation of methionine biosynthesis in the enterobacteriaceae. *Prog. Biophys. Molec. Biol.* **56,** 145–185.
8. Ryan, P. C., Lu, M., and Draper, D. E. (1991) Recognition of the highly conserved GTPase center of 23S ribosomal RNA by ribosomal protein L11 and the antibiotic thiostrepton. *J. Mol. Biol.* **221,** 1257–1268.
9. Wyman, J. and Gill, S. J. (1990) in *Binding and Linkage: Functional Chemistry of Biological Macromolecules,* Chapter 2, University Science Books, Mill Valley, CA.
10. Siebenlist, U. and Gilbert, W. (1980) Contacts between *Escherichia coli* RNA polymerase and an early promoter of phage T7. *Proc. Natl. Acad. Sci. USA* **77,** 122–126.
11. Hayes, J. J. and Tullius, T. D. (1989) The missing nucleoside experiment: a new technique to study recognition of DNA by protein. *Biochemistry* **28,** 9521–9527.
12. Maxam, A. M. and Gilbert, W. K (1980) Sequencing end-labelled DNA with base-specific chemical cleavages. *Meth. Enzymol.* **65,** 499–560.

CHAPTER 21

The Gel Shift Assay for the Analysis of DNA–Protein Interactions

John D. Taylor, Alison J. Ackroyd, and Stephen E. Halford

1. Introduction

The gel shift assay is one of the most powerful methods for the analysis of DNA–protein interactions *(1,2)*. The assay itself is simple. DNA and protein are mixed together, the solution subjected to electrophoresis through polyacrylamide, and the gel is then analyzed for DNA, usually by autoradiography of radiolabeled DNA *(3,4)*. Binding of the protein to the DNA can result in a complex that has a different electrophoretic mobility from the free DNA. In general, the mobility of the complex is retarded relative to the unbound DNA and thus the assay is often called gel retardation. However, with circular DNA substrates (typically, minicircles of 200–400 bp), the DNA–protein complex can migrate faster than the free DNA *(5,6)*. The separation of the complex from the free DNA, and therefore the detection of the complex, is dependent on a variety of factors. These must be determined experimentally for each system. However, the ease with which the assay can be performed means that the optimal conditions can be discovered quickly. Factors that influence the electrophoretic mobility of DNA–protein complexes include the molecular weight of the protein and the DNA *(7,8)*, the ionic strength and the pH of the electrophoresis buffer *(9)*, the concentration of the gel matrix, and the temperature. Particularly useful accounts of how modifications to the assay can affect the mobility of DNA–protein complexes have been published *(2,10)*.

The principle of the gel shift assay is that the entry of the mixture of free DNA and DNA–protein complex into the gel matrix results in the physical separation of the two species. In the subsequent electrophoresis, the protein can make no difference to the mobility of the free DNA and, provided that the bound DNA remains associated with the protein, it too will have a characteristic mobility. The gel matrix may stabilize the complex by hindering the diffusion of the protein away from the DNA *(4,7,8)*, although it has been shown that complexes reversibly dissociate and reassociate within the gel *(11)*. However, even if the bound DNA dissociates from the protein during electrophoresis, it can never "catch up" with the DNA that was free at the start of the run. Thus, the method has the potential of "freezing" the equilibrium between bound and free DNA at the moment of entry into the gel. The concentration of each species can then be determined. Assays of this type can yield the equilibrium constant for the binding of the protein to its DNA ligand and also the kinetics of the interaction, the latter by analyzing samples at different times after mixing the DNA with the protein *(3,6,7,12)*. An important parameter in kinetic experiments is the "electrophoretic dead time," the time taken for the complex to migrate from the solution loaded in the well to the gel itself. This needs to be made as short as possible *(8)*.

If the DNA substrate has one specific site that binds the protein much more tightly than any other site, a gel shift experiment should reveal just one DNA–protein complex (Fig. 1). However, the appearance of a single retarded band does not necessarily mean that a specific complex has been formed. One test for specificity is to add unlabeled competitor DNA to a binding reaction containing [^{32}P]-labeled substrate DNA *(3,7,13)*. On addition of competitor DNA, all nonspecific complexes with the labeled DNA should be titrated away. Any remaining complexes should represent the specific DNA–protein association, and this can confirmed by observing their disappearance on further titrations with unlabeled substrate DNA. Indeed, the interaction of a protein with a specific DNA sequence is best analyzed in the presence of nonspecific competitor DNA as shown in Fig. 1. Moreover, by using unlabeled competitor with labeled substrate, the gel shift assay can be applied to unfractionated nuclear extracts, to detect

Gel Shift Assay

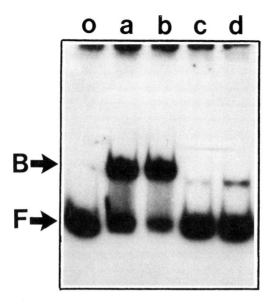

Fig. 1. Autoradiograph from a gel shift assay, showing binding to one specific site. The reactions contained both a 203-bp DNA fragment ($[^{32}P]$-labeled, 5 nM) with one binding site for the resolvase from Tn3 and unlabeled calf thymus DNA. These were mixed with either no additional protein (lane 0), the resolvase from Tn3 (in the lanes marked 3 above the gel; 100 nM in a, 200 nM in b), or helix-swap resolvase (in the lanes marked H; 100 nM in a, 200 nM in b). The samples were then analyzed by electrophoresis through polyacrylamide. F and B, on the left of the autoradiograph, mark the positions of the 203 bp DNA, either free or bound to resolvase. Resolvase possesses a helix-turn-helix motif and helix-swap resolvase was constructed from Tn21 resolvase by replacing all of the amino acids in its recognition helix with the equivalent sequence from Tn3 resolvase. Native Tn21 resolvase has virtually no affinity for this segment of DNA from Tn3 but the replacement of its recognition helix allows for some binding to Tn3 DNA, although the yield of the complex is much lower than with the cognate Tn3 resolvase. (Data from ref. *13*: Reprinted with permission from *J. Mol. Biol.* Copyright [1990] Academic).

proteins that bind specifically to the substrate DNA *(14)*. Which proteins in the crude extract are bound to the substrate can then be determined immunologically *(15)*.

In addition to separating bound from free DNA, the gel shift assay can also resolve DNA–protein complexes with different stoichiometries; that is, the DNA with 1, 2, 3, or more molecules of protein bound to it *(4,16,17)*. The gel shift experiment then reveals a series of bands of DNA with progressively reduced electrophoretic mobili-

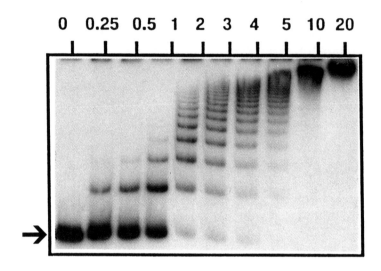

Fig. 2. Autoradiograph from a gel shift assay, showing binding to multiple nonspecific sites. The reactions contained a 381-bp DNA fragment ([^{32}P]-labeled, 0.1 nM), with either no added protein (lane 0; the arrow on the left of the gel marks the free DNA), or with *Eco*RV restriction endonuclease at the concentrations (nM) indicated above each lane: The binding buffer contained EDTA. The DNA contains one copy of the *Eco*RV recognition sequence but, instead of observing a single DNA-protein complex, the *Eco*RV restriction enzyme produces a series of complexes. The first retarded band is the DNA bound to 1 molecule of enzyme, the second with 2 molecules of enzyme, and so forth. In the absence of Mg^{2+} ions, to prevent cleavage of the DNA, this restriction enzyme binds all DNA sequences, including its recognition site, with the same equilibrium constant. (Data from ref. *17*: Reprinted with permission from *Biochemistry*. Copyright [1991] American Chemical Society).

ties (*see* Fig. 2). The stoichiometry of each complex can be determined: The change in mobility from one complex to the next is usually proportional to the change in molecular weight caused by adding one more molecule of protein *(2,16,17)*. A more rigorous, although technically demanding, approach to determining stoichiometries involves labeling the protein with ^{14}C and the DNA with ^{32}P so that the molar ratios, in each band on the gel, can be measured directly *(9,18)*. The gel can also be stained for protein, with Coomaasie blue or silver, and another approach is quantitative Western blotting *(19)*. The ability of gel shifts to resolve DNA species multiply bound by one protein (as in Fig. 2), or DNA bound by two or more different pro-

Gel Shift Assay 267

teins *(3,7),* is one of the major advantages of this technique over many other methods for DNA–protein interactions. For example, in filter-binding *(see* Chapter 20), retention can be established by just one molecule of protein binding to the DNA. Moreover, gel shift assays can also resolve complexes having the same stoichiometry but different conformations, in which the DNA in the complex is either looped *(5–7,20)* or bent *(21,22; see also* Chapter 22).

1.1. Choice and Preparation of DNA Fragment

If specific binding is to be investigated and the potential binding sequence is known, the first requirement is a DNA fragment of suitable length that contains the site. It is not essential to radiolabel the DNA. After a gel shift experiment, the DNA can be visualized by staining the gel with ethidium bromide *(4).* However, it is preferable to use low DNA concentrations, lower than can be detected with ethidium. This necessitates radiolabeled DNA followed by autoradiography. There is no minimum length of DNA below which it cannot be used for gel shifts, provided that it contains the complete binding site, although there is effectively a maximum length. If the DNA is >500 bp, only a small shift in mobility is caused by binding one molecule of protein, unless the protein is very large (such as DNA gyrase or RNA polymerase). However, some DNA–protein interactions involve several proteins associating with an array of binding sites and, in these cases, longer DNA fragments can be used in gel shift experiments. For the electrophoresis of large nucleoprotein assemblies, agarose, rather than polyacrylamide, is the preferred matrix *(23).* Mixtures of polyacrylamide and agarose have also been used as the matrix *(24).*

The minimum length of DNA will need to be determined experimentally, but usually it will be the target sequence and a few bp on either side. Short DNA molecules can be readily made in an oligonucleotide synthesizer and the synthetic oligonucleotides should be purified by electrophoresis or HPLC. Alternatively, if the target sequence is not known, it can be identified by gel shift experiments on oligonucleotides of random sequence *(25; see also* Chapter 23). Larger DNA fragments, 100–500 bp, are generally prepared by either restriction digests or PCR reactions on plasmid vectors. They are then isolated from the vector by electrophoresis through polyacrylamide

(26). If a restriction fragment is to be used, it is best prepared from a plasmid that has been purified on CsCl/ethidium gradients and it is advantageous to use restriction enzymes that leave 5'-extensions. These ends can be filled in by the Klenow fragment of DNA polymerase I in the presence of the appropriate α-[^{32}P]dNTP, with the other dNTPs as required *(26)*. A less attractive alternative is to dephosphorylate the DNA with alkaline phosphatase and then label it with polynucleotide kinase and γ-[^{32}P]ATP. The labeled DNA *must* be separated from the unincorporated label by gel filtration or by repeated precipitations of the DNA with ethanol *(26)*.

Virtually all gel shift experiments require not only the substrate DNA but also other DNA samples. As noted above, the specificity of a DNA–protein interaction can be examined by adding unlabeled competitor DNA to the radiolabeled substrate *(3,4,14)*. The competitor should not contain any copies of the specific binding site for the protein. A nonspecific DNA of known sequence is one possibility: Phage λ DNA and many plasmids may meet this requirement. Sheared *E. coli* or calf thymus DNA have also been used, as have polynucleotides and heparin *(12,27)*. Unnatural polynucleotides with simple repeat sequences, such as poly d(I-C), are particularly useful *(27)*. The amount of competitor DNA that needs to be added to remove nonspecific bands will depend on the specificity of the interaction. On the addition of a large excess of cold DNA, even specific complexes will dissociate because the sum of the affinities for all of the nonspecific binding sites will then be greater than the affinity for the specific site *(28)*. Another test for specificity is to use an isogenic DNA fragment that differs from the original fragment by 1 or 2 bp *(9,17)*.

Certain experiments demand further variations in the DNA fragments. For example, in the analysis of DNA bending by gel shift, a permuted series of DNA fragments will be needed, all of the same length but with the binding site for the protein at different locations relative to the center of the DNA molecule (*see* Chapter 22). Similarly, in the study of DNA looping interactions, where one protein binds simultaneously to two separate sites on the DNA, a series of DNA fragments may be needed with different lengths of DNA between the two sites *(5,20)*.

2. Materials

1. Standard apparatus for the electrophoresis of polyacrylamide gels and an electrophoresis dc power supply capable of 200 V. If possible, do not use apparatus that has been used previously for SDS-PAGE on proteins (*see* Note 1).
2. Electrophoresis running buffer (TBE): $0.089M$ Tris base, $0.089M$ boric acid, and 2 mM EDTA. Make this up at 5 times the required strength from 54 g of Tris base, 27.5 g of boric acid, 20 mL of $0.5M$ EDTA, pH 8.0, and water to a final volume of 1 L: Dilute fivefold with water before use. (All water throughout this procedure is double-distilled; *see* Note 2.)
3. Acrylamide (29/1 acrylamide / N',N'-methylene *bis*-acrylamide). Make up a stock solution at 30% (w/v) from 290 g of acrylamide and 10 g of *bis*-acrylamide with water to bring the final volume to 1 L, then stir until dissolved. Add a couple of spoonfuls of a mixed-bed resin (such as Amberlite MBI, BDH), leave stirring for 30 min, filter, and store at 4°C (*see* Note 3). Also needed for the polyacrylamide gel are TEMED (N,N,N',N'-tetramethylene diamine), and a freshly prepared solution of ammonium persulfate (10% w/v in water).
4. ^{32}P-radiolabeled substrate DNA, and additional DNA samples as required for the type of gel shift experiment being carried out (*see* Section 1.1.). Depending on the nature of the substrate, the method chosen for its preparation and for the incorporation of the label, a wide range of the standard reagents and equipment for molecular biology will be needed at this stage (either the vector DNA with the fragment of interest and restriction endonucleases, or synthetic oligonucleotides, Klenow polymerase, both unlabeled and [^{32}P]-labeled nucleotide triphosphates, and so on). Standard procedures *(26)* can be followed in the preparation of both the substrate and control DNA samples.
5. The DNA-binding protein. *(See* Note 4.)
6. Binding buffer: For example, 50 mM Tris-HCl, pH 7.5, 100 mM NaCl, 10 mM β-mercaptoethanol, 0.1 mM EDTA, and 500 µg/mL of BSA (bovine serum albumin). To prepare this, make up a solution of Tris-HCl, NaCl, and EDTA, all at twice the required concentrations, and sterilize it by autoclaving. Once cool, add the β-mercaptoethanol to again give twice the required concentration: This solution is described below as 2X binding buffer. Prepare separately a stock solution of nuclease-free BSA at 5 mg/mL by first dissolving the protein (from any supplier) in 10 mM Tris-HCl, 0.1 mM EDTA, pH 7.5; readjust the pH to 7.5 and heat the solution to 67°C for at least 2 h; store at 4°C (*see* Note 5).

7. Loading buffer: The buffer added to samples prior to loading the gel is the same as binding buffer but supplemented with 40% (w/v) sucrose and 100 µg/mL bromophenol blue (*see* Note 6).
8. Gel fixer solution (10% v/v glacial acetic acid, 10% v/v methanol, 80% water), 3MM chromatography paper (Whatman), and a gel drier.
9. Autoradiography cassettes, preflashed X-ray film (Hyperfilm MP, Amersham International, Arlington Heights, IL), and developer for the autoradiograph (*see* Note 7).
10. If the autoradiographs are to be assessed quantitatively, a scanning densitometer may be required.

3. Methods
3.1. The Gel Shift Assay

1. Make the gel well in advance of performing the actual assay. First, both the gel tank and its component parts (glass plates, spacers, and comb) must be rigorously cleaned and dried, especially if the equipment has been used previously for SDS-PAGE. The gel plates should be cleaned with 95% ethanol and one plate should be "silanized" by coating it with a solution of 1% v/v dichlorodimethlysilane in $CHCl_3$ and allowing this to evaporate in the fume chamber.
2. Prepare a polyacrylamide gel in the appropriate buffer (*see* Note 2) by the standard procedure *(26)* (*see* Note 8). Use a comb that creates wells that are substantially larger than the volume of the sample so that the sample, once loaded in the well, occupies the minimal height above the gel matrix, that is, wide wells are better than narrow ones. The reason for this is that it facilitates the transfer of the DNA-protein complexes from the sample solution into the gel matrix, and it also minimizes streaking from the edges of the bands.
3. Allow the polyacrylamide gel to set for at least 2 h. However, it is advisable to remove the comb as soon as the gel has set, and thoroughly wash the wells with buffer. Then mount the gel in the gel tank and fill the chambers with electrophoresis buffer.
4. Prerun the gel at constant voltage, at the same voltage as will be used in the subsequent assay. The minimum length of time for prerunning is until the current becomes invariant with time. The gels are prerun first to remove all traces of the ammonium persulfate that was used in the polymerization of the gel; second, to distribute evenly throughout the gel any cofactors that may have been added to the electrophoresis buffer (especially if these were reagents that could not be added to the buffer used to make up the gel); and third, and most important, to ensure that the gel is at a constant temperature. The latter is absolutely essential for

gels run in the coldroom: There is no point in working in the coldroom unless both the gel and the buffers are at thermal equilibrium before the samples are loaded. Just before loading the samples (*see below*, step 8), layer a solution of binding buffer across the wells in order to minimize the change in conditions that the sample undergoes when loaded on the gel.

5. Use a series of sterile 0.5-mL Eppendorf tubes for the binding reactions. Each reaction will have a final volume of 20 µL. In each tube, place 10 µL of 2X binding buffer, 2 µL of nuclease-free BSA (stock at 5 mg/mL), the desired amount of the [^{32}P]-labeled DNA substrate, and, if needed, the competitor DNA. In general, the amount of the radiolabeled DNA should be the minimum that gives a band that can be observed readily on the autoradiograph (this can be as little as 0.1 ng/tube). Before use, the stocks of both substrate and competitor DNA should be diluted with buffer so that the addition of the DNA causes the minimal change to the composition of the binding reaction: 10 mM Tris-HCl, 0.1 mM EDTA, pH 7.5, can be used for this dilution. Add sterile water so that the volume in each tube, after the subsequent addition of the protein, is 20 µL. Incubate the tubes in a water bath at the required temperature for a few minutes.

6. Add the DNA-binding protein to all but one of the tubes. Ideally, this should be the same small volume to each tube (typically, <10% of the assay volume), but at different dilutions of the protein (*see* Note 9). To the tube to which no protein was added, add the same volume of the buffer used for diluting the protein. The protein should obviously be diluted in a buffer in which it remains stable. However, less obvious is the fact that some DNA-binding proteins seem to be inactivated by "dilution shock" on the addition of a small volume of protein to a large volume of buffer. In these cases, use serial dilutions with small steps between each stage.

7. Thoroughly mix the solution of DNA and protein. This can be done by vortexing, but sometimes DNA-protein complexes that are observed by gel shift after gentle mixing cannot be detected after vortex mixing *(4)*. An alternative is to flush the sample repeatedly in a micropipet tip. The binding reaction is then allowed to equilibrate at the required temperature (*see* Note 10).

8. Add 8 µL of loading buffer to each sample and immediately load an aliquot (15-µL) from each on the polyacrylamide gel. The additional loading buffer, the loading itself, and starting the electrophoresis all need to be done as quickly as possible. One procedure is to start by gently placing the loading buffer on top of all of the separate samples. The buffer is viscous and will maintain its position above each sample for a short time. The individual tubes are then tapped to mix their con-

tents immediately before loading. The loading is best done with plastic tips on a micropipet, using special (flexible) gel-loading tips if necessary to ensure an even layer across the bottom of each well. To minimize the time that the complexes take to enter the gel, some workers load the gel while it is running at a very low voltage. However, you can be electrocuted if you follow their procedure (see Note 11).

9. Turn up the electrophoresis power supply as quickly as possible after loading the samples on the gel. Run the gel, typically at 10 V/cm, until the free DNA is close to the bottom of the gel (see Note 12). The best voltage for a particular application, and the time of the run, are determined by trial and error. However, running the gel at too high a voltage will heat the gel excessively and denature the protein. Recirculation of the electrophoresis buffer can be advantageous, especially with buffers of low capacity, and it can be essential if cofactors are included in the buffer (see Note 2). The buffer should be recirculated at a flow rate of 50% of the total buffer vol/h.
10. After the gel is run, dismantle the apparatus. Remove the gel from the plates, place it in a bath of gel fixer for 30 min, then layer it onto Whatman 3MM paper and dry it on the gel dryer. The gel is finally autoradiographed with preflashed X-ray film.
11. If quantitative data are required, scan the autoradiograph in a densitometer. Determine the proportion of the free DNA and the bound complexes, relative to the total amount of DNA in each sample, by measuring the area of each peak in the densitometric scans. See Note 13.

3.2. Analysis of Results

If the gel shift assay reveals a single DNA–protein complex as a result of the binding of the protein to one specific site on the DNA (as in Fig. 1), the binding reaction between DNA (D) and protein (P) to form the complex (DP) can be described by:

$$D + P \leftrightarrow DP$$

The concentrations of bound and free DNA that are measured directly by the gel shift assay, $[D_b]$ and $[D_f]$ respectively, can then be related to an equilibrium constant:

$$K = [D_b]/\{[D_f] \times [P_o - D_b]\}$$

where $[P_o]$ is the total concentration of protein in the assay. Provided that the protein concentration is much higher than the concentration of the DNA substrate, so that $[P_o - D_b]$ is effectively equal to $[P_o]$, the increase in the amount of bound DNA with increasing levels of protein should follow a rectangular hyperbola. Hence, K can be evaluated from the variation in $[D_b]$ with $[P_o]$ either graphically (by an Eadie plot; $[D_b]$ against $[D_b]/[P_o]$) or, better, by using a computer program that fits data directly to rectangular hyperbolas (for example, ENZFITTER from Biosoft, Cambridge, UK). However, if any dissociation of the DNA-protein complex occurs during electrophoresis, the value for $[D_b]$ that is measured from the amount of DNA in the retarded complex (i.e., band B in Fig. 1) will be less than the true value for the complex. However, the value for $[D_f]$ measured from the amount of free DNA (i.e., band F in Fig. 1), will still be the true value: DNA that dissociates from the protein during the run cannot "catch up" with the DNA that was free at the start of the run. In this situation, it is better to evaluate K from the decrease in $[D_f]$ rather than the increase in $[D_b]$.

If the gel shift assay reveals multiple DNA-protein complexes (as in Fig. 2), it may be a result of nonspecific binding of the protein anywhere along the DNA. One test for this type of binding is to repeat the gel shift assay with DNA fragments of different lengths: The number of complexes resolved by gel shift should increase as the length of the DNA is increased. For example, the *Eco*RV restriction enzyme produces three complexes with a 55 bp DNA (the DNA bound to 1, 2, or 3 molecules of protein), six complexes with a 100 bp DNA, and >12 complexes with a 235 bp DNA *(17)*. The binding equilibria are then given by the equation:

$$D + P \leftrightarrow DP_1 \leftrightarrow DP_2 \leftrightarrow DP_3 \ldots DP_n$$

where n is the maximum number of protein molecules that can fit on the DNA. However, even though the concentrations of D_f, DP_1, DP_2, through to DP_n can all be measured by gel shift assays, the intrinsic equilibrium constant for the binding of the protein to an individual site on the DNA cannot be determined by the procedures used above for analyzing specific binding.

The analysis of nonspecific binding must take account of the following *(28)*: First, the DNA will contain a large number of binding sites for the protein (each bp is in effect the start of another binding site); second, since each molecule of protein will cover a certain number of bp, the sites overlap one another and only a fraction of them can be filled at any one time; third, the number of sites that are left after the binding of one or more molecules of protein depends on the location of the protein on the DNA (i.e., is the distance between the protein and the end of the DNA, or the distance between two protein molecules, long enough to accommodate another molecule of protein?); and fourth, if there are cooperative interactions between protein molecules on the DNA, the protein may bind preferentially to a site immediately adjacent to one already occupied by protein. One procedure for evaluating equilibrium constants from the concentrations of all of the complexes separated by gel shift has recently been published *(17).*

4. Notes

1. Almost any apparatus for polyacrylamide gel electrophoresis, either homemade or commercial, can be used for gel shift assays. However, it is best to have one where the gel plates are virtually immersed in running buffer (for example, the Mini-Protean II system from Bio-Rad, Richmond, CA), to prevent localized overheating in the gel. Additional features that are required for some, but not all, gel shift experiments are buffer recirculation and thermostatic control of the gel surface. However, the latter is not essential even in the analysis of complexes that are stable only at low temperatures. Running the gel in a cold room is often an adequate substitute.

2. The electrophoresis running buffer is a key factor in successful gel shift assays. Ideally, one wants a buffer in which the DNA–protein complex is so stable that none of it dissociates during electrophoresis. In many cases, the transfer of the DNA–protein complex from a binding buffer containing NaCl to a gel matrix in TBE achieves this goal. Other systems need electrophoresis buffers with lower ionic strengths, for example, 0.04M Tris-acetate, 1 mM EDTA, pH 8.0. Moreover, the separation of the DNA–protein complex from the free DNA can depend on the pH of the electrophoresis buffer *(9).* In addition, the buffer may require additional components to maintain the stability of the complex. For instance, a stable complex with a restriction enzyme demands the absence of Mg^{2+} ions, which can be ensured by having EDTA in the buffer *(17),* whereas complexes with DNA gyrase are seen by gel shift only if the buffer contains Mg^{2+} *(29).* Many proteins bind a cofactor before they bind specifically to DNA (for example, tryptophan with the *trp* repres-

sor *[9]*), and these require their cofactor in the electrophoresis buffer. Buffer recirculation is then needed to maintain a constant concentration of cofactor throughout the gel.
3. *Acrylamide is a potent neurotoxin.* Always wear gloves when working with acrylamide and, when handling the dry powder, a face mask is also essential.
4. For certain types of gel shift experiments, it is essential that the protein is purified to homogeneity. One example is the analysis of the DNA sequence-specificity of a DNA-binding protein by using gel shifts to measure its binding to a series of DNA fragments that differ from each other by only 1 or 2 bp within the target sequence. Major artifacts can be generated by using this approach with partially purified preparations of the protein (*see* ref. *30* for the elimination of one such artifact). But for other experiments, such as the identification of a protein that binds to a given DNA sequence, it is unnecessary to purify the protein. Indeed, as noted above (Sections 1. and 1.1.), the latter experiments can be carried out with unfractionated nuclear extracts *(14,27),* provided that the target DNA is mixed with a suitable competitor, such as poly d(I-C).
5. The buffer for the binding reaction will obviously vary from one DNA-binding protein to another. Nearly all DNA–protein interactions are highly dependent on the salt concentration *(31);* they can also vary with pH, and they often require some cofactor that must be added to the binding buffer as well as the electrophoresis buffer (*see above,* Note 2). The stability of the protein in different buffers will also influence the choice of binding buffer; in some cases, glycerol or spermidine must be added as well as BSA, in order to stabilize the protein.
6. The purpose of adding sucrose to binding buffer, to create loading buffer, is simply that this is the minimal perturbation to increase the density of the solution. This is to facilitate loading into the wells of the polyacrylamide gel and to prevent the binding reaction from mixing with electrophoresis buffer before the sample has entered the gel. The sucrose can be replaced by either glycerol or Ficoll400 (Pharmacia-LKB, Piscataway, NJ), and it is worthwhile using these to check that the binding equilibrium is not perturbed by sucrose. Moreover, if the binding buffer contains glycerol (*see* Note 5), it may be possible to dispense with loading buffer and load the samples directly on to the gel. Bromophenol blue is just a marker for electrophoresis and is not really necessary: In gel shift assays, electrophoresis should be continued until the free DNA is close to the end of the gel, and this must determined experimentally. The size of the DNA that comigrates with bromophenol blue will vary with the percentage of polyacrylamide in the gel *(26).*

7. If a storage phosphor system *(32)* is available, use it in place of film autoradiography. It will provide quantitative data directly, by measuring the amount of radioactivity in each band on the gel, and it avoids the problems associated with nonlinear film response (*see* Note 13).
8. The gel strength (% acrylamide) for optimal resolution of complexes can only be found by trial and error. It will be a function of the sizes of both the DNA and the protein. For DNA fragments of >200 bp, initial experiments should be done with 5 or 6% gels. With short synthetic oligonucleotides, gels at 10 or 15% may be better, provided that the protein is not so large that the complex then cannot enter the gel. Avoid situations in which the complex sticks in the well—it should migrate a measurable distance into the gel. Another factor that can affect the mobility of DNA–protein complexes is the ratio of acrylamide to *bis*-acrylamide, and it may be necessary to alter the 29/1 ratio given here (*see* Section 2., step 3).
9. If the binding assay is carried out as described here, by titrating a fixed amount of DNA with varied amounts of protein, then the range of protein concentrations should extend from the situation in which very little of the DNA is complexed with protein to that in which essentially all of the DNA is complexed (for example, Fig. 2). As a first experiment, it is best to assay over a wide range of protein concentrations. Subsequent assays, once shifts have been observed, can then be carried out over a narrower range.
10. For kinetic experiments, the samples are subjected to electrophoresis as soon as the relevant time has elapsed after mixing the DNA with the protein. Generally, the association rates of proteins to specific sites on DNA are extremely rapid *(31)*, and, for equilibrium experiments, an incubation time of 15 min is usually sufficient. However, it is important to test that the system has reached equilibrium by carrying out assays with different incubation times. In some cases, it may be helpful to first incubate the sample at a higher temperature (e.g., 37°C) before reequilibrating it at a lower temperature.
11. If you chose to load the gel while it is running at low voltage, *take every possible precaution to avoid electrocution.* If it is to be attempted at all, wear dry plastic gloves and use only plastic tips. The voltage across the gel *must* be <1V/cm. *Do not, under any circumstances, attempt to load a gel at too high a voltage.*
12. The sample to which no protein has been added gives one marker for the free DNA. But in preliminary experiments, it is also advisable to run in a separate lane some DNA size markers to act as standards. Observed shifts can then be compared between different experiments.

13. If the autoradiographs are to be analyzed by densitometry, it is essential to work in the range where the film gives a linear response to the amount of radioactivity. This range can be established by running on a gel different amounts of a known sample of [^{32}P]-labeled DNA. On preflashed film, the areas under the peaks in the densitometric scans increase linearly with increasing amounts of DNA, but only up to a certain amount of DNA. Beyond that amount, successive increments of DNA produce proportionally smaller increments in peak areas. If the film is not preflashed, the scale will also be nonlinear with very low amounts of DNA. Only those band intensities that lie within the linear range should be used to obtain quantitative data. One way to ensure that the data are within the linear range is to autoradiograph the same gel for a number of different exposure times. Strong black bands from a long exposure allow for easy visualization, but densitometry generally requires faint bands from a short exposure.

References

1. Revzin, A. (1987) Gel electrophoresis assays for DNA-protein interactions. *BioTechniques* **7**, 346–355.
2. Fried, M. (1989) Measurement of protein-DNA interaction parameters by electrophoresis mobility shift assay. *Electrophoresis* **10**, 366–376.
3. Garner, M. and Revzin, A. (1981) A gel electrophoresis method for quantifying the binding of proteins to specific DNA regions: application to components of the *Escherichia coli* lactose operon regulatory system. *Nucleic Acids Res.* **9**, 3047–3060.
4. Fried, M. and Crothers, D. M. (1981) Equilibria and kinetics of lac repressor-operator interactions by polyacrylamide gel electrophoresis. *Nucleic Acids Res.* **9**, 6505–6525.
5. Krämer, H., Amouyal, M., Nordheim, A., and Müller-Hill, B. (1988) DNA supercoiling changes the spacing requirement of two *lac* operators for DNA loop formation with lac repressor. *EMBO J.* **7**, 547–556.
6. Lobell, R. B. and Schleif, R. F. (1990) DNA looping and unlooping by *ara*C protein. *Science (Washington, DC)* **250**, 528–532.
7. Fried, M. and Crothers, D. M. (1983) CAP and RNA polymerase interactions with the *lac* promoter: binding stoichiometry and long range effects. *Nucleic Acids Res.* **11**, 141–148.
8. Fried, M. and Crothers, D. M. (1984) Kinetics and mechanism in the reaction of gene regulatory proteins with DNA. *J. Mol. Biol.* **172**, 263–282.
9. Carey, J. (1988) Gel retardation at low pH resolves *trp* repressor-DNA complexes for quantitative study. *Proc. Natl. Acad. Sci. USA* **85**, 957–979.
10. Ceglarek, J. A. and Revzin, A. (1989) Studies of DNA-protein interactions by gel electrophoresis. *Electrophoresis* **10**, 360–365.
11. Revzin, A., Ceglarek, J., and Garner, M. (1986) Comparison of nucleic-acid interactions in solution and in polyacrylamide gels. *Anal. Biochem.* **153**, 172–177.

12. Shanblatt, S. and Revzin, A. (1984) Kinetics of RNA polymerase-promoter complex formation: effects of nonspecific DNA-protein interactions. *Nucleic Acids Res.* **12**, 5287–5306.
13. Avila, P., Ackroyd, A. J., and Halford, S. E. (1990) DNA binding by mutants of Tn21 resolvase with DNA recognition functions from Tn3 resolvase. *J. Mol. Biol.* **216**, 645–655.
14. Strauss, F. and Varshavsky, A. (1984) A protein binds to a satellite DNA repeat at three specific sites that would be brought into mutual proximity by DNA folding in the nucleosome. *Cell* **37**, 889–901.
15. Kristie, T. M. and Roizman, B. (1986) α4, the major regulatory protein of herpes simplex virus type 1, is stably and specifically associated with promoter-regulatory domains of α genes and of selected other viral genes. *Proc. Natl. Acad. Sci. USA* **83**, 3218–3222.
16. Hudson, J. M., Crowe, L., and Fried, M. (1990) A new DNA binding mode for CAP. *J. Biol. Chem.* **265**, 3219–3225.
17. Taylor, J. D., Badcoe, I. G., Clarke, A. R., and Halford, S. E. (1991) *Eco*RV restriction endonuclease binds all DNA sequences with equal affinity. *Biochemistry* **30**, 8743–8753.
18. Hendrickson, W. and Schleif, R. (1985). A dimer of *ara*C protein contacts three adjacent major groove regions of the *ara*I DNA site. *Proc. Natl. Acad. Sci. USA* **82**, 3129–3133.
19. Hoess, R., Abremski, K., Irwin, S., Kendall, M., and Mack, A. (1990). DNA specificity of the *cre* recombinase resides in the 25 kDa carboxyl domain of the protein. *J. Mol. Biol.* **216**, 873–882.
20. Krämer, H., Niemöller, M., Amouyal, M., Revet, B., von Wilcken-Bergmann, B., and Müller-Hill, B (1987) *Lac* repressor forms loops with linear DNA carrying two suitably spaced *lac* operators. *EMBO J.* **6**, 1481–1491.
21. Wu, H-M. and Crothers, D. M. (1984) The locus of sequence directed and protein-induced DNA bending. *Nature* **308**, 509–513.
22. Liu-Johnson, H-N., Gartenberg, M. R., and Crothers, D. M. (1986) The DNA binding domain and bending angle of *E. coli* CAP protein. *Cell* **47**, 995–1005.
23. Berman, J., Eisenberg, S., and Tye, B. K. (1987) An agarose gel electrophoresis assay for the detection of DNA-binding activities in yeast cell extracts. *Meth. Enzymol.* **155**, 528–537.
24. Topol, J., Ruben, D. M., and Parker, C. S. (1985) Sequences required for in vitro transcriptional activation of a *Drosophila* hsp 70 gene. *Cell* **42**, 527–537.
25. Pollock, R. and Treisman, R. (1990) A sensitive method for the determination of protein-DNA binding specificity. *Nucleic Acids Res.* **18**, 6197–6204.
26. Sambrook, J., Fritsch, E. F., and Maniatis, T. (1989) *Molecular Cloning, A Laboratory Manual*, 2nd Ed. Cold Spring Harbor Laboratory, Cold Spring Harbor, NY.
27. Varshavsky, A. (1987) Electrophoretic assay for DNA-binding proteins. *Meth. Enzymol.* **151**, 551–565.
28. McGhee, J. D. and von Hippel, P. H. (1972) Theoretical aspects of DNA-protein interactions: cooperative and noncooperative binding of large ligands to a one-dimensional homogeneous lattice. *J. Mol. Biol.* **86**, 469–489.

29. Maxwell, A. and Gellert, M. (1984) The DNA dependence of the ATPase activity of DNA gyrase. *J. Biol. Chem.* **259,** 14,472–14,480.
30. Carey, J., Lewis, D. E. A., Lavoie, T. A., and Yang, J. (1991) How does *trp* repressor bind to its operator? *J. Biol. Chem.* **266,** 24,509–24,513.
31. Lohman, T. M. (1986) Kinetics of protein-nucleic acid interactions: use of salt effects to probe mechanisms of interaction. *CRC Crit. Rev. Biochem.* **19,** 191–245.
32. Johnston, R. F., Pickett, S. C., and Barker, D. L. (1990) Autoradiography using storage phosphor technology. *Electrophoresis* **11,** 355–360.

CHAPTER 22

Improved Plasmid Vectors for the Analysis of Protein-Induced DNA Bending

Christian Zwieb and Sankar Adhya

1. Introduction

Bending of DNA by proteins plays an important role in transcription initiation, DNA replication, and recombination. The degree of protein-induced DNA bending is conveniently determined by combining gel retardation techniques with the use of so-called bending vectors *(1,2)*. Bending vectors contain duplicated circular permuted restriction sites and cloning sites for insertion of protein binding sequences. Restriction enzyme digestion readily generates fragments that are identical in size, but differ in the location of the binding site (Fig. 1).

The mobility of a DNA fragment is less when a bend is located in the middle, compared to complexes with a bend closer to the ends *(3,4)*. The bending angle α is defined as the angle by which a segment of the rod-like DNA duplex departs from linearity; therefore, 0° is the value for a straight DNA fragment. It is possible to estimate α by measuring μ_M (mobility of the complex with the protein bound at the *m*iddle of the fragment) and μ_E (mobility of the complex with the protein bound near the *e*nd of a DNA fragment) using an empirical relationship $\mu_M/\mu_E = \cos(\alpha/2)$ *(4,5)*.

To carry out a bending experiment, the protein binding site is inserted into the bending vector. Next, DNA fragments with the binding site located either in the middle or at the end are generated by digestion of

From: *Methods in Molecular Biology, Vol. 30: DNA–Protein Interactions: Principles and Protocols*
Edited by: G. G. Kneale Copyright ©1994 Humana Press Inc., Totowa, NJ

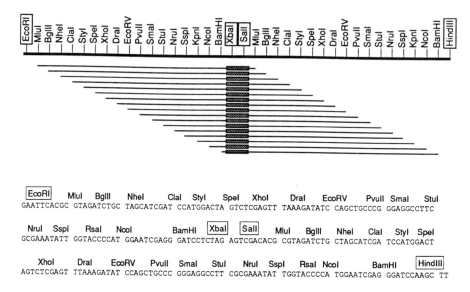

Fig. 1. Schematic representation of the pBend3-insert between the EcoRI–HindIII sites. pBend3 was constructed by cloning of the 236 bp EcoRI HindIII fragment of pBend2 (4) into pBluescript SK- (Stratagene). pBluescript is a high copy number plasmid and generates a large amount of DNA on plasmid extraction. The EcoRI–HindIII fragment contains 17 duplicated restriction sites. The duplicated sites can be used to generate DNA fragments of identical length, but in which the protein binding sequence (gray rectangle in the upper part) is shifted. The sites XbaI and SalI (in boxes) are unique and suitable for cloning of the protein binding sequence. Restriction sites are not drawn to scale. The sequence of the insert is shown in the lower part.

the vector with different restriction enzymes. Finally, DNA–protein complexes are analyzed by gel electrophoresis to determine the bending angle α. As an example, we illustrate bending at *gal* operator sites by Gal repressor *(2,6)*. However, the method is easily adopted to other protein–DNA complexes if they can be analyzed by the gel retardation technique. General considerations for successful complex formation and for the selection of restriction sites are discussed in the Notes.

pBend3, 4, and 5 (Figs. 1 and 2), are improved versions of the previously described bending vector pBend2 *(4)*. pBend2 and pBend3 contain the same 236-bp *EcoRI–HindIII*-fragment with 17 duplicated restriction sites. Because of the higher copy number of pBend3, preparation of plasmid DNA is more efficient. The digested vector is less likely to comigrate

DNA-Bending Vectors

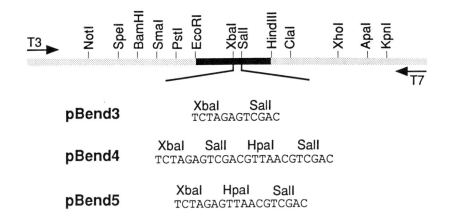

Fig. 2. Restriction- and cloning sites of pBend3, 4, and 5. Portions of pBluescript SK- (Stratagene) are shown in gray; the XbaI and SalI sites of pBluescript are abolished by partial digestion with XbaI or SalI, filling-in with DNA polymerase (Klenow) and ligation with T4 DNA ligase. Promoters for T3 and T7 polymerase are indicated by the arrows. They can be used to study the bending by proteins that bind to double-stranded RNA or to RNA/DNA hybrids. The region between the EcoRI- and HindIII-sites (indicated in black) is identical to the one shown in Fig. 1. pBend4 and pBend5 contain additional SalI- and HpaI-sites.

with protein–DNA complexes; therefore, larger DNA binding proteins with a high degree of bending can be analyzed without tedious purification of individual DNA fragments. pBend4 and pBend5 contain an additional HpaI cloning site to facilitate the insertion of DNAs with blunt ends. The promoters for T3- and T7-polymerase have potential use for analyzing the bending of double-stranded RNA or RNA/DNA hybrids by proteins, such as transcription factor TFIIIA (7). Details of the construction of pBend3, 4, and 5 are described in the legends to Figs. 1 and 2.

2. Materials

2.1. Insertion of the Protein Binding Site into pBend

1. pBend-vector (provided by the authors on request).
2. Oligonucleotides or DNA fragments containing the investigated protein binding site, compatible with the cloning sites of pBend.
3. Restriction enzymes: HpaI, EcoRI, HindIII, and MluI at about 10 U/μL with 10X concentrated digestion buffers as specified or provided by the vendors.

4. 500 mM EDTA, pH 8.0: Mix 93 g of disodium ethylene diamine tetraacetate–2H$_2$O in 400 mL water, adjust pH to 8.0 by adding about 20 g of NaOH pellets (only then will EDTA dissolve completely), adjust volume to 500 mL, and autoclave.
5. TE: 10 mM Tris-HCl, pH 7.5, 1 mM EDTA.
6. 7.5M Ammonium acetate.
7. 80% Ethanol. Store at 4°C.
8. Oligonucleotide annealing buffer: 10 mM Tris-HCl, pH 7.5, 1 mM EDTA, 100 mM NaCl.
9. 5X Ligation buffer: 500 mM Tris-HCl, pH 7.6, 100 mM MgCl$_2$, 100 mM DTT. Store at –20°C.
10. 5 mM ATP: Dissolve 3 mg of ATP (disodium salt) in 1 mL of TE.
11. T4 polynucleotide kinase: 10 U/µL.
12. T4 DNA ligase: 1 U/µL.
13. Competent *E. coli* cells (e.g., strain DH5α).
14. LB-amp plates: Suspend 10 g of LB powder and 7.5 g of Bacto-agar in 500 mL of water; autoclave and dissolve agar by swirling, cool solution to 55°C in a waterbath, add 50 mg of ampicillin (for final concentration of 100 µg/mL), dissolve by swirling, and pour plates. Store plates in a plastic bag at 4°C.
15. LB-media: Suspend 10 g of LB powder in 500 mL water and autoclave. Store at room temperature. For the ampicillin-containing LB, add and dissolve the specified amount to the media at room temperature.
16. Tris-sucrose: 5mM Tris-HCl, pH 7.8, containing 10% (w/v) sucrose (prepare fresh).
17. Lysozyme solution: 10 mg/mL in 250 mM Tris-HCl, pH 8.0, keep frozen aliquots at –20°C. Thaw once immediately before use and discard unused portions.
18. 200 mM EDTA, pH 8.0.
19. TLM: 150 mM Tris-HCl, pH 8.0, 200 mM EDTA, 0.3% (v/v) Triton X-100.
20. Phenol: Add 62.5 mL water to bottle with 250 g phenol, mix, and warm as little as possible. Add 300 mg 8-hydroxyquinoline and fill 20-mL aliquots in 30-mL Falcon tubes. On use, thaw and add 1 mL of 1M Tris-base to one aliquot. Keep refrigerated not longer than 1 mo.
21. RNaseA: 250 µg/mL in 10 mM Tris-HCl, pH 7.5.
22. 20X Tris-acetate: 96.8 g of Tris-base, 22.84 mL glacial acetic acid, 40 mL of EDTA (500 mM, pH 8.0), dissolved in 1 L of water (final volume).
23. 2% Agarose gel, about 5 mm thick, 7 cm long, and 10 cm wide: Mix 0.6 g agarose (electrophoresis grade), 1.5 mL of 20X concentrated Tris-acetate, and 30 mL of water in a 100-mL reagent bottle. Be sure that the

DNA-Bending Vectors

cap of the bottle is loose before the agarose in the mixture is melted in a microwave oven. Swirl the mixture occasionally and dissolve the agarose completely. Adjust the volume with water to 30 mL and pour the gel. Insert a comb for about 3-mm wide slots and let the agarose solidify. Cover the gel with Tris-acetate electrophoresis buffer containing 1 µg/mL ethidium bromide (CAUTION: Ethidium bromide is mutagenic).

24. 10 mg/mL Ethidium bromide solution: Dissolve 1 g of ethidium bromide in 100 mL of water by stirring for several hours. Store in the dark at 4°C.
25. Agarose loading buffer: Mix 1 vol of 50% glycerol, 1 vol of Tris-acetate electrophoresis buffer, and 1/10 vol of a 2.5% (w/v) bromophenol blue solution in TE.
26. DNA molecular weight markers in the range of 100–1000 bp, e.g., HaeIII-digest of ΦX174.
27. UV transilluminator.
28. Horizontal electrophoresis apparatus for agarose gel, approximate dimensions 7 × 10 cm.

2.2. Purification of pBend DNA

1. *E. coli* suspension buffer: 50 mM Tris-HCl, pH 8.0, 100 mM NaCl.
2. Tris-glucose-EDTA: 25 mM Tris-HCl, pH 8.0, 50 mM glucose, 10 mM EDTA. Prepare freshly by mixing and dissolving 2.5 mL 1M Tris-HCl, pH 8.0, 1 g glucose, and 2 mL 500 mM EDTA, pH 8.0, in a total volume of 100 mL of water.
3. NaOH-SDS: 200 mM NaOH, 1% SDS. Prepare by mixing 10 mL of 10% SDS and 2 mL of 10N NaOH in a total volume of 100 mL of water.
4. 3M Sodium acetate: pH 4.8 (adjust pH with glacial acetic acid).
5. 2M Ammonium acetate. Store at 4°C.
6. Isopropanol.
7. Cheesecloth.
8. Quick-Seal polyallomer centrifuge tubes (e.g., Beckman [Fullerton, CA] No. 342413).
9. Cesium chloride.
10. 1M NaCl: Dissolve 58.44 g of NaCl in 1 L of water, autoclave, and store at room temperature.
11. CsCl mix: Dissolve 122.1 g of cesium chloride (DNA grade) in 128 mL TE and add 4.13 mL of ethidium bromide (10 mg/mL). The final volume is 165 mL.
12. *n*-Butanol: Mix *n*-butanol with an equal volume of water and an amount of cesium chloride that leaves some undissolved. Use only the upper (*n*-butanol) phase.

2.3. Analysis of DNA–Protein Complexes

1. Vertical electrophoresis apparatus for polyacrylamide gel, approximate dimensions 15 × 15 cm.
2. Acrylamide/*bis*-acrylamide: Add 30 g acrylamide and 0.8 g *bis*-acrylamide dissolved in 100 mL of water.
3. 10X TBE: 108 g of Tris-base, 55 g of boric acid, and 9.3 g of EDTA (Ethylenediaminetetraacetic acid, disodium salt) dissolved in water. Make up to a total volume of 1 L.
4. 8% Polyacrylamide slab gel (about 15 × 15 cm, 1-mm thick): Mix 31 mL of water, 13.3 mL of 30% acrylamide/*bis*-acrylamide, 5 mL of 10X TBE, 600 µL of 10% APS (dissolve 1 g of ammonium persulfate in 10 mL of water), and 60 µL of TEMED (*N,N,N',N'*-Tetra-methylethylenediamide). Pour solution between the glass plates of the assembled vertical electrophoresis apparatus. Insert a comb for about 1-cm wide slots. Let the gel polymerize for several hours, preferably overnight.
5. 5X Gal repressor protein binding buffer: 50 mM KCl, 50 mM Tris-HCl, pH 7.6, 50 mM MgSO$_4$, 0.5 mM EDTA, 0.5 mM DTT, 250 µg/mL BSA, 50% glycerol.
6. Ethidium bromide stain: Add 100 µL of stock ethidium bromide solution (10 mg/mL) to 1 L of TE.

3. Methods

3.1. Insertion of the Protein-Binding Site into pBend

1. Add 25 µL of pBend–DNA (1 mg/mL), 10 µL of 10X *Hpa*I digestion buffer, 55 µL of water, and 10 µL (100 U) of *Hpa*I restriction endonuclease to a 1.5-mL Eppendorf tube. Mix and incubate at 37°C for 2 h, or overnight.
2. To precipitate the restriction enzyme, place the digest on ice, add 3 µL of 500 mM EDTA, pH 8.0, 100 µL of ice-cold TE, and 100 µL of ice-cold 7.5M ammonium acetate. Keep the sample at 4°C for 10 min. Centrifuge in a tabletop centrifuge for 10 min. Remove the supernatant (containing the DNA) and add it to an Eppendorf tube filled with 600 µL ice-cold ethanol. Incubate at –70°C for 20 min, and centrifuge in a tabletop centrifuge for 10 min. Discard the supernatant, add 500 µL of ice-cold 80% ethanol, centrifuge for 5 min, and discard the supernatant. Dry the DNA pellet in a vacuum centrifuge and dissolve the sample in 250 µL of TE. Store the linearized pBend–DNA at –20°C.
3. Synthesize two complementary oligonucleotides that when annealed to each other form the protein binding site (*see* Note 1 for the design of

DNA-Bending Vectors

DNA inserts). The purification of the oligonucleotides is likely to be unnecessary if they are shorter than 30 nucleotides. Dissolve each oligonucleotide in autoclaved distilled water at a concentration of 200 µg/mL.

4. Add 10 µL of each oligonucleotide and 180 µL of oligonucleotide annealing buffer to a 1.5-mL Eppendorf tube. Incubate the sample for 3 min in a 300-mL beaker with about 150 mL of boiling water. Place the beaker with the tubes in the coldroom at 4°C to allow for annealing of the oligonucleotides over a period of several hours. Store the DNA at −20°C.
5. Mix in a 1.5-mL Eppendorf tube: 2 µL of annealed oligonucleotides, 3 µL of 5X concentrated ligation buffer, 1 µL of 5 mM ATP, 8.5 µL of water, and 0.5 µL of T4 polynucleotide kinase. Incubate for 10 min at 37°C. Place the sample on ice and add 3 µL of 5X concentrated ligation buffer, 1 µL of linearized vector DNA (from step 2 above), 1 µL of 5 mM ATP, 9 µL of water, and 1 µL of T4 DNA ligase. Incubate at 15°C for several hours, or overnight. The samples can be stored in the refrigerator for several days and aliquots can be used for several transformations.
6. Transform competent *E. coli* cells according to the protocol provided by the vendor and streak on LB-amp plates. Incubate the plates at 37°C overnight or until the colonies appear.
7. For preparation of the plasmid DNA on a small scale, use sterile tooth picks to transfer individual colonies to 15-mL tubes containing 5 mL of LB with 200 µg/mL ampicillin; also, streak cells from each transformant onto a LB-amp plate. Incubate this master plate at 37°C and shake the liquid cultures at 37°C overnight.
8. Pellet the cells by centrifugation for 15 min at about 700 g at 4°C (e.g., at 3000 rpm in a Sorvall RT6000B refrigerated centrifuge with a H1000B rotor). Decant the supernatant, add 200 µL of Tris-sucrose, and transfer to 1.5-mL Eppendorf tubes. Add 25 µL of lysozyme solution. Mix and add 130 µL of 200 mM EDTA, pH 8.0, and 130 µL of TLM. Mix and place at 65°C until lysis occurs (which usually takes a few minutes). Vortex briefly and centrifuge for 15 min in a tabletop centrifuge. Remove the pellet with a sterile toothpick; add 0.5 vol of the prepared phenol and 0.5 vol of chloroform. Vortex for 10 s, centrifuge for 10 min, and carefully remove about 200 µL of the aqueous (upper) phase while staying clear of the interface. Add 400 µL of ice-cold ethanol, mix, and centrifuge for 5 min, decant the supernatant, add 1 mL of 80% ethanol, centrifuge for 2 min, carefully decant the supernatant, and dry the pellet in a vacuum centrifuge. Dissolve the pellet in 30 µL of TE with occasional mixing. Store the samples at −20°C.
9. To verify successful insertion of the protein binding site, digest an aliquot of the DNA with *Eco*RI and *Hind*III. To a 5-µL aliquot of the

plasmid preparation add 2 µL of water, 1 µL of 10X EcoRI digestion buffer, 1 µL of EcoRI, 1 µL of HindIII, and 1 µL of RNase. As a control, digest 1 µg of pBend-DNA. Incubate all samples at 37°C for several hours or overnight. Place digests on ice, add 1 µL of 200 mM EDTA, pH 8.0, 90 µL of ice-cold TE, and 50 µL of ice-cold 7.5M ammonium acetate. Keep on ice for 10 min. Centrifuge in tabletop centrifuge for 10 min. Collect the supernatant and add it to an Eppendorf tube containing 300 µL ice-cold ethanol. Mix and incubate at −70°C for 20 min. Centrifuge in tabletop centrifuge for 10 min. Remove supernatant, add 300 µL ice-cold 80% ethanol to the pellet, centrifuge for 5 min, and discard supernatant. Carefully dry the pelleted DNA in a vacuum centrifuge and dissolve it in 5 µL of TE. Add 5 µL of Tris-acetate loading buffer and mix briefly.

10. Prepare a 2% agarose gel. Load the samples from step 9 in parallel with DNA molecular weight markers. Electrophorese at 80 V until the bromophenol blue has migrated about 4 cm. Examine the DNA under a UV-transilluminator and take a picture with a Polaroid camera (film type 57 or 55). Successful insertion is indicated by an EcoRI–HindIII fragment of the expected mobility (242 bp plus insert). (Note 2 has recommendations for ensuring cloning of the insert.) Electrophoresis can be continued to discover minor mobility differences, but then fresh electrophoresis buffer should be used. Eventually, the nature of the positive clone must be verified by DNA-sequencing, which also reveals the orientation of the inserted binding site (*see* Note 3 for the selection of suitable sequencing primers).

3.2. Purification of pBend DNA

1. To obtain pure DNA of the positive pBend-derivative, set up a 5 mL culture of the positive clone in LB with 500 µg/mL of ampicillin starting from an individual colony of the master plate (step 7 of Section 3.1.). Shake at 37°C for several hours, until the culture becomes turbid. Transfer the cells to a 2-L sterile Erlenmeyer containing 400 mL of LB with ampicillin. Shake overnight at 37°C.
2. Place the culture on ice and transfer the cells into centrifuge bottles. Pellet the cells by centrifugation at 4°C at about 1600g (e.g., 3500 rpm in a H6000A rotor of a Sorvall RC3C centrifuge). Decant the supernatant, resuspend the pellet in 20 mL of *E. coli* suspension buffer, and transfer the cells to a 50-mL centrifuge tube (preferably Nalgene, cat. no. 3131-0024). Centrifuge at 4°C for 10 min at about 5000g (e.g., in a Sorvall SS34-rotor at 10,000 rpm). Freeze the pellet completely by placing the sample on dry ice or in a −80°C freezer.

3. Thaw the pellet and resuspend the cells in 6 mL of Tris-glucose-EDTA. Add 12 mg of lysozyme powder, mix, and keep on ice for 30 min. Bring to room temperature and add 12 mL of NaOH-SDS. Mix and place on ice for 5 min.
4. Add 9 mL of $3M$ sodium-acetate, pH 4.8, shake, and leave on ice for 30 min. Centrifuge at about $12,000g$ at 4°C for 15 min (e.g., in an SS34 rotor at 15,000 rpm).
5. Transfer the supernatant to a new centrifuge tube by filtering through a cheesecloth. Add 0.5 vol of isopropanol to the transfered solution, leave 5 min at room temperature, and centrifuge at 4°C for 10 min at $5000g$ (e.g., in the SS34 rotor at 10,000 rpm).
6. Discard the supernatant and add 6 mL of ice-cold $2M$ ammonium-acetate to the pellet. Vortex repeatedly to dissolve the plasmid DNA until only small particles are visible. Centrifuge at 4°C for 10 min at $5000g$.
7. Transfer the supernatant (containing the plasmid DNA) to a new centrifuge tube, add 4 mL of isopropanol, mix, and centrifuge at 4°C for 10 min at $5000g$.
8. Discard the supernatant and completely dissolve the pellet in 7 mL of TE with occasional shaking. Add 8 g of cesium chloride, 400 µL of $1M$ NaCl, 400 µL of $1M$ Tris-HCl, pH 8.0, 160 µL of ethidium bromide (10 mg/mL), and 2.5 mL of CsCl mix (*see* Section 2.2., item 11).
9. After the CsCl is dissolved, draw the solution into a 20-mL syringe and transfer it into a Quick-Seal centrifuge tube. Fill a second tube with CsCl mix and make sure that the two tubes are balanced. Seal the tubes and centrifuge overnight at about $200,000g$ at 20°C (e.g., at 50,000 rpm in a Beckman NTV65 rotor).
10. Remove the tubes from the rotor, puncture the top of the tube, then collect the lower of the two visible bands with a syringe by puncturing the side of the tube. Transfer the DNA into a 15-mL Corex glass centrifuge tube.
11. Extract the ethidium bromide by adding 1 mL of *n*-butanol that has been saturated with water and cesium chloride. Vortex and remove the upper phase with a glass pipet. Repeat this process beyond the point where the color becomes invisible (usually about six times).
12. Add 2.5 mL of water and 7 mL of ethanol. Mix and incubate at −70°C for 15 min. (Do not leave too long, otherwise CsCl will precipitate). Centrifuge for 15 min at 4°C at about $7000g$, preferably in a swinging-bucket rotor (e.g., Sorvall HB4 rotor at 10,000 rpm).
13. Pour off the supernatant, add 5 mL of ice-cold 80% ethanol to the pellet, repeat the centrifugation, discard the supernatant, and evaporate excess ethanol under vacuum.

14. Dissolve the DNA in 500 μL of TE and determine the absorbance at 260 nm. Add the appropriate amount of TE to adjust the concentration of the plasmid DNA to 1 mg/mL (1 A_{260} is equivalent to 50 μg/mL). Store the DNA at 4°C.

3.3. Analysis of DNA–Protein Complexes

1. To generate restriction fragments with the protein binding site located at the end or in the middle, digest the pBend-construct separately with *Mlu*I (end) and *Eco*RV (middle) (*see* Note 4). One digestion contains 100 μL (100 μg) of plasmid DNA from step 14 of Section 3.2., 30 μL of 10X *Mlu*I- (or *Eco*RV) digestion buffer, 160 μL of water, and 10 μL of *Mlu*I or *Eco*RV restriction enzyme (100 U). Incubate the samples at 37°C for several hours, or overnight. Place the samples on ice, add 15 μL of 500 m*M* EDTA, pH 8.0, 150 μL of ice-cold 7.5*M* ammonium acetate, and leave at 4°C for 10 min. Centrifuge in a tabletop centrifuge for 10 min. Remove and add the supernatant (containing the DNA) to a new Eppendorf tube filled with 900 μL of ice-cold ethanol. Mix and incubate at –70°C for 20 min. Centrifuge in a tabletop centrifuge for 10 min. Discard the supernatant, add 500 μL of ice-cold 80% ethanol, centrifuge for 5 min, and decant the supernatant. Carefully dry the pellet in a vacuum centrifuge and dissolve the DNA in 50 μL of water. Verify the success of the digestion by electrophoresis of an aliquot on a 2% agarose gel (described in step 23 of Section 2.1.).
2. Pour a vertical 8% polyacrylamide slab gel. Assemble the electrophoresis apparatus and preelectrophorese at room temperature for 1 h at 100 V with TBE buffer in the reservoirs.
3. Mix at room temperature: 2 μL of *Mlu*I- or *Eco*RV-digested DNA (from step 1, above), 1.6 μL of 5X *Gal* repressor protein binding buffer, 4.4 μL of water, 2 μL of diluted protein. Keep the time between diluting the protein and addition to the DNA as short as possible. Do not vortex; mix gently with the tip of the pipet. Prepare a control without added protein. Incubate all samples for 10 min at room temperature.
4. Flush the wells of the polyacrylamide gel with reservoir buffer and load samples without the addition of loading buffer and tracking dyes. The glycerol in the binding buffer gives the sample sufficient density. A long plastic microcapillary tip is helpful to deliver the sample to the bottom of the slot. Glass capillaries should be avoided because proteins tend to stick to glass. In a separate slot, load DNA-molecular weight markers with bromophenol blue. Electrophorese at room temperature for 5 h at 200 V. Separate the glass plates and immerse the gel in ethidium bromide stain for visualization of the DNA under UV light.

Take a picture with a Polaroid camera. (Consult Notes 5 and 6 if complexes cannot be tetected.)
5. Measure the distances between the the slot and the position of the protein–DNA complexe of the *Mlu*I- (μ_E) and the *Eco*RV-digest (μ_M). Also, examine the mobilities of the free DNA (Fig. 3) to make sure that the DNA fragments contain no intrinsic bending. Calculate the bending angle α using the empirical formula $\mu_M/\mu_E = \cos(\alpha/2)$. (*See* Note 7 for help on determining the bending angle.)

4. Notes

1. Protein binding sites can be inserted into the pBend-vectors using restriction fragments or synthetic oligonucleotides. Restriction fragments should not be considerably larger than the protein binding site to be tested. Oligonucleotides are normally available with blunt ends, therefore the *Hpa*I-sites of pBend4 or pBend5 (Fig. 2) should be used. Newly synthesized oligonucleotides might be designed with "sticky" ends, so that they are compatible with the *Xba*I- and the *Sal*I-site; they can be cloned efficiently and insert into the DNA in a single orientation.
2. If many transformants are obtained, but none contains the protein binding site, the pBend–DNA might not be fully linearized. Alter the DNA–enzyme ratio in favor of the enzyme and confirm complete digestion of an aliquot by electrophoresis on an agarose gel.
3. Insertion of the protein binding site may not occur if the oligonucleotides are of poor quality. In this case, they should be purified and checked by polyacrylamide gel electrophoresis. If the transformation efficiency with supercoil control DNA is high, yet very few transformants are obtained with the annealed oligonucleotides, reduce the amount of insert DNA. Multiple insertion of the binding site can occur and is detected by gel electrophoresis and sequencing. For sequencing use primers named T3, T7, M13–20 or reverse primer (Stratagene, La Jolla, CA). Do not use the SK- and KS-primers (Stratagene) because they are not fully complementary to pBend3, 4, and 5.
4. In the initial bending experiment, it is advisable to restrict only with *Mlu*I or *Bam*HI (to place the binding site close to the ends) and *Eco*RV or *Pvu*II (to place the binding site in the middle). Do not select restriction sites that also occur in the protein binding sequence. When exploiting the 17 circular permutated restriction sites attention must be paid to the property of some of the restriction enzymes as follows: *Cla*I-sites are methylated in most *E. coli* strains; its use is therefore limited to prior growth of the plasmid in a methylation-defective (*dam-*) host. *Sty*I will also cut at *Nco*I of the repeat; for the purpose of a bending experiment,

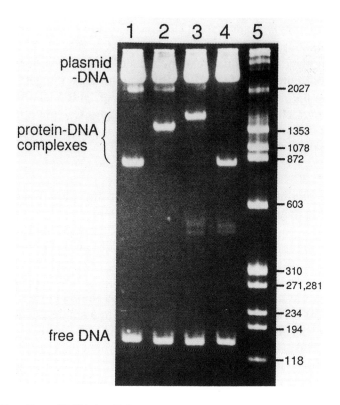

Fig. 3. Bending of DNA by Gal-repressor. Analysis by electrophoresis of complexes on a 8% polyacrylamide gel followed by staining with ethidium bromide. The protein concentration was chosen such that about half of the DNA fragment forms a complex and the other half remains as free DNA. The linear remainder of the vector DNA (pBend5 in this case) stays close to the top of the gel. Lane 1: Complex with the *Mlu*I-digested plasmid containing the *gal*-operator O_E; lane 2: Same as lane 1, but with the *Eco*RV-digested plasmid; lane 3: complex with the *Eco*RV-digested plasmid containing the *gal*-operator O_I; lane 4: Same as lane 3, but with the *Mlu*I-digested plasmid; lane 5: molecular weight markers; the size of the fragments in bp is indicated on the right. Note the large mobility differences between complexes formed with DNA fragments containing the binding site at the ends (lanes 1 and 4) and in the middle (lanes 2 and 3); a comparison of the mobilities of the complexes in lanes 2 and 3 shows that when interacting with Gal repressor, O_I is bent more than O_E.

it can only be used under partial digestion conditions. An additional *Spe*I-site is present in the vicinity of the single *Eco*RI-site as part of pBluescript SK- (*see* Fig. 2). *Spe*I-digestion generates an additional small fragment, which contains no protein binding site and does not interfere

with the bending assay. Three additional *Dra*I-sites are located in the plasmid corresponding to pBluescript coordinates 1912, 1931, and 2623. Depending on their electrophoretic property, some of the vector-derived DNA fragments might comigrate with certain protein–DNA complexes. Make sure to include a control without added protein. Likewise, two additional *Pvu*II-sites correspond to pBluescript coordinates 529 and 977. A *Sma*I-site is present close to *Eco*RI (*see* Fig. 2). Two additional *Ssp*I-sites correspond to pBluescript coordinates 442 and 2850, and two additional *Rsa*I-sites correspond to pBluescript coordinates 665 and 2526. *Nco*I will also cut at *Sty*I of the repeat and can only be used under partial digestion conditions. An additional *Bam*HI-site is present close to the *Eco*RI-site (*see* Fig. 2).
5. One of the frustrating aspects of conducting a bending experiment can be the inability to detect a complex on the polyacrylamide gel. Even if the binding and electrophoresis conditions are known one should be careful to avoid solutions and equipment that has been in contact with SDS. If possible, dedicate one electrophoresis set-up to "gel-shift" experiments. Many DNA binding proteins are insoluble in the low salt concentration of the electrophoresis buffer and must be stored at high ionic strength. Limit the time between dilution and addition to the DNA. Avoid vortexing during complex formation and do not add tracking dyes because they interact with the complex and might change its mobility. Larger protein–DNA complexes (e.g., *lac* repressor, *[8]*) behave better in low percentage polyacrylamide gels (e.g., 4%) with a high acrylamide/*bis*-acrylamide ratio (80:1). The protein concentration for obtaining about equal amounts of free and complexed DNA should always be determined in a preliminary experiment.
6. Bending experiments with radioactively labeled DNA are particularly useful if the protein has not been purified or if its availability is limited. Labeling can be accomplished with T4 polynucleotide kinase and [^{32}P]-γATP. Often, the DNA ends generated by the various restriction enzymes are labeled to different degrees. This problem can be overcome by loading the gel with aliquots of the binding reaction adjusted for the efficiency of fragment labeling. It is best to purify and isolate the protein-binding fragments because the radioactively labeled plasmid-DNA might obscure the region where the complexes are located. Another potential problem (which is also the case with unlabeled DNA) might be that bands appear that represent minor digestion products. In order to identify those, make sure to include controls without added protein.
7. The bending angle α assumes a value of 0° for a straight duplex. Since the mobility of a rigid DNA fragment is related to its end-to-end dis-

tance, the latter equals L · cos(α/2), with L being the length of the unbent DNA. The end-to-end distance of a fragment bent at the end will be virtually the same as L. Thus μ_M/μ_E = L · cos(α/2)/L = cos (α/2), where μ_M is the mobility of the complex with the protein bound centrally and μ_E the mobility of the complex with the protein bound at the end of the DNA fragment. The apparent bending angles for two *gal* operators, O_E and O_I, induced by Gal repressor are 95 and 112°, respectively (Fig. 3, ref. 2). We measure the distance between the top of the gel and the front of the band representing the protein-DNA complex. Whatever method is used, one must be consistent. Possible intrinsic bending in the free DNA must be considered in the calculation of the bending angle. It should be noted that the calculated values may be different from absolute bending angles, since factors other than the end-to-end distance influence the mobility of protein-bound and unbound DNA fragments. The method measures the net bend and cannot distinguish between a single sharp bend at one position and a smooth curving over a larger DNA region. For precise determination of bending angles, at least three independent experiments should be carried out. If possible, control lanes should be added of a similar size complex in which the DNA is bent to a known degree.

References

1. Crothers, D. M. and Fried, M. G. (1983) Transmission of long-range effects in DNA. *Cold Spring Harbor Symp. Quant. Biol.* **47,** 263–269.
2. Zwieb, C., Kim, J., and Adhya, S. (1989) DNA bending by negative regulatory proteins: Gal and Lac repressors. *Genes Dev.* **3,** 606–611.
3. Wu, H.-M. and Crothers, D. M. (1986) The locus of sequence-directed and protein-induced DNA bending. *Nature (London)* **308,** 509–513.
4. Kim, J., Zwieb, C., Wu, C., and Adhya, S. (1989) Bending of DNA by gene-regulatory proteins: construction and use of a DNA bending vector. *Gene* **85,** 15–23.
5. Thompson, J. F. and Landy, A. (1988) Empirical estimation of protein-induced DNA bending angles: application to site-specific recombination complexes. *Nucleic Acids Res.* **20,** 9687–9705.
6. Majumar, A. and Adhya, S. (1984) Demonstration of two operator elements in *gal*: in vitro repressor binding studies. *Proc. Natl. Acad. Sci. USA* **81,** 6100–6104.
7. Zwieb, C. and Brown, R. S. (1990) Absence of substantial bending in the *Xenopus laevis* transcription factor IIIA-DNA complex. *Nucleic Acids Res.* **18,** 583–587.
8. Fried, M. G. and Crothers, D. M. (1983) CAP and RNA polymerase interaction with the *lac* promoter: binding stoichiometry and long-range effects. *Nucleic Acids Res.* **11,** 141–185.

CHAPTER 23

Determination of Sequence Preferences of DNA Binding Proteins Using Pooled Solid-Phase Sequencing of Low Degeneracy Oligonucleotide Mixtures

Joseph A. Gogos and Fotis C. Kafatos

1. Introduction

Mutational analysis via synthetic oligonucleotides, combined with a number of biochemical approaches, has provided a wealth of information on the nature of protein–DNA interactions. Although in vitro mutational studies of the DNA target site were based initially on individual oligonucleotides, recently their power has been greatly enhanced by procedures involving degenerate oligonucleotide mixtures, in which all possible sequence combinations can be represented and from which particular sequences can be selected by protein binding. Strategies based on this approach, however, typically are quite laborious: They involve cloning of the oligonucleotides after the selection step and sequencing of a great number of individual clones to yield statistically significant information.

To make selection procedures convenient for everyday practice, we have experimented with direct sequencing of *low degeneracy* oligonucleotide mixtures. In this manner, the "acceptability" of a mutant base to the protein might be evaluated from the intensity of the corresponding band in the autoradiogram of a single sequencing gel, rather than from the statistics of binary representation (presence or absence) among multiple individual clones. This is, in theory, possible with the low degeneracy oligonucleotides, because the protein is expected to always bind in the same phase, using the nonvariable positions as anchoring points. Indeed, we were successful in developing a strategy for binding site selection analysis (BSSA), which starts with a low affinity site and detects in the protein-bound oligonucleotide mixture the increased abundance of bases that improve the binding affinity *(1)*. Here we describe a protocol for one selection cycle and pooled sequencing, which is greatly facilitated by immobilizing the selected and nonselected oligonucleotides on a solid support compatible with direct chemical sequencing (Fig. 1). The double-stranded DNA substrate for the binding reaction is obtained by synthesis of a template strand with certain bases degenerate, and a primer from which the second strand is formed by Klenow extension. The bound DNA sequences are selected by gel retardation, immobilized, and then sequenced. Sequencing of immobilized DNA fragments is fast and convenient, since it obviates both DNA extractions from polyacrylamide gels and the time-consuming ethanol precipitation steps of chemical DNA sequencing. This assay can be an informative complement to well-established approaches, such as methylation interference or missing contact assays, for rapid and detailed analysis of protein–DNA interactions. In addition, the convenience and sensitivity of these latter approaches can benefit from the solid phase reaction protocol that we have developed.

2. Materials

1. Custom-made oligonucleotides, synthesized on an automated oligonucleotide synthesizer using nucleoside phosphoramidite precursors.
2. 30% fresh ammonium hydroxide. To ensure a concentrated solution, buy only small bottles of ammonium hydroxide and keep them at 4°C tightly capped and sealed with parafilm.
3. Water bath at 55°C.
4. Screw-capped microcentrifuge tubes.
5. Parafilm.
6. Savant SpeedVac with connections to a vacuum pump and to a water aspirator.
7. TE buffer: 10 mM Tris-HCl, pH 7.4, 1 mM EDTA, pH 8.0.
8. Spectrophotometer.
9. Stock 10X TBE buffer: 0.09M Tris-borate, 0.002M EDTA, pH 8.0.

Binding Site Selection Analysis

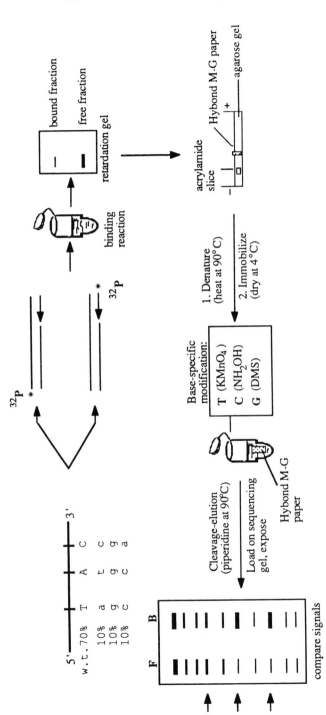

Fig. 1. Schematic representation of experimental strategy. One protocol utilizes synthetic oligonucleotides of a low affinity sequence that are mutagenized in three positions by including low concentrations (10% each) of the other three (nonwild-type) bases at the corresponding steps of the synthesis (bases shown in lower case letters). The complementary strands are then synthesized by primer extension using the Klenow fragment of polymerase I. Single end labeling is achieved using end-labeled primer or template (asterisk). The mixture of double-stranded products is used in excess as a substrate in a binding reaction with a protein of interest. Protein-bound and free oligonucleotides are separated by nondenaturing electrophoresis, and gel slices that include the bands of interest are cut out and embedded in a layer of agarose (1% 1X TBE). The DNA from each gel slice is electrophoretically transferred (30 min to 1 h) to a piece of Hybond M-G paper. Double-stranded DNA is denatured in situ by placing the still wet piece of paper over boiling water vapor for 2 min, immobilized by drying at 4°C and subjected directly to base specific modification. The consequences of each base substitution are evaluated after gel electrophoresis and autoradiography, by comparing band intensities in the bound and free fractions.

10. Materials for denaturing polyacrylamide gel electrophoresis (19:1 acrylamide: N,N'-methylene bis-acrylamide [40% w/v]; ultrapure grade urea; stock 10X TBE buffer).
11. T4 polynucleotide kinase.
12. 10X Kinase reaction buffer: 0.5M Tris-HCl, pH 7.6, 0.1M MgCl$_2$, 50 mM dithreitol, 1 mM spermidine-HCl, 1 mM EDTA, pH 8.0.
13. [^{32}P]-γ-ATP(>6000 Ci/mmol).
14. Materials for ethanol precipitation: 96% ethanol; 3M sodium acetate; glycogen 2 μg/μL (=100X).
15. Klenow fragment of DNA polymerase.
16. 3X Polymerase buffer: 30 mM Tris-HCl, pH 7.4, 150 mM NaCl, 15 mM DTT.
17. Heating block.
18. Water bath at 37°C.
19. dNTPs mix, stock solution: 10 mM each dNTP.
20. Materials for nondenaturing polyacrylamide gel electrophoresis (29:1 acrylamide: N,N'-methylene bis-acrylamide [30% w/v]; glycerol 50%, stock; TBE buffer).
21. Materials for binding reactions: protein (crude or purified bacterial extracts containing recombinant protein, crude nuclear extracts), bulk competitor DNA (poly[dI.dC]poly[dI.dC],poly[dA.dT]poly[dA.dT] [Pharmacia, Piscataway, NJ], salmon sperm DNA [Sigma, St. Louis, MO], E. coli DNA), NaCl or KCl, 5X binding buffer (e.g., 60 mM HEPES NaOH, pH 7.9, 20 mM Tris-HCl, pH 7.9, 5 mM EDTA, pH 8.0, 25% glycerol, 2.5 mM DTT).
22. X-ray film.
23. Plastic wrap.
24. Razor blades and forceps.
25. Agarose.
26. Oven at 65°C.
27. Hybond M-G paper (Amersham, Arlington Heights, IL).
28. Potassium permanganate (KMnO$_4$), 99% pure (Aldrich, Milwaukee, WI): Prepare stock solution 2 mg/mL in water. Make a fresh stock solution every 2 wk. Store at 4°C, protected from light.
29. Hydroxylamine (NH$_2$OH), 99% pure (Aldrich): Prepare stock solution 275 mg/mL in water. Adjust the pH to 6.0 with di- or triethylamine. Store at 4°C. Make a fresh stock solution every 2 wk (although solutions kept for several months are still active).
30. Dimethyl sulfate (DMS), 99% pure (Aldrich). Store it as small aliquots at 4°C. DMS is toxic, is a mutagen, and a suspected carcinogen. Small amounts of dilute DMS solutions can be inactivated by 5M NaOH.

Binding Site Selection Analysis

31. Ammonium formate buffer: Dissolve 31.5 mg of ammonium formate in 5 mL of water and adjust the pH to 3.5 with formic acid. Finally add water up to 10 mL. Store at 4°C for several months.
32. Whatman absorbent paper.
33. Eppendorf tubes, 0.5 mL.
34. $10M$ Piperidine (Aldrich); store at 4°C.
35. 0.1 N NaOH.
36. Formamide dye solution: 1 mL formamide, 10 µL 0.1% bromophenol blue, 10 µL 0.1% xylene cyanol.

3. Method
3.1. Synthesis and Purification of the Oligonucleotides

3.1.1. Synthesis of Oligonucleotides and Primers with an Automated Oligonucleotide Synthesizer

Since the technique is best suited for identifying DNA sequences that bind to a certain protein with higher affinity relative to other sequence combinations, the scheme we follow most frequently begins with a sequence encompassing a low affinity binding site. However, high affinity sites can also be used. Synthetic oligonucleotides of the starting sequence are prepared in which mutations are introduced, usually at three or four positions (*see* Note 1), by including low concentrations of the three nonwild-type bases at the corresponding step of synthesis (*see* Note 2). A general procedure for synthesizing such low degeneracy mixed oligonucleotides is as follows: At positions where mutations are not desired, such as those including the primer complementary sequence (*see* Note 3), add a single nucleoside phosphoramidite precursor. In contrast, at positions where mutations are desired, add a defined mixture of nucleoside phosphoramidite. In general, the frequency of addition of particular nucleotides depends on the relative molarities of the precursors. Sometimes this is not the case for guanines, presumably because of the instability of the G precursors, especially if freshly prepared solutions are not used. The most accurate and reproducible way to achieve a desired mixture is to combine appropriate amounts of solid nucleoside phosphoramidite prior to solubilization in acetonitrile.

3.1.2. Deprotection of the Oligonucleotides After Synthesis

1. Add 1 mL of fresh 30% ammonium hydroxide to the resin of synthesis (typically 0.2 µmol) inside a screw cap tube. Seal the tube with parafilm and place it in a 55°C waterbath for 5–12 h (*see* Note 4).

2. Cool the tube to room temperature and open it in the hood. Transfer the supernatant into a clean Eppendorf tube, leaving behind the resin. Rinse the resin with 200 µL ddH$_2$O and combine the supernatants.
3. Draw off the ammonium hydroxide using the Savant SpeedVac connected to a water aspirator for vacuum. When the volume has decreased by about one half to one third, the tubes can be moved to a Savant SpeedVac connected to a vacuum pump. NEVER use a vacuum pump to evaporate ammonia.
4. When the sample is dry, a pellet is usually visible, which should be dissolved in a suitable volume of water or TE buffer.

3.1.3. Purification of Oligomers of the Expected Length from the Shorter Synthetic Products

Use denaturing polyacrylamide gel electrophoresis with appropriate percentage gels (*see* Chapter 16). Quantitate the DNA by measuring absorbance at 260 nm and adjust the concentration to 1 pmole/µL.

3.2. End-Labeling and Duplex Formation

3.2.1. Phosphorylation of Template Oligonucleotides and Primers at High Specific Activity with T4 Polynucleotide Kinase

Phosphorylate in separate reactions ~1–2 pmole of template (30–60 ng of 45-mer) and an equimolar amount of primer using 10 µL of [^{32}P]-γ-ATP (6000 µCi/mmole) in a 15 µL reaction. This should be enough for two binding reactions (*see also below*). Have in mind that you want to use ~1–2 × 10^5 cpm per binding reaction that using minimal protein to DNA ratios can usually give you enough retarded signal for one or two sequencing reactions.

3.2.2. Removal of Unincorporated Nucleotides (Optional)

Use ethanol precipitation in the presence of carrier (glycogen): Add 75 µL of H$_2$O, 10 µL of 3M sodium acetate, 14 µL of glycogen (2 µg/µL), and 250 µL of ethanol. Incubate at –70°C for 30 min, centrifuge, wash the pellet with 70% ethanol, and dry.

3.2.3. Conversion of the Mixture to Perfectly Matched Duplexes

This is done by second strand synthesis using a primer and the Klenow fragment of DNA polymerase I. Anneal equimolar amounts of primer and template in 15 µL of 3X polymerase buffer (30 mM

Tris-HCl, pH 7.4, 150 mM NaCl, 15 mM DTT) by heating at 65°C for 10 min and slow cooling to room temperature. This can be conveniently done in a heating block that is set at 65°C and then turned off. Perform extension for ~1 h in 45 µL of 1X buffer at 37°C after adding dNTPs, at a final concentration of 0.2 mM, along with Klenow fragment (10 U).

3.2.4. Separation of Double-Stranded Products from Excess Template or Unextended Primer

Use nondenaturing gel electrophoresis (12% acrylamide, 1X TBE, 5% glycerol) and purify the duplex from acrylamide by crush elution following the standard protocol *(2)*.

3.3. Binding Reaction and Electrophoresis

3.3.1. Protein-Binding Reactions

This includes optimization and scaling up of the reaction:

1. Optimization aims, on the one hand, at determining the amount of probe necessary for the preparative retardation experiments and, on the other hand, at reducing or eliminating any irrelevant binding (specific or nonspecific), most importantly that which produces retarded bands of electrophoretic mobility similar or identical to the band of interest. The latter step is quite critical when crude bacterial extracts containing recombinant protein are being used (*see* Note 5). Optimization should involve the following parameters:
    a. DNA probe and protein concentration. Since the proportion of the oligonucleotides able to bind protein can vary widely in different mixtures, depending on the number of positions that have been varied, the proportion of mutant bases in each, and the tolerance of the protein, it is helpful to do an initial analysis on the amount of probe required for each mixture. However, do not use excess protein to compensate for reduced binding because low protein to DNA ratios are associated with lower background and highest sensitivity. For the initial analytical experiments ~1–3 ng (0.1 pmole of 45-mer) at ~20,000 cpm of end-labeled DNA probe and ~1–2 fmol of protein per 20 µL reaction are generally good starting conditions. More protein might be necessary if it is not purified. In practical terms use the minimum amount of protein that on scaling up will give you retarded signal enough for one or two sequencing reactions (~10^3–10^4 cpm).
    b. Competitor DNA concentration. Synthetic alternate copolymers are most commonly used as nonspecific DNA competitors. The choice

of the particular copolymer depends on the sequence of the binding site. Generally poly[dI.dC] poly[dI.dC] (Pharmacia) at a final concentration of 3 μg/20 μL reaction is a good starting point. Salmon sperm and/or *E. coli* DNA may also be helpful, especially if the irrelevant binding is specific, but tend to reduce the binding of interest as well.

  c. Salt concentration. NaCl or KCl should be added at a final concentration that decreases or eliminates nonspecific binding, but does not disrupt specific binding. Concentrations between 75 and 100 mM work generally well. Do not forget to correct for the salt contribution of the protein extract.

  d. Additional cofactors. This can include, for example, $ZnSO_4$ for zinc finger proteins, especially of the C_2H_2 type.

2. Scaling up of the reaction 5- to 10–fold is usually enough. Note that $\sim 1-2 \times 10^5$ cpm can usually give enough retarded signal for one or two sequencing reactions (~1000–3000 cpm) and that you do not want to load more than 2×10^5 cpm per lane of the retardation gel since the smearing of the free probe will increase the background in the retarded band. In a 0.5-mL tube combine the following for a 100 mL reaction:

  a. $1-2 \times 10^5$ cpm of DNA probe.

  b. 15 μg of bulk competitor (poly[dI.dC]poly[dI.dC], poly[dA.dT]poly[dA.dT], *E. coli* DNA, and so forth, or a combination).

  c. Reaction buffer: a typical composition (5X) is 60 mM HEPES NaOH, pH 7.9, 20 mM Tris-HCl, pH 7.9, 5 mM EDTA, pH 8.0, 25% glycerol, 2.5 mM DTT.

  d. Any additional cofactor.

  e. Mix all of the above components prior to adding the protein. After addition of the protein mix gently by tapping the side of the tube.

Incubation at room temperature (~22°C) is preferred, if acceptable, since it tends to decrease nonspecific binding. Incubation time can vary. Since the selection is performed under equilibrium conditions, a time course of the reaction should be done beforehand. For most interactions equilibrium is attained between 20 and 45 min.

3.3.2. Mobility Shift Analysis Using Nondenaturing Gel Electrophoresis

Use low ionic strength gels of the composition 6–8% acrylamide (29:1), 5% glycerol, and 0.25X TBE. The gel buffer should be 0.25X TBE. Prerun the gel at 180 V for ~1 h in the cold room (4°C). Load the samples on the gel at 40 V without addition of any dye/sample buffer since the glycerol will weigh down the sample. In an adjacent well load a blank sample with dye/sample buffer to monitor electrophoresis (*see also* Chapter 21). It is important to squirt the wells clean of glycerol prior to loading your sample, otherwise your bands will be distorted. If more than 200,000 cpm are needed to get a reasonable sequencing signal, then load the binding reaction in more than one lane (*see* Section 3.4. step 3). Run the gel at 160–200 V.

3.4. Transfer of DNA

1. After the gel run, carefully remove one glass plate so that the gel is resting on one plate with its other side exposed. Cover the exposed side with plasticwrap. In the darkroom place a film on the gel and tape it to the plasticwrap. Before taping align one corner of the glass plate with one corner of the film and mark that corner by notching it. Place gel and film in a light tight cassette and wrap it with aluminium foil. Put metal plates as a weight over the film, but do not clamp, because you may distort the bands. Expose for 6–12 h at 4°C and develop the film.
2. Using a razor blade, cut out from the X-ray film small rectangles containing the autoradiographic images of the bound and free DNA. Align the film over the gel and cut out the segments of polyacrylamide underneath the rectangular holes in the film. Do not attempt to remove the plasticwrap from the gel before cutting, but instead cut through both gel and plasticwrap.
3. Remove with the razor blade or forceps the acrylamide strips containing free and shifted probe, peel off the plasticwrap, and place the strips on the surface of a thin layer of agarose prepared on the bottom of a minigel apparatus (if more than one lane is loaded with binding reaction then the strips can be stacked one over the other). In addition, remove a strip containing the dye that migrates closest to the uncomplexed DNA in order to monitor the transfer. Because low percentage (29:1) acrylamide gels are sticky, the strips may be difficult to handle. However, prechilling the gel at 4°C make the strips easier to manipulate.
4. Pour 1% agarose/1X TBE over the strips, without disturbing them, and allow it to set. Cover the agarose with plasticwrap and put the mini gel apparatus in a 65°C oven for 10 min. Remove the plasticwrap and make an incision with a razor blade just ahead of each strip in the direction of electrophoretic migration. Prewet in 1X TBE a small piece of Hybond M-G paper (Amersham). Place the piece of Hybond M-G paper in the slot ahead of the DNA. Resume electrophoresis in 1X TBE for 30 min to 1 h (enough for the monitoring dye to move out of the strip). Avoid excessive electrophoresis.
5. Following electrophoretic transfer, remove the pieces of paper, steam denature DNA *in situ* by exposure of the *still wet paper* to boiling water

vapor for 2 min, and leave the pieces to air dry at 4°C. After drying, wash the pieces successively for 30 s by immersing in 50 mL of double-distilled water to remove the TBE buffer and blot them on Whatman paper to remove excess liquid. Dehydrate them for another 30 s in 50 mL of 95% ethanol and allow them to air dry for 2–3 min. Finally, cut the paper into three pieces (usually we perform three sequencing reactions per binding reaction), and transfer each piece into a separate tube.

3.5. Sequencing Reactions

3.5.1. Solid-Phase Chemical Sequencing

These reactions are performed at room temperature directly on the free and bound DNAs immobilized on Hybond M-G paper, using $KMnO_4$ for T, NH_2OH for C, and DMS for G according to the published Hybond M-G protocol supplied by Amersham *(3) (see* Notes 7, 8, and 11). In our experience the activity of each chemical can vary from batch to batch; accordingly, if too much or too little cleavage is observed concentration adjustment must be done. This is preferred to reaction time adjustments. Perform the following three reactions for the free and bound DNA.

1. Potassium permanganate reaction: Add 1 mL of double-distilled water to the tube along with 30 µL of a 2 mg/mL $KMnO_4$ stock solution. After 10 min replenish the liquid and continue the reaction for an additional 10 min.
2. Hydroxylamine reaction: Add 1 mL of a 275 mg/mL hydroxylamine stock solution, pH 6.0 (prepared fresh each week), to the tube and incubate for 20 min.
3. DMS reaction: Add 1 mL of 50 m*M* ammonium formate buffer, pH 3.5, to the tube along with 7 µL of DMS and leave the reaction to proceed for 6 min (*see* Notes 8 and 9).

3.5.2. Termination of Reactions

This is done simply by removing the paper with forceps and rinsing it for 30 s with double-distilled water. Blot the excess liquid on Whatman paper. Rinse once more in water, blot, rinse in ethanol for 30 s, and blot again.

3.5.3. Cleavage and Recovery of Modified DNA

Cut the dried paper into 3–4 smaller pieces, place them in the bottom of a 500-µL Eppendorf tube, and cover them with 100 µL of piperidine (cutting prevents the paper from floating on the piperidine solution). Seal the tubes with parafilm and incubate at 90°C for 30 min. At the end of the incubation remove the pieces of paper with forceps, freeze the samples, and lyophilize in a SpeedVac evaporator until dry. Add 100 µL of distilled water, freeze, and lyophilize again. Repeat the same cycle once more (*see* Note 11). Measure the samples for Cerenkov counts to determine cpm.

3.5.4. Analysis of the Products on Sequencing Gels

Dissolve the pellets in 2 µL of 0.1*N* NaOH and add 4 µL of formamide including dyes. Heat for 3 min at 94°C and quick chill on ice. Load the samples on a 15–20% polyacrylamide/urea sequencing gel, run, dry, and expose the gel for autoradiography (*see* Chapter 10 in vol. 4 of this series for details of running DNA sequencing gels).

3.6. Interpretation of the Results

At the positions of fixed sequence, intense bands must be seen at the expected positions, depending on the sequence. At the positions of degeneracy, bands appear in all four lanes of the free, unretarded fraction with intensities related to their proportions. In the retarded fraction, however, if selection is at work, a change in the relative intensity of these latter bands is observed. If one starts with a low affinity site and low proportions of mutant bases, selection is evident as an increase in intensity of a minor band(s), depending on the contribution of the particular base to the affinity of binding. Usually, no decrease in the intensity of the autoradiographic bands corresponding to minority bases selected against is expected (*see* Note 12). However, a decrease in the intensity of the wild-type majority base is expected and this serves as an *internal control* against any sequencing artifact. If equimolar amounts of wild/nonwild-type bases are used, then in theory it is possible to observe a decrease in the band intensity in the nonselected positions. However, since the magnitude of this decrease is expected to be small (at most 25 to 10%) it might be difficult to observe it in practice. Further discussion of the method and some examples of its application can be found in Notes 13–18.

4. Notes

1. This binding site selection analysis combined with pooled sequencing has its limitations on the number of degenerate position because of phasing problems in protein binding. Therefore, the number of degenerate positions cannot be high and certainly not higher than the actual length

of the binding site. In this latter case a phasing problem arises because the protein binding site can begin at different positions along the random sequence and direct sequencing cannot give a unique sequence pattern.
2. For the degenerate positions the mixtures of phosphoramidites we use usually contain 70% wild-type base and 10% each of the three other bases. However, this procedure can also be applied using equimolar mixtures of bases. The proportions of nonwild-type bases are dictated by two kinds of considerations. The upper limit of these proportions (and the limit of three or four mutated positions per oligonucleotide) is imposed mainly by the desire to have a high frequency of singly mutated oligonucleotides in the mixture, in order to simplify the final analysis. The lower limit is dictated by the background noise of the sequencing reactions as explained below.
3. Primers as small as 6 nucleotides are acceptable, maximizing the proportion of the oligonucleotide available for mutation.
4. Deprotection can also be accomplished at room temperature for 24–48 h in ammonium hydroxide. However, 55°C for at least 5 h ensures complete deprotection of all side chains. If the volume of the ammonium hydroxide drops significantly at 55°C, the deprotection may not be complete. In that case add more ammonium hydroxide and repeat the deprotection procedure.
5. Crude bacterial extracts containing recombinant protein can be used; in this case, as has been explained, a control binding reaction, using nonrecombinant bacterial extracts, must be run to verify the absence of nonspecific complexes with the same mobility as the specific complex. If such interfering complexes cannot be eliminated, either further purification of the protein is necessary or control BSSA must be run using only control extract *(4).* In addition, as mentioned below, this procedure is well suited to the mutational analysis of proteins that are present in small amounts; in fact crude nuclear extracts can also be used in the binding reaction, to analyze the binding specificity of complexes that have not as yet been cloned, but are prominent in mobility shift experiments.
6. During the electrophoretic transfer the DNA is retained quantitatively on the Hybond M-G paper.
7. The sequencing reaction conditions described are preferred because of their convenience (all are at room temperature), and because background considerations are not crucial when looking only for enrichment of minor bases (*see* Note 12 below). If desired, slightly lower background can be obtained with less convenient, but optimized reaction conditions *(5).*
8. In theory two sequencing reactions, one specific for a given purine and the other for the noncomplementary pyrimidine, should suffice for follow-

ing the fate of all four bases when applied to both strands. In practice we use three different reactions for both labeled strands. For crosschecking the results it would be desirable to have available four reactions, but no A-specific protocol has yet been described with Hybond M-G.
9. Formate buffer should be at room temperature before addition of DMS to avoid precipitation of the chemical.
10. During piperidine incubation DNA is eluted by the piperidine, which also cleaves a certain proportion of the modified bases.
11. The size of the Hybond M-G paper must be kept to a minimum to minimize the amount of residue that is formed by minor paper destruction during the piperidine reaction and persists after lyophilization. When larger paper size cannot be avoided and particulate materials are observed in the piperidine solution after the incubation, centrifuge the solution for 5 min, remove the supernatant in a separate tube, and precipitate with ethanol (in 100 µL of piperidine add 100 µL of 0.6M sodium acetate, 2 µg of glycogen, and 2.5 vol of ethanol). This is to remove any residual quarternary ammonium salts, freed from the membrane during destruction, because they interfere with electrophoresis. Phenol extraction at this stage is detrimental because DNA is partitioned mainly in phenol in the presence of these salts.
12. We have investigated the inherent background noise of sequencing reactions. This noise makes the use of a low affinity binding site convenient as the starting point for the procedure. The reason, as we have found, is that it is practical to detect selection reliably both for and against the major (wild-type) base, but only for and *not against* the minor (nonwild-type) bases, if the latter are each present at less than approx 10–15% of the total. Thus, if at a degenerate position one starts with, for example, wt T 85% and mutants C 5%, A 5%, G 5%, and C is unacceptable by the protein (that is C→0%), one is unable to follow that change by direct Maxam-Gilbert sequencing. Therefore, this approach cannot be used for identification of "down" mutations. The lowest threshold for detecting negative selection depends on the nature of the major base, whose nonspecific modification causes most of the background, and also on the sequence context.
13. The technique is most discriminant when the DNA probe is in excess over the protein. Under DNA excess, the relative proportions of the various sequences in the bound fraction are independent of the amount of protein and depend only on the concentrations of the sequences and their respective affinity constants. If saturating amounts of protein are used, no detectable differences between bound and nonbound DNAs are expected, especially since the major DNA sequence permits binding (albeit

weakly). Minimal protein to DNA ratios are also the conditions for highest sensitivity with respect to protein availability; this feature makes the procedure well suited for studying the specificity of prominent binding activities in crude nuclear extracts, before protein purification or cloning.
14. The sensitivity of this approach (especially for detection of *minor* differences in affinity) can be enhanced by using the polymerase chain reaction, as was first demonstrated by Blackwell and Weintraub *(4)*, who developed independently a similar selection/pooled sequencing protocol involving reiterative protein binding of oligonucleotide mixtures and PCR amplification. In this case it is also practical to perform single-strand dideoxynucleotide sequencing. Note however that under these experimental conditions (repeated selection rounds, decreasing protein concentration) only the highest affinity sequences are selected eventually; by overcoming even minor affinity differences relative affinity contributions among bases at a certain position may appear "distorted." In our hands Maxam-Gilbert sequencing has lower background and is more reliable (sequence independent) than the Mn^{2+}/dideoxynucleotide method, when used to sequence oligonucleotide mixtures. Apart from the advantages for sensitivity, the PCR method is also useful when higher redundancy oligonucleotides are to be used, for example in an exploratory search for an unknown binding site *(6)*.
15. The protocol described here is attractive because of its extreme simplicity and convenience. It is suited for checking the behavior of variants of a known DNA motif, such as may be encountered in diverse members of a multigene family. Since it is more direct, involving only one selection cycle without the intervention of PCR and cloning, it is more accurate in estimating relative affinity differences among bases at a certain position. Furthermore, it is an informative and convenient complement to methylation interference or missing contact analysis. All of these techniques use similar methods (Maxam-Gilbert sequencing, potentially on the same solid support), and thus can be incorporated in an integrated experimental approach, providing different types of information on protein–DNA interactions at the same time.
16. As an example of the possibilities of this method, we will summarize our findings on the interaction of the *Drosophila melanogaster* CF2 (chorion factor 2) with a low affinity site on the *s15* gene promoter *(1)*. Our analysis had two purposes: first, to identify positions important for binding within this site and the base preference of each position and second to analyze the relation of the CF2 complex with another complex that is formed on the same site by a yet unidentified nuclear factor. This latter complex is prominent in retardation experiments using nuclear

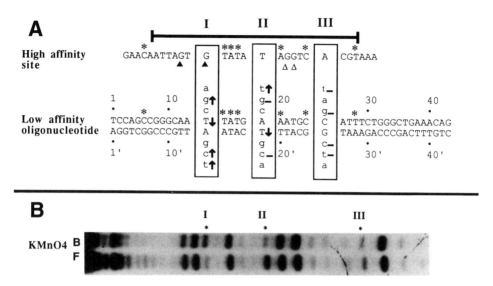

Fig. 2. **A.** Summary of the design and the conclusions of a BSSA experiment (1). Sequence alignment between a high affinity CF2-binding site and the degenerate oligonucleotide used for BSS analysis. Methylation interference and footprint results on the high affinity site (7,9) are indicated as follows: bar (footprint), filled triangle (complete interference), open triangle (partial interference). Asterisks indicate identity between high and low affinity sites. Although alternative alignments are possible, this alignment maximizes identities in the center of the footprint. The low affinity oligonucleotide is shown in double-stranded form, and minority bases at degenerate positions (numbered I, II, III) are indicated in lower case. Up and down arrows indicate observed selection for or against the corresponding bases. Absence of detectable selection is indicated by (–). **B.** Autoradiogram of a sequencing gel. The results from probing the lower strand with KMnO$_4$ are shown as an example. Bases are numbered on the top, exactly as in **A**. F is free DNA and B is protein-bound DNA.

extracts. CF2 is a zinc finger-containing protein (7–9). Gel retardation, DNase I protection, and methylation interference studies have revealed that CF2 expressed in bacteria has multiple binding sites in the *Drosophila s15* and *s36* chorion genes, as well as in several moth chorion gene promoters that are capable of expression in transgenic *Drosophila*. This behavior is by no means specific for CF2, but characterizes a number of transcription factors. An oligonucleotide was designed based on a low affinity segment of *s15*, with degenerate mutations at three positions that differ from another high affinity CF2 site. Figure 2 summarizes the design and the conclusions of the experiment.

At position I, where the high affinity site has a G in the plus strand scored as a contact residue by methylation intereference *(7)* and the low affinity site has T, clear selection was evident in favor of the minor G base and against the major T. Selection was also scored in favor of the minor C and T bases in the negative strand. At position II, where the high affinity site has T and the low affinity one has A in the plus strand, selection was noted for the minor T base, but not for G; in the negative strand selection was evident against the major T base, but C appeared to be indifferent. Note that in both positions the decreased intensity of the wild-type base served as an internal control against potential sequencing artifacts, and confirmed the positive selection of certain minor bases. It is notable that in both cases mutations that make the low affinity site more similar to the high affinity sequence were selected positively. At position III, no selection was evident within the sensitivity of the assay; either this position is indifferent in terms of binding affinity, or selection operates only against one or two minor bases. In conclusion, using our selection protocol we identified two positions that are important for CF2 binding and a third one that appears not to be important. In the former two positions increased similarity to the high affinity site resulted in greater binding, as expected, thus validating the technique. In one of the important sites an additional favorable substitution was identified. Note that these results could not have been obtained with typical methylation interference experiments using the low affinity oligonucleotide. The validity of these results was confirmed by further characterization of the CF2 binding site using highly degenerate oligonucleotides and reiterative cycles of binding and PCR amplification *(8)*.

Complex X is prominent in mobility shift experiments and is formed on the same oligonucleotide that binds CF2 with low affinity, by a *Drosophila* factor that has not as yet been cloned. We investigated how the two binding sites are related on the oligonucleotide. In this case our results indicated that, in contrast to the apparent indifference of CF2 to mutations in position III, complex X selects for the minor T base in the same position. We performed followup experiments testing for complex X formation with singly mutated oligonucleotides, and verified that both T and C are acceptable at position III, but A is not. Thus, even before cloning the complex X factor, this series of experiments permitted us to document that the CF2 and X binding sites are nonidentical, although they may be overlapping.

17. A similar approach has been used independently by Ueda and Hirose *(10)* to determine the sequence recognized by BmFTZ-F1, the *Bombyx*

mori homolog of the *Drosophila melanogaster* factor FTZ-F1. These authors used three 28-bp degenerate oligonucleotides (with 11-bp primer sequence) that were based on the sequence of a DmFTZ-F1 binding site upstream of the *fushi taratsu* gene. Each oligonucleotide had four variable positions with equimolar base proportions at each position. This set scanned the region corresponding to a previously identified 12-bp consensus. It was found that a combination of direct chemical sequencing after one selection round and gel mobility shift competition analysis using a limited number of oligonucleotides was quite effective in identifying the binding specificity of the moth homolog. The sequence of the oligonucleotides used in the competition analysis was based on the initial results of the direct chemical sequencing and their use served two purposes: to verify clear base preferences obtained with direct sequencing and to examine the contribution in specificity of positions that did not show clear-cut base preferences. Their results identified a 9 bp sequence involved in specific binding that did not include the last 3 bp of the previously proposed consensus. Notably, these last 3 bp included a G previously identified as important by methylation interference experiments. Consistent with our experience, their analysis showed that relatively high differences in affinity (at least three- to fivefold, best if more than 10-fold, compared to other possible sequence combinations) are able to produce clear-cut results with direct sequencing after one round of selection.

18. Alternative BSSA schemes are also feasible. For example, a possible scheme is to start with a consensus sequence and design degenerate oligonucleotides with the most frequently used bases constant and several surrounding positions varied randomly. The feasibility of such a scheme after one round of selection was established by Blackwell and Weintraub *(4)*.

References

1. Gogos, J. A., Tzertzinis, G., and Kafatos, F. C. (1991) Binding site selection analysis of protein-DNA interactions via solid phase sequencing of oligonucleotide mixtures. *Nucleic Acids Res.* **19,** 1449–1453.
2. Sambrook, J., Fritsch, E. F., and Maniatis, T. (1989) *Molecular Cloning: A Laboratory Manual* (2nd Ed.), vol. 1, Cold Spring Harbor Laboratory, Cold Spring Harbor, NY, pp. 6.46–6.47.
3. Hybond M-G protocol (supplied with the product). Amersham, Arlington Heights, IL.
4. Blackwell, T. K. and Weintraub, H. (1990) Differences and similarities in DNA binding preferences of Myo-D and E2A protein complexes revealed by binding site selection. *Science* **250,** 1104–1110.

5. Gogos, J. A., Karayiorgou, M., Aburatani, H., and Kafatos, F. C. (1990) Detection of single base mismatches of thymine and cytosine residues by potassium permanganate and hydroxylamine in the presence of tetralkylammonium salts. *Nucleic Acids Res.* **18,** 6807–6814.
6. Blackwell, T. K., Kretzner, L., Blackwood, E. M., Eisenman, R., and Weintraub, H. (1990) Sequence specific DNA-binding by the c-Myc protein. *Science* **250,** 1149–1151.
7. Gogos, J. A., Hsu, T., Bolton, J., and Kafatos, F. C. (1992) Sequence discrimination by alternatively spliced isoforms of a DNA binding zinc finger domain. *Science* **257,** 1951–1955.
8. Hsu, T., Gogos, J. A., Kirch, S., and Kafatos, F. C. (1992) Multiple zinc finger forms resulting from developmentally regulated alternative splicing of a transcription factor gene. *Science* **257,** 1946–1950.
9. Shea, M. J., King, D. L., Conboy, M. J., Mariani, B. D., and Kafatos, F. C. (1990) Proteins that bind to *Drosophila chorion* cis-regulatory elements: a new C_2H_2 zinc finger protein and a C_2C_2 steroid receptor like component. *Genes Dev.* **4,** 1128–1140.
10. Ueda, H. and Hirose, S. (1991) Defining the sequence recognized with BmFTZ-F1, a sequence specific DNA factor in the silkworm, *Bombyx mori,* as revealed by direct sequencing of bound oligonucleotides and gel mobility shift competition analysis. *Nucleic Acids Res.* **19,** 3689–3693.

Chapter 24

Analysis of DNA–Protein Interactions by Intrinsic Fluorescence

Mark L. Carpenter and G. Geoff Kneale

1. Introduction

Changes in the fluorescence emission spectrum of a protein on binding to DNA can often be used to determine the stoichiometry of binding and equilibrium binding constants; in some cases the data can also give an indication of the location of particular residues within the protein. The experiments are generally quick and easy to perform, requiring only small quantities of material (1). Spectroscopic techniques allow one to measure binding equilibria (unlike, for example, gel retardation assays and other separation techniques, which are strictly nonequilibrium methods). Fluorescence is one of the most sensitive of spectroscopic techniques, allowing the low concentrations (typically in the μM range) required for estimation of binding constants for many protein–DNA interactions. Considerable care, however, needs to be exercised in conducting the experiment itself and in the interpretation of results. The fundamental principles of fluorescence are discussed briefly below.

A molecule that has been electronically excited with UV/visible light can lose some of the excess energy gained and return to its ground state by a number of processes. In two of these, fluorescence and phosphorescence, this is achieved by emission of light. Phosphorescence is rarely observed from molecules at room temperature and will not be considered further. Although electrons can be excited to a number of higher energy states, fluorescence emission in most cases only occurs from the first vibrational level of the first excited state. This has two implications for the measurement of fluorescence emission spectra. First, some of the energy initially absorbed is lost prior to emission, which means that the light emitted will be of longer wavelength (i.e., lower energy) than that absorbed. This is known as the Stoke's shift. Second, the emission spectrum and therefore the wavelength of maximum fluorescence will be independent of the precise wavelength used to excite the molecule. Thus, for tyrosine, the wavelength of the fluorescence maximum is observed around 305 nm, regardless of whether excitation is at the absorption maximum (ca. 278 nm) or elsewhere in the absorption band. Of course, the fluorescence intensity will change as a consequence of the difference in the amount of light absorbed at these two wavelengths.

The fraction of light emitted as fluorescence compared to that initially absorbed is termed the quantum yield. The value of the quantum yield for a particular fluorophore will depend on a number of environmental factors, such as temperature, solvent, and the presence of other molecules that may enhance or diminish the probability of other processes deactivating the excited state. The deactivation or quenching of fluorescence by another molecule, either through collisional encounters or the formation of excited state complexes, forms the basis of many of the fluorescence studies on protein–DNA interactions.

The study of protein–nucleic acid interactions is greatly simplified by the fact that all detectable fluorescence arises from the protein, all four of the naturally occurring DNA bases being nonfluorescent by comparison. Tyrosine and tryptophan residues account for almost all the fluorescence found in proteins. As a general rule, when both residues are present the emission spectrum will be dominated by tryptophan, unless the ratio of tyrosines to tryptophans is very high. The quantum yield of a tyrosine residue in a protein compared to that observed in free solution is generally very low, illustrating the susceptibility of tyrosine to quenching. Tryptophan residues are highly sensitive to the polarity of the surrounding solvent, which affects the energy levels of the first excited state with the result that the emission maximum for tryptophan can range from 330 nm in a hydrophobic environment to 355 nm in water. Thus, in proteins containing only one tryptophan, the general environment surrounding the residue can be ascertained. Tryptophan fluorescence, like that of tyrosine,

Intrinsic Fluorescence

can be quenched by a number of molecules including DNA. Unlike tyrosine, the emission maximum can also change if tryptophan is involved in the interaction and this can also be used to monitor DNA binding *(2)*.

The extent to which the fluorescence of a protein is quenched by DNA is proportional to the concentration of quencher. Since quenching is caused by the formation of a complex between the protein and the DNA, the extent of quenching is proportional to the amount of bound protein. Thus, by determining the extent to which the protein fluorescence is quenched when fully bound to DNA (i.e., at saturation), the fraction of bound and free protein at any point in a titration can be determined. From this data, the stoichiometry and binding constant of the interaction can often be obtained. Note that to establish an accurate stoichiometry, a high concentration of protein is preferred when titrating with DNA (i.e., well above the K_d of the complex) to ensure stoichiometric binding. To establish the binding constant itself, one should be working at much lower concentrations of protein so that at the stoichiometric point, there is a measurable concentration of unbound protein. In the case of protein–DNA interactions having a low K_d, this may not be possible.

If fluorescence quenching is being used to follow DNA binding, it is vital to take account of sample dilution, as well as the increased absorption of the sample as DNA is titrated in. The latter effect is known as the inner filter effect, and arises from the absorption of the excitation beam (and generally to a lesser extent, the emission beam) on passing through the sample (*see* Fig. 1). One should aim to keep the absorption of the sample (at the excitation wavelength) as low as possible, although absorbances up to 0.2 can normally be corrected for without too much difficulty. A small pathlength cell will also help (if rectangular, the excitation beam should pass through the smallest path). Ideally the absorption of the sample at the excitation and emission wavelengths (A_{ex} and A_{em}) should be measured for each point in the titration (if not, one can calculate these values from the known concentrations of protein and nucleic acid at each point). For normal right angled geometry of observation, the corrected fluorescence F_{corr} can be obtained from the observed fluorescence F_{obs} by the formula:

$$F_{corr} = F_{obs} \cdot 10^{(A_{ex}/2 + A_{em}/2)} \qquad (1)$$

Often the value of A_{em} is small enough to ignore (for a detailed treatment of the inner filter correction, *see* ref. *3*). Note that it is equally

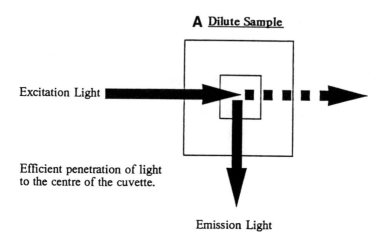
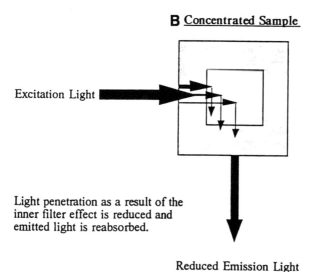

Fig. 1. Schematic representation of the inner filter effect in fluorescence, showing the effect of high concentration on the absorbance of the excitation beam.

important to correct for the inner filter effect whether titrating protein into DNA or vice versa.

The following method deals only with the determination of DNA binding curves by intrinsic fluorescence quenching. However, fluo-

Intrinsic Fluorescence 317

rescence anisotropy can also be used if the molecular size of the complex is sufficiently different from the free protein—for example to investigate proteins that bind cooperatively to DNA *(4)*. Time resolved fluorescence techniques are also advantageous (for measurements of fluorescent lifetimes or rotational correlation times) but require sophisticated instrumentation *(5)*.

2. Materials

1. Reagents used in buffer solutions should be of the highest purity available and the solutions prepared in double-distilled water. The buffer should have negligible absorbance in the 260–300 nm wavelength range and should not be used if it shows any fluorescence in the region 290–400 nm. High-mol wt ions should be avoided. Phosphate buffer should not be used if tyrosine fluorescence is being monitored.
2. Stock solutions of protein and DNA should be divided into small aliquots and stored at −20°C, assuming this is not detrimental to the protein.
3. High quality quartz cuvets with all four faces polished (*see* Notes 1 and 2).
4. Most commercially available fluorimeters allow scanning by both the excitation and emission monochromators and are suitable for use in these studies. Cell compartments that can be thermostatically controlled and contain a magnetic stirrer are preferable but not essential (*see* Notes 3 and 4). We routinely use a Perkin-Elmer (Norwalk, CT) LS5B fluorimeter.
5. To record fluorescence spectra an X-Y recorder or on-line computer will need to be linked to the fluorimeter. We generally control the fluorimeter and record data with an Apple Macintosh computer, using the software package "Bearphoton" written by Jeff Foote.

3. Methods

The method described in this section assumes that nothing is known concerning the fluorescence properties of the protein or its complex with DNA. Consequently the initial steps described in Section 3.1. are concerned with characterizing some of the fluorescence properties of the two species, such that the optimal conditions for obtaining accurate and reliable data can be obtained. Section 3.2. describes the procedure for obtaining data for a protein that is quenched by DNA, which results only in a decrease in fluorescence intensity, and how this data can be used to obtain information on binding. Several variations of the method are mentioned in the Notes.

3.1. Preliminary Experiments

1. Switch on the fluorimeter and allow 10 min for the components to stabilize. Set the excitation and emission slits to intermediate values (e.g., 10 nm bandpass).
2. Fill the cuvet with protein solution (~1 mM but *see* Note 5). Allow time for the solution to equilibrate to the temperature of the compartment. To prevent local heating of the solution or possible photodecomposition, the excitation shutter should be kept closed except when taking measurements.
3. If the absorption spectrum of the protein is known set the excitation wavelength to that corresponding to the absorption maximum between 265 and 285 nm; if no peak exists the protein does not contain tyrosine and tryptophan residues and will not fluoresce. If the absorption spectrum is unknown set the excitation wavelength to 280 nm.
4. Open the excitation shutter and quickly scan the emission between 285 and 400 nm looking for the wavelength at which a maximum value for the intensity is given on the readout. Return the emission monochromator to this wavelength.
5. Find the excitation wavelength maximum between 265 and 285 nm in the same manner, with the emission monochromator set at the wavelength of maximum fluorescence.
6. With both the excitation and emission wavelengths set at their peak values, adjust the instrument to give a reading corresponding to about 90% of the full scale. Narrow slit widths and a lower amplification (expansion factor, gain) are preferred and a compromise between the two may have to be found (*see* Note 6).
7. Determine the emission spectrum by scanning the emission monochromator over the entire wavelength range over which fluorescence occurs. A scan speed of 60 nm/min is generally suitable.
8. Add a small aliquot of a concentrated DNA solution to the cuvet, such that the concentration of DNA is in excess. Mix and immediately check the fluorescence emission at the emission maximum of the protein. Check several times over the next few minutes until a consistent reading is obtained. Allow this time for equilibration in subsequent experiments. Do not adjust slit widths or the amplification.
9. Obtain an emission spectrum and compare with that obtained for the protein only. If fluorescence quenching is suspected, make sure that allowance for sample dilution has been made. If an inner filter correction is required, measure the absorbance of the sample in a spectrophotometer (in the same cuvet) and correct the observed fluorescence as discussed in Section 1.

Intrinsic Fluorescence

10. Add aliquots of DNA until there is no further change in fluorescence intensity in the emission spectrum (*see* Notes 7 and 8).

3.2. Protein–DNA Titrations

1. Examine the emission spectrum of the free protein. If it is characteristic of tyrosine fluorescence check for interference from the Raman band (*see* Note 9). If tryptophan-like check for tyrosine contributions that may be masked (*see* Note 10).
2. Examine the emission spectrum of the protein bound to DNA. The titration method described in the following passage is particularly applicable when the only change in the spectrum is a change in fluorescence intensity. Several variations of this method are described briefly in the Notes, including an example in which the emission spectrum of the protein shifts on binding DNA (*see* Notes 11 and 12).
3. Accurately determine the concentration of protein and DNA solutions (e.g., by UV spectroscopy). Since we are titrating DNA into protein try to use a stock concentration of DNA that is at least 20 times the concentration of protein used in the experiment multiplied by the estimated stoichiometric ratio; for example, if the protein concentration used is 10 μM and the estimated stoichiometry is 5 bases per protein then the DNA concentration should be at least $20 \times 10\,\mu M \times 5 = 1000\,\mu M$ (1 mM). This would mean that the dilution of the original protein solution will be only 5% at the stoichiometric point.
4. Using the protein solution set up the instrument as described in steps 1–6 of Section 3.1. If measuring tyrosine fluorescence use an excitation wavelength near the maximum. This wavelength can also be used for tryptophan excitation if tyrosine fluorescence is insignificant, otherwise use an excitation wavelength of 295 nm.
5. Run a buffer blank and check that the profile of the emission spectrum is consistent with that previously obtained. Subtract this spectrum from subsequent spectra, if this can be done automatically.
6. Set the emission monochromator to the emission wavelength maximum and ensure that the readout is about 90% of its maximum value. Note down the value.
7. To begin the titration add a small aliquot from the stock DNA solution to the protein in the cuvet. Mix and allow the sample to equilibrate (use the time period determined earlier) before taking a reading. The aliquots should be sufficiently small, such that the protein is still greatly in excess and a linear change is observed as more DNA is added.
8. Continue to add the same quantity of DNA for 8–10 points. If changes are still approximately linear at this stage gradually increase the volume of the DNA added, noting the total amount added at each point.

9. When the change in intensity begins to deviate significantly from linearity decrease the size of the aliquot so that more data points are obtained in this region.
10. As quenching approaches the maximum larger aliquots of DNA can be added. Continue until no change in quenching is observed for several points.
11. After the last point check that the emission spectrum of the complex is consistent with that previously obtained for the bound protein.
12. Remove the sample, wash the cuvet thoroughly, and run a blank spectrum consisting of cell plus buffer. This should have negligible or no fluorescence. Subtract any value at the emission wavelength maximum from the data points, if not already done automatically (see Note 9).
13. For each data point calculate the fluorescence quenching ($Q = 100 [Fo - F]/Fo$, where F = the measured fluorescence and Fo is the fluorescence in the absence of DNA), having made any corrections for inner filter effects and dilution of the sample. Also calculate the nucleotide concentration at each point ($N = ny/[n + x]$, where n = total volume of DNA added up to that point, x = initial volume of sample in cuvet, and y = molar concentration of DNA stock solution). From the DNA concentration calculate R, which is the ratio of the concentration of DNA to that of protein. (For polynucleotides it is usual to express N as the concentration of nucleotides; for short synthetic duplexes, it is more usual to use the molar concentration of duplex.)
14. Plot a graph of Q against R (or N). If only one mode of binding is occurring the graph should look like one of the curves shown in Fig. 2. If the "break point" in the titration is sharp, it indicates a high value for the binding constant (i.e., a small dissociation constant compared with the protein concentration used). Conversely, too weak a binding constant (or too dilute a protein solution) will give rise to a smoothly rising curve with no apparent break point.
15. The stoichiometry of binding is the value of R at which the slope obtained from the initial linear range of the titration crosses the horizontal line defined by Q_{max} at which no further change in intensity occurs.
16. Further information can be extracted from the binding curve by fitting it to an appropriate model. In the simplest case of a bimolecular interaction ($P + N = PN$), then a useful expression to estimate the binding constant is $K = \theta/(1 - \theta)^2 [P_o]$ where θ is the fraction of bound protein at the stoichiometric point and $[P_o]$ is the total protein concentration in the cuvet. This expression also applies to more complex cooperative binding along a linear DNA lattice (6), assuming the cooperativity is sufficiently high, when K becomes equal to the apparent binding con-

Intrinsic Fluorescence

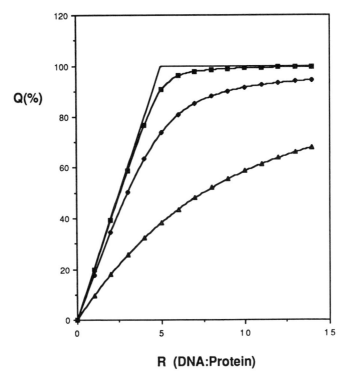

Fig. 2. A graph of fluorescence quenching against DNA:protein ratio (R). The curves illustrate the addition of DNA (in this case, a polynucleotide) to protein (10 μM). The theoretical binding curve for infinitely strong binding (upper curve) shows a stoichiometry of 5 nucleotides bound per protein. Typical curves are shown for binding constants of 10^7 (■), 10^6 (♦), and 10^5 (Δ) M^{-1}.

stant (and approximates to the product of the cooperativity factor and the intrinsic binding constant for one site). For a more extensive discussion of complex DNA binding equilibria, *see* ref. 7.

4. Notes

1. Most fluorimeter cell compartments are designed to take cuvets with a 1 cm path length (distance between opposite faces) and usually require 2.5–3.0 cm^3 of sample for measurements. If smaller sample volumes are required then reduced volume cuvets similar to those used in absorption studies but suitable for fluorescence work can usually be obtained from most cuvet suppliers. Also, most suppliers are prepared to construct cuvets to your own specifications (at a cost). The major requirement is that the sample is located in the center of the cell

(assuming standard right angle observation). Cuvets holding as little as 300 μL have been successfully used by the authors.
2. Care should be taken when handling fluorescence cuvets since both fingerprints and scratches can introduce significant artifacts into the experiment. After use, cuvets should be thoroughly washed with distilled water and a mild detergent if necessary. If greatly contaminated, then immerse the cells in a 50:50 mix of ethanol with sulfuric acid ($4M$) for several hours and then rinse thoroughly with water.
3. If the fluorimeter is not equipped with a magnetic stirrer unit, adequate mixing can usually be achieved by gently drawing the solution in and out through a plastic pipet tip. Avoid introducing bubbles into the sample since this can both denature the protein and cause light scattering. For accurate measurements, temperature control is essential because fluorescence is highly sensitive to temperature.
4. If the use of a reduced volume cell prevents the use of commercially available magnetic fleas then substitutes can be made as follows:
   a. Seal the narrow end of a Pasteur pipet or micropipet by heating it in the flame of a Bunsen burner.
   b. Insert a small length of iron wire (cut up a paperclip) and shake it down to the sealed end.
   c. Cut the pipet just above the wire using a glass cutter and seal the open end in the flame. These fleas should only be used once because some rusting occurs with time.
5. Fluorescence intensity is only proportional to concentration when the absorbance is no greater than 0.1 absorbance units at the excitation wavelength selected. If a molar extinction coefficient for the protein is known, use this to calculate a suitable protein concentration. Remember that the absorption bands for proteins and nucleic acids overlap and in titrations the contribution of the nucleic acid to the overall absorption must be considered.
6. Although it is preferable to have narrow excitation and emission slit widths and a low amplification factor, there may be a need to compromise in order to obtain a stable reading. For proteins displaying tyrosine fluorescence, the small wavelength difference between the excitation and emission maxima suggests that it would be better to maintain narrow slits and increase the signal amplification. For proteins dominated by tryptophan fluorescence the greater the difference between the excitation and emission wavelengths, the greater the feasibility of increasing the slit widths and maintaining a lower amplification. In general, when measuring emission spectra it is better to use a narrow emission slit width and widen the excitation slit width. For

broad banded spectra, such as that seen with tryptophan, both slits can be widened.

7. If no changes in the emission spectrum of the protein are observed when DNA is added (after inner filter and dilution corrections if necessary), then either the protein is not binding to DNA, or binding cannot be detected by this procedure and will need to be assessed by another method, such as fluorescence anisotropy or the use of an extrinsic probe (*see* Chapter 25).
8. Note the molar ratio of DNA: protein at which no further changes occur. This will provide a rough guide for future experiments.
9. Tyrosine emission can often be confused with Raman scattering of light that occurs around 305 nm when an excitation wavelength of 280 nm is used. The presence of the Raman band can be assessed by measuring the emission spectrum using a different excitation wavelength. The fluorescence emission spectrum is independent of excitation wavelength whereas Raman scattering occurs at a constant wavenumber (= $1/\lambda$ in cm^{-1}) from that used for excitation and will shift in the same direction as the change in excitation wavelength. The contribution of the Raman band to the overall intensity of the signal can be assessed by running an emission spectrum of a buffer blank. Automatically subtract out this spectrum from subsequent spectra where possible.
10. To check the contribution tyrosine may make to a fluorescence emission spectrum dominated by tryptophan, run an emission spectrum using an excitation wavelength of 295 nm. At this wavelength only tryptophan emission will be observed. If the emission spectrum is unchanged then it can be concluded that the contribution from tyrosine residues is negligible. (Of course the intensity will be lower since tryptophan absorption is greater at 280 nm than it is at 295 nm.)
11. In cases where both the emission maximum shifts and the fluorescence intensity is quenched, the method described can be used provided that an emission wavelength is chosen outside the wavelength region overlapped by the emission spectra of the free and bound protein. Alternatively, the ratio of the intensity of the emission maxima of the free and bound protein can be followed (for an example *see* ref. 2). The use of a ratio method means that the dilution factor and inner filter correction can usually be ignored, although strictly speaking the ratio is not a linear function of degree of binding.
12. In some cases it may be preferable to titrate DNA with protein (for an example of this method, *see* ref. 8). The procedure is similar to that given here, but in this case the experiment should be repeated by adding protein to the buffer in the absence of DNA as a reference. Subtrac-

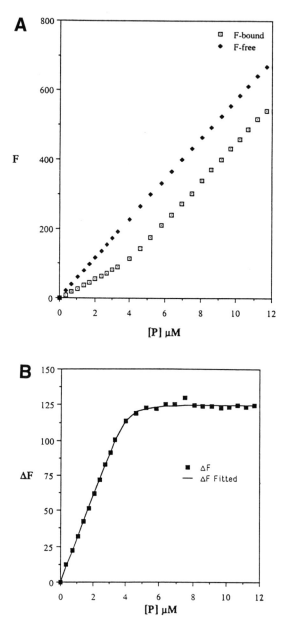

Fig. 3. (**A**) Titration of an oligonucleotide with fd gene 5 protein. Fluorescence (305 nm) of increasing concentrations of protein (P) is measured in the presence (lower curve) and absence (upper curve) of DNA. The difference between these two curves is plotted in (**B**) along with the theoretical binding curve. In this experiment, the starting concentration of DNA was 16.7 μM.

tion of the two curves should yield a clear binding curve (*see* Fig. 3). (For a discussion of the merits of whether to titrate protein with DNA or vice versa, *see* ref. *4*.) We have found that in some cases different results can be found dependant on the direction of the titration; this can occur when the fluorescence changes observed include contributions from protein-protein interactions accompanying DNA binding in addition to (or instead of) contributions from the interaction with the DNA itself.

References

1. Harris, D. A. and Bashford, C. L., eds. (1987) *Spectrophotometry and Spectrofluorimetry: A Practical Approach.* IRL, Oxford, UK. (Chapters 1 and 4 are particularly relevant).
2. Kneale, G. G. and Wijnaendts van Resandt, R. W. (1985) Time resolved fluorescence of the Pf1 bacteriophage DNA binding protein: determination of oligo- and polynucleotide binding parameters. *Eur. J. Biochem.* **149,** 85–93.
3. Birdsall, B., King, R. W., Wheeler, M. R., Lewis, C. A., Goode, S. R., Dunlap, R. B., and Roberts, G. C. K. (1983) *Anal. Biochem.* **132,** 353–361.
4. Carpenter, M. L. and Kneale, G. G. (1991) Circular dichroism and fluorescence analysis of the interaction of Pf1 gene 5 protein with poly(dT). *J. Mol. Biol.* **217,** 681–689.
5. Greulich, K. O., Wijnaendts van Resandt, R. W., and Kneale G. G. (1985) Time resolved fluorescence of bacteriophage Pf1 DNA binding protein and its complex with DNA. *Eur. Biophys. J.* **11,** 195–201.
6. Kelly, R. C., Jensen, D. E., and von Hippel, P. H. (1976) Fluorescence measurements of binding parameters for bacteriophage T4 gene 32 protein to mono-, oligo- , and polynucleotides. *J. Biol. Chem.* **251,** 7240–7250.
7. McGhee, J. D. and von Hippel, P. H. (1974) Theoretical aspects of DNA-protein interactions: cooperative and noncooperative binding of large ligands to a one-dimensional homogeneous lattice. *J. Mol. Biol.* **86,** 469–489.
8. Alma, N. C. M., Harmsen, B. J. M., de Jong, E. A. M., Ven, J. V. D., and Hilbers, C. W. (1983) Fluorescence studies of the complex formation between the gene 5 protein of bacteriophage M13 and polynucleotides. *J. Mol. Biol.* **163,** 47–62.

CHAPTER 25

A Competition Assay for DNA Binding Using the Fluorescent Probe ANS

Ian Taylor and G. Geoff Kneale

1. Introduction

Fluoresence spectroscopy is a useful technique for investigating the interaction of DNA binding proteins with DNA. Generally, use is made of the intrinsic fluorescence of the protein arising from the aromatic amino acids, which is frequently perturbed in a DNA–protein complex (*see* Chapter 24). In some cases, however, changes in the intrinsic fluorescence emission of the protein arising from its interaction with nucleic acid may not be detectable. For example, if tryptophans and/or tyrosines are not located in the proximity of the DNA binding site, the emission spectrum may not be perturbed by the interaction. Furthermore, if the protein contains a large number of tryptophan and tyrosine residues, any effects may be masked because of the large number of aromatic residues whose emission spectrum is unperturbed by the binding of nucleic acid.

To overcome this problem an alternative approach to the measurement of DNA binding is to add an extrinsic fluorescence probe to the system that competes with DNA for the binding site of the protein. One can then measure the change in the fluorescence emission spectrum of the fluorescent probe as DNA is added. If the fluorescence characteristics of the free and bound probe differ, then displacement of the probe by DNA can be observed.

From: *Methods in Molecular Biology, Vol. 30: DNA–Protein Interactions: Principles and Protocols*
Edited by: G. G. Kneale Copyright ©1994 Humana Press Inc., Totowa, NJ

The fluorescent probe 1-anilinonapthalene-8-sulfonic acid (ANS) and its derivatives have long been used to study protein structure *(1)* and more recently to study protein-nucleic acid interactions *(2–4)*. ANS has the property that its fluorescence is enhanced approx 100-fold with a 50 nm blue wavelength shift when transferred from an aqueous environment to a less polar solvent, such as methanol (*see* Fig. 1). ANS will bind weakly to hydrophobic patches on protein molecules with an average dissociation constant K_d in the region of 100 µM *(3)*. When molecules of ANS are bound to protein, enhancement and shifting of the fluorescence spectrum similar to that observed in apolar media often occurs (*see* Fig. 2). Thus bound molecules of ANS fluoresce much more strongly and at a shorter wavelength than ANS molecules in an aqueous solvent.

The precise reason why ANS should bind at DNA binding sites is not entirely clear, since such sites are not particularly hydrophobic. Nevertheless, ANS is a planar aromatic molecule that will have some properties in common with the DNA bases despite the lack of hydrogen bonding capacity. Furthermore, the negatively charged sulfonate group of ANS may mimic the phosphate group of the DNA backbone.

The protocol given below was used successfully to investigate DNA recognition a type I modification enzyme (*Eco*R124 DNA methylase *[4]*). It was possible to demonstrate differential binding affinity for an oligonucleotide containing the canonical recognition sequence, compared with one that differs by just one base pair in the nonspecific sequence. It remains to be seen just how general the method is, since there have been very few instances of its application to DNA binding proteins. Since it can be established fairly rapidly whether ANS binds to a given protein, and whether there is some release of the bound ANS by the addition of DNA, it is a technique worth investigating. Sections 3.1. and 3.2. deal with these preliminary experiments. If these are encouraging, then accurate fluorescence titrations can be done to investigate the DNA binding characteristics of the protein in more detail (Section 3.3.).

2. Materials

1. A fluorimeter is required that is capable of scanning with both the emission and excitation monochromators, and in which the emission and excitation slit widths can be varied. In our laboratory we routinely use a Perkin-Elmer (Norwalk, CT) LS5B. The fluorimeter is controlled and

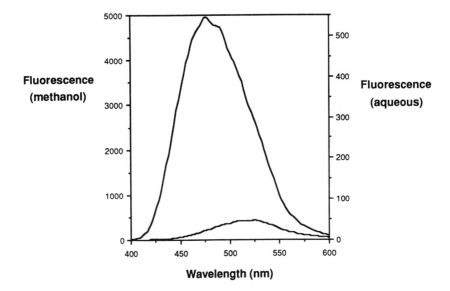

Fig. 1. Fluorescence emission spectra of ANS in methanol (upper curve) and aqueous buffer (lower curve), using an excitation wavelength of 370 nm. The buffer composition is given in Section 2.

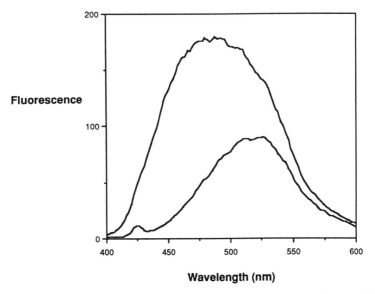

Fig. 2. Fluorescence shift in the spectrum of ANS (100 µM) induced by the addition of protein (2.24 µM *Eco*R124 methylase). Conditions as in Fig. 1.

data collected by an Apple Macintosh computer using the software package "Bearphoton" (written by J. Foote).
2. It is also necessary to be able to thermostatically control the cell holder compartment. The temperature should be controlled to 0.1°C by circulating water through the cell holder block, the temperature of which is controlled from a programmable circulating waterbath (e.g., Neslab [Newington, NH] RTE-100).
3. All buffer solutions should be prepared from the higest quality reagents and ultrapure water. The buffer should be degassed and filtered to remove any particulate contaminants. In the example provided, the assay was carried out in 10 mM Tris-HCl, pH 8.2, 100 mM NaCl, 5 mM MgCl$_2$ (*see* Note 1).
4. High purity 1-anilino-napthalene-8-sulfonic acid (ANS) free of contaminating *bis*-ANS is obtained from Molecular Probes (Eugene, OR). Solutions of ANS should always be made fresh just prior to use in a titration.
5. A 50 µM stock solution of purified DNA binding protein.
6. A 100 µM stock solution of synthetic oligonucleotide duplex (*see* Note 2).
7. Good quality quartz cuvets (preferably stoppered), with all four faces polished. A 1 × 0.4 cm (semimicro) cuvet is preferred, since this minimizes the inner filter effect compared to the standard 1 × 1 cm (3 mL) cuvets. The excitation beam should pass through the 0.4 cm path, since absorption of the emission beam is negligible.

3. Methods

3.1. Titration of Protein with ANS

To find optimal conditions for the use of ANS in a DNA binding experiment it is advisable to first titrate the protein of interest with ANS to check the extent to which the protein binds the fluorescent probe. The precise concentrations of reagents and composition of the buffer used here work well for the *Eco*R124 methylase, and its subsequent binding to a 30-bp DNA fragment. It may be necessary to vary the conditions for other systems.

1. Prepare 250 mL of degassed and filtered buffer—e.g., 10 mM Tris-HCl, pH 8.2, 100 mM NaCl, 5 mM MgCl$_2$.
2. Prepare 1 mL of a 1 µM solution of protein in the same buffer.
3. Prepare 50 mL of a 1 mM ANS solution in buffer. The concentration of ANS can be determined from its UV absorption spectrum ($\varepsilon_{370\,aqueous}$ = 5500M^{-1}cm^{-1}; *see* Note 3).
4. Adjust the excitation and emission slits on the fluorimeter to 2.5 nm (wider slits can be used if the signal is weak). Place 1 mL of buffer in a

Fluorescence Competition Assay

quartz fluorimeter cuvet and record the fluorescence emission spectrum between 400 and 600 nm, using an excitation wavlength, λ_{ex}, of 370 nm.

5. Add 2-µL aliquots (2 µM ANS per addition) of the 1 mM ANS solution to the cell, mix gently by inversion, and after each addition record the fluorescence emission spectrum as above. For the final point in the titration, measure the absorbance (at 370 nm) of the sample in the same cuvet, making sure the pathlength of the cuvet is the same as that used for the excitation beam in the fluorimeter.
6. In a fluorimeter cuvet prepare 1 mL of 1 µM protein solution in buffer. Record the fluorescence emission spectrum between 400 and 600 nm (using $\lambda_{ex} = 370$ nm).
7. Repeat the ANS titration (as in step 5, but with the protein now present). Again, for the final point in the titration, measure the absorbance of the sample at 370 nm.
8. Choose the wavelength in the emission spectrum that shows the largest difference between the two titrations. This will probably be in the vicinity of 480 nm, although shorter wavelengths can be used to minimize the background from the free ANS (*see* Fig. 2).
9. Correct the observed fluoresence intensity values at the emission wavelength chosen (F_{obs}) at each individual ANS concentration for inner filter effects using Eq. 1.

$$F_{corr} = F_{obs} \cdot 10^{(A_{ex}/2)} \qquad (Eq.\ 1)$$

where A_{ex} is the absorbance of the sample at the excitation wavelength (370 nm) for the appropriate path length. The absorbance can be calculated from the ANS extinction coefficient and the known ANS concentration at each point, but this should be checked against the measured absorbance at the end point of the titration (*see* Note 4).

10. Plot the corrected fluorescence at the chosen wavelength for each titration against ANS concentration (corrected for dilution if significant; *see* Note 5). A typical case is shown in Fig. 3. The titration curve in the absence of protein should be linear if the corrections for inner filter and dilution have been correctly applied.
11. Subtraction of the curve of ANS added to buffer from the curve of ANS added to the protein solution yields the binding curve of ANS to the protein as illustrated in Fig. 4. The shape of this curve will depend on the ANS binding properties of the particular protein under study (*see* Note 6). In the case shown in Fig. 4, this curve is fairly representative of several ANS molecules associated weakly with a protein.

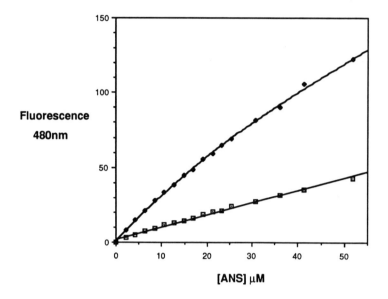

Fig. 3. Fluorescence titration of *Eco*R124 methylase (1.5 μ*M*) with ANS (upper curve). The lower curve shows the fluorescence in the absence of the protein. Both spectra were corrected for the inner filter effect.

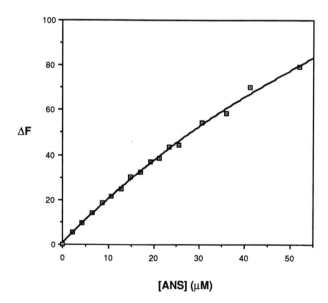

Fig. 4. Protein-ANS binding curve, generated by the subtraction of the two curves in Fig. 3.

3.2. Preliminary Investigation of the Displacement of ANS by DNA

Once satisfied that the protein under investigation binds ANS one must ascertain if any of the bound ANS molecules are located in the DNA binding site of the protein. If so, their fluorescence will change on being displaced by the bound DNA.

1. Make up a solution of 100 μM ANS in buffer (*see* Note 7). Measure its fluorescence emission spectrum between 400 and 600 nm (using λ_{ex} = 370 nm).
2. Make up an identical solution of ANS but containing 1 μM DNA. Measure the emission spectrum as in step 1. These two spectra should be effectively identical, that is, there should be no observable interaction between the nucleic acid and ANS.
3. Make up a 100 μM ANS solution containing 1 μM protein, and an identical solution containing, in addition, 1 μM DNA. Measure the fluorescence emission spectra of these two samples as above.

The presence of protein in the ANS solution should cause a change in the shape and intensity of the emission spectrum. An increase in quantum yield accompanied by a blue shifted spectrum should be observed. When the nucleic acid is present and bound to the protein, any ANS (which is weakly bound) in the DNA binding site of the protein will be displaced and change the form of the spectrum toward that of free ANS (*see* Fig. 5).

3.3. Fluorescence Competition Assay

Once it has been established that the DNA fragment of interest shows a measurable displacement of ANS, further investigation of the DNA binding characteristics of the protein can be carried out. Titrations can be done in a number of different ways but we have found it more reproducible to titrate the protein into of ANS in the presence and absence of the DNA. The difference in fluorescence at each point in the titration then represents the amount of ANS displaced, that is, the amount of DNA bound. The concentrations used should be those found to be optimal from the earlier experiments. Since the concentration of ANS is constant throughout, the absorption at 370 nm should remain unchanged and there is no need for inner filter corrections.

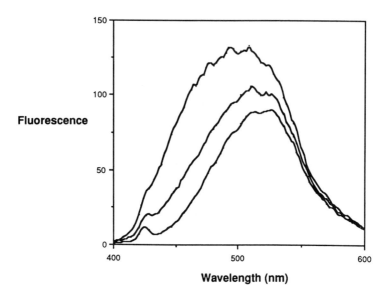

Fig. 5. Fluorescence emission spectra of ANS alone (100 µM; lower curve) and with the addition of 1 µM EcoR124 methylase (upper curve). The addition of 1 µM DNA (a 30-bp synthetic duplex containing the recognition sequence) in the presence of the protein displaces a substantial fraction of the bound ANS, as judged by the resemblance of the resulting spectrum (middle curve) to that of the free ANS.

1. Place 1 mL of a 100 µM ANS solution (in buffer) in the cuvet and record the emission spectrum between 400 and 600 nm (using λ_{ex} = 370 nm).
2. Add 10-µL aliquots of the 50 µM stock protein solution (500 nM per addition) to the cell up to a final protein concentration of 3 µM. After each addition record the fluorescence intensity at the chosen wavelength (e.g., 480 nm; see Section 3.1.8.).
3. Make up a 1 mL solution of 1 µM DNA/100 µM ANS (in buffer) in the cuvet. Again record the fluorescence emission intensity.
4. Again add 10-µL aliquots of the 50 µM stock protein to the cuvet and record the intensity after each addition.
5. Plot the fluorescence at each point of the titration against protein concentration for both experiments (see Fig. 6).
6. To obtain a binding curve for the protein–nucleic acid interaction, subtract the fluorescence intensities at each point in the two experiments (with and without DNA) and replot against protein concentration (see Fig. 7). This difference represents the amount of bound DNA, since DNA binding is solely responsible for the decrease in fluorescence, through displacement of ANS from the binding site. For a high affinity

Fluorescence Competition Assay

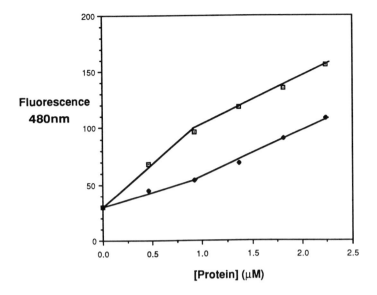

Fig. 6. ANS competition assay. The *Eco*R124 methylase protein was added to ANS (100 μ*M*) in the presence (lower curve) and absence (upper curve) of the DNA duplex (1 μ*M*).

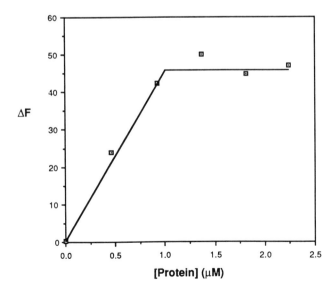

Fig. 7. Binding curve for the interaction of *Eco*R124 methylase with its cognate DNA sequence, generated by subtraction of the two curves shown in Fig. 6.

DNA protein interaction (with K_d substantially less than the concentration of DNA used in the titration) competition from the ANS will be negligible and a stoichiometric binding curve will be produced; the sharp break in the curve at the stoichiometric point indicates the point at which all the available DNA is bound. Curvature of the plot around the stoichiometric point represents a lower affinity interaction. *See* Chapter 24 for further discussion of the analysis of DNA binding curves.

4. Notes

1. The exact composition of the binding buffer will be dependent on the DNA binding protein being studied. If the interaction has been studied by other methods (for example gel retardation) it is advisable to carry out the fluorescence experiment in the binding buffer used in previous studies. However, one should avoid the presence of quenchers.
2. In our laboratory titrations are carried out using short synthetic oligonucleotide duplexes (30-mers), which contain the recognition sequence of the protein of interest, or variations of this sequence. However, there is no reason why experiments cannot be conducted using longer nucleic acids (restriction fragments, polynucleotides, and so forth).
3. The value of $\varepsilon_{370\,aqueous}$ was derived for ANS in the buffer used by comparison of the A_{370} of two equimolar solutions, one in methanol for which ε_{370} is known ($6800 M^{-1}cm^{-1}$; Molecular Probes) and the other in buffer. If the buffer to be used is significantly different to that used in the above procedure, then it is advisable to recalculate an ε_{370}.
4. *See* Chapter 24 for a more detailed discussion of the inner filter effect. As a guide, for excitation at 370 nm in an aqueous buffer 52 µM ANS has an absorption of 0.11 in a 0.4 cm pathlength cell (giving an inner filter correction of 1.14).
5. If the stock solutions of ANS and protein are not available at high concentration then the volume added by the end of the titration will significantly dilute the sample and thus decrease the fluorescence, and this must be corrected for. Unlike absorbance, fluorescence is not strictly proportional to concentration because of inner filter effects. However, it is approximately so if the absorbance is low and the dilution factor is kept small.
6. The binding curve generated for ANS can in principle take many forms depending on the exact affinity of ANS for sites on the protein. If the protein contains high affinity sites the curve may be biphasic and allow the stoichiometry of the interaction to be determined. A more likely situation is that there will be many weak sites for ANS binding (K_d >100 µM) with different affinities, resulting in a curved plot.
7. The concentration of the ANS solution used in the titration must be determined empirically from the previous experiments. It should be high

enough to ensure that a good fraction of the ANS binding sites on the protein are occupied (as determined in Section 3.1.), but not so high as to swamp the signal arising from the displaced ANS (Section 3.2.).
8. As well as using the direct excitation of the fluorescent probe (i.e., with an excitation wavelength of 370 nm for ANS), it may be possible to investigate energy transfer effects between the aromatic amino acid residues of the protein and the bound probe. If excitation is performed at 280 nm to excite both tyrosine and tryptophan, energy transfer to ANS will be apparent from the emission spectrum from 400 to 600 nm. In principal this effect could also be used to follow displacement of ANS in a titration with DNA.
9. A further extension of the fluorescent probe approach is to employ the covalent probe I-AEDANS *(5)*. This reagent reacts with accessible cysteine residues in a protein, and has a higher quantum yield than ANS in aqueous solution. One can look at the emission spectrum of the bound probe, or it may be possible to observe energy transfer from aromatic residues in the protein. Any of these fluorescence characteristics could change when DNA is bound, if the probe is located near the DNA binding site. We have used this technique to study the interaction of *Eco*R124 methylase with DNA, and found that energy transfer from the protein to the bound probe decreased by over 30% when DNA was bound. As long as the presence of this probe does not inhibit DNA binding, then titrations with DNA can be used to produce DNA binding curves. If the probe does inhibit DNA binding, this can also be informative; the labeled cysteine(s) can be identified by peptide mapping by analogy with the method reported in Chapter 12.

References

1. Brand, L. and Golke, J. R. (1972) Fluorescent probes. *Annu. Rev. Biochem.* **41,** 843–863.
2. Secnik, J., Wang, Q., Chang., C. M., and Jentoft, J. E. (1990) Interactions at the nucleic acid binding site of the avian retroviral nucleocapsid protein: studies utilizing the fluorescent probe 4,4'- bis (phenylamino) (1,1'-binapthalene)-5,5'-disulfonic acid. *Biochemistry* **29,** 7991–7997.
3. York, S. S., Lawson, R. C. Jr., and Worah, D. M. (1987) Binding of recrystallized and chromatographically purified 8-anilino–napthalene-sulfonate to *Escherichia coli* lac repressor. *Biochemistry* **17,** 4480–4486.
4. Taylor, I., Patel, J., Firman, K., and Kneale, G. (1992) Purification and biochemical characterisation of the *Eco*R124 type I modification methylase. *Nucleic Acids Res.* **20,** 179–186.
5. Kelsey, D. E., Rounds, T. C., and York, S. S. (1979) Lac repressor changes conformation on binding to poly [d(A-T)]. *Proc. Natl. Acad. Sci. USA* **76,** 2649–2653.

CHAPTER 26

Circular Dichroism for the Analysis of Protein–DNA Interactions

Mark L. Carpenter and G. Geoff Kneale

1. Introduction

The asymmetric carbon atoms present in the sugars of nucleotides and in all the amino acids (with the exception of glycine) results in nucleic acids and proteins displaying optical activity. Further contributions to the optical activity of the polymers result from their ability to form well-defined secondary structures, in particular helices, which themselves possess asymmetry. As a consequence, circular dichroism (CD) has found widespread use in secondary structure prediction of proteins *(1)*. Similar studies, though less widespread, have sought to correlate structural parameters of DNA with their CD spectrum *(2)*. It follows that the disruption of secondary structure by, for example, denaturation or ligand binding can be usefully followed by circular dichroism.

Plane polarized light can be resolved into left- and right-handed circularly polarized components. Circular dichroism measures the difference in the absorption of these two components,

$$\Delta \varepsilon = \varepsilon_L - \varepsilon_R \qquad (1)$$

where ε is the molar extinction coefficient ($M^{-1}\text{cm}^{-1}$) for the left (L) and right (R) components (*see* Note 1). When passing through an optically active sample the plane of polarized light is also rotated, which means that the emerging beam is elliptically polarized. Thus CD is often expressed in terms of ellipticity ($[\theta]_\lambda$ in degrees) or molar ellipticity ($[\theta]_\lambda$ in degrees cm^{2}dm^{-1}).

$$[\theta]_\lambda = 100 \cdot \theta_\lambda/c.l \qquad (2)$$

where c is the molar concentration and l is the pathlength in cm. The two expressions are interconvertible with the expression $[\theta]_\lambda = 3300 \, \Delta\varepsilon$.

The overlap of the UV absorption bands of nucleic acids and proteins means that CD studies of protein–DNA interactions can be complicated by the contributions observed from both components. This is particularly true for wavelengths <250 nm. In practice CD spectra between 250 and 300 nm are dominated by that of the nucleic acid, the contribution arising from the aromatic chromophores of the protein being weak by comparison. Changes in conformation can usually be attributed to the polynucleotide because the random distribution of aromatic amino acids means a large conformational change throughout the protein would be required to cause a significant change in the CD spectrum.

The low molar ellipticity of polynucleotides means that for accurate CD measurements high concentrations (10^{-4}–$10^{-5}M$) of nucleotide are required. For this reason circular dichroism is not generally used to determine binding constants of protein–DNA interactions. However, circular dichroism can be used to obtain accurate values for the stoichiometry of protein–nucleic acid interactions *(3)* and in the case of the fd gene 5 protein was used to show the existence of two distinct binding modes *(4)*. Circular dichroism has also been used to show that conformational changes induced by bound *Lac* repressor are different for operator DNA and for random sequence DNA *(5)*. Similar studies on *Gal* repressor demonstrated the involvement of the central G-C base pairs of the operator sequence in repressor induced conformational changes *(6)*. Similar studies on the interaction of *cro* protein of bacteriophage λ have also revealed different conformational changes for specific and nonspecific DNA binding *(7)*. Despite the apparent lack of any direct interaction of the central base pair of the operator sequence with *cro* protein, base substitution at this site was shown to affect the CD spectrum considerably.

It should be emphasized that circular dichroism provides complementary data to other spectroscopic techniques, such as fluorescence, since such techniques can monitor different components of the interaction. Indeed, even the apparent stoichiometry of binding can be significantly different for this reason, even when measured under the same solution conditions *(3)*.

2. Materials

1. A high quality quartz cell with low strain is required for accurate measurements. The cell pathlength will depend on the absorption properties and concentration of the sample. Cells with pathlengths between 0.05 and 1 cm are often used, depending on the CD signal to be measured and the absorption of the sample (including the buffer). A pathlength of 1 cm is usually recommended for measurement of the DNA signal in the vicinity of 275 nm, where the signal is weak and buffer absorption is negligible.
2. Buffers should be prepared using high quality reagents and water. Use buffers that have low absorbance in the wavelength region of interest. Tris-HCl, perchlorate, and phosphate are routinely used.
3. Stock solutions of appropriate protein and nucleic acid solutions in the same buffer: The protein should be as concentrated as possible to minimize dilution during the titration. If a synthetic DNA fragment containing the recognition sequence is to be used, it should be close to the minimum size required for binding to maximize the change in CD signal. To avoid denaturation and degradation keep concentrated solutions of protein and DNA frozen in small aliquots, if it has been established that this procedure does not damage the protein.
4. A supply of dry nitrogen (oxygen free).
5. (+)10-camphor-sulfonic acid at a concentration of 0.5 mg/mL, to be used as a calibration standard. Check the accurate concentration by UV spectroscopy using a molar extinction coefficient of 34.5 at 285 nm *(1)*.
6. A circular dichroism spectrometer (spectropolarimeter). In the procedure given below, we have made use of a Jasco J-20 spectrometer that, although an older model, is capable of giving reasonable results. However, more modern machines (for example, the J600 or J700 series from Jasco) offer clear advantages in terms of sensitivity, stability, and general convenience of use, and will minimize potential problems because of baseline drift.

3. Methods

For most proteins there is no significant CD spectrum between 250 and 300 nm compared to that seen for nucleic acids. Experiments involving the addition of protein can thus be conveniently carried out in this wavelength range, as described below. Below 250 nm both proteins and DNA have optical activity and any experiments here may require resolution of the spectrum into protein and DNA components.

1. To prevent damage to the optics, flush the instrument with nitrogen for 10–15 min before switching the lamp on. Continue to purge the instrument for the duration of the experiment (*see* Note 2).
2. Switch the lamp on and allow the instrument to stabilize for 30 min before making any measurements.
3. While waiting for the instrument to warm up, measure the UV spectrum of both the DNA and protein and calculate the concentration of the stock solutions from their extinction coefficients. The stock solution of protein should be at as high a concentration as possible, to minimize corrections for dilution in subsequent titrations.
4. Measure the UV absorbance of the cell to be used in the CD experiment against an air blank (*see* Note 3).
5. Using the data from steps 3 and 4 determine the concentrations of DNA and protein that can be used in the experiment such that the total absorbance of all components including the cell is <1.0.
6. Once the CD instrument has warmed up use a 1-cm cell filled with water to determine the baseline between 250 and 320 nm. This should be flat.
7. Calibrate the instrument by replacing the water with the previously prepared solution of camphor sulfonic acid and measure the CD between 250 and 320 nm. A 0.5 mg/mL solution in a 1-cm pathlength cell has ellipticity of 168 millidegrees at 290.5 nm.
8. Take the cell set aside for the experiment and fill with buffer. Place the cell in the instrument taking care to note the orientation of its faces in the beam. If using a cylindrical cell place it so that the neck of the cell rests against the side of the cell holder. Run a baseline between 250 and 320 nm.
9. Replace the buffer with the DNA solution and run the spectrum under the same conditions. If you remove the cell to do this remember to place the cell back in the holder with the same face toward the light source.
10. Accurately pipet a small aliquot of the stock protein solution into the DNA in the cell and mix. Allow time for equilibration and measure the CD spectrum. The volume of the protein solution added will be determined by its concentration, the DNA concentration in the cuvet, and a rough idea of the expected stoichiometry. For initial experiments, the molar quantity of protein added at each step should be perhaps 10% of that of the DNA.

11. Repeat the addition of protein to the DNA until no further changes in the CD are observed.
12. Thoroughly wash the cell and refill with buffer. Rerun the baseline. If the baseline has drifted from its initial value this will have to be considered when analyzing your results. If drifting is severe, *see* Note 4.
13. Plot the measured CD parameter at a particular wavelength against the concentration of protein added. The stoichiometry can be determined from the point at which a line drawn along the initial slope intersects that of the titration end point, which should be horizontal assuming dilution (if significant) has been corrected for (*see* Fig. 1).
14. Once the spectral changes and stoichiometry have been established, it is often useful to repeat the experiment with more titration points, using smaller aliquots of protein. This can be done at a single wavelength, without the need for scanning. Although the maximum CD signal from DNA is normally obtained around 275 nm, one should work at the wavelength that corresponds to the largest difference between free and bound DNA, which may well be different.

The CD spectrum of the DNA at saturation can be used to assess conformational changes that result from protein binding to a given sequence. Although analysis of the spectrum in terms of molecular structure is not straightforward *(2)*, it may be possible to interpret CD spectra of double-stranded DNA in terms of changes in helical twist angle (underwinding or overwinding of the helix). The spectral changes that accompany protein binding to different DNA sequences or with a variety of cofactors can also be informative.

4. Notes

1. For proteins and polynucleotides molarity is usually expressed in terms of moles of amino acid or nucleotide residue respectively for such calculations.
2. High intensity UV radiation converts oxygen to ozone, which damages the optics. Failure to purge will lead to deterioration in instrument performance.
3. For most experiments at these wavelengths, cells with pathlengths of 0.5 or 1 cm can usually be used. Longer pathlength cells for work with more dilute solutions are available at a price.
4. If significant drifting of the baseline has occurred the experiment may have to be repeated. To minimize the effect of drifting use two cells, one as a baseline reference and one containing sample. The baseline can then be standardized at each point in the titration. Remember that when swapping cells it is vitally important to present the same section of the cell to the beam each time.

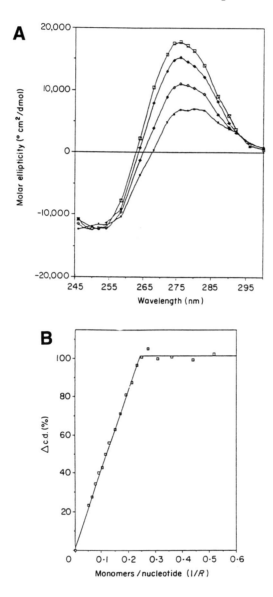

Fig. 1. Circular dichroism titrations of Pf1 gene 5 protein to poly dT. In this example, the protein is a single-stranded DNA binding protein that binds cooperatively to DNA with little sequence preference. (**A**) Successive CD spectra during the titration of poly dT with Pf1 gene 5 protein. The top curve corresponds to free DNA; below are increasing concentrations of protein corresponding to molar ratios of nucleotide to protein of 19.2, 6.7, and 4.0 (which represents saturation). (**B**) The normalized change in fluorescence signal at 276 nm is plotted against the ratio of gene 5 protein:poly dT (expressed as subunit concentration:nucleotide). The data show clear stoichiometric binding with an end-point of 0.25 (i.e., 4.0 nucleotides bound per protein subunit).

General Reading

Bayley, P. M. (1973) The analysis of circular dichroism of biomolecules. *Prog. Biophys.* **27,** 1–76.

Bush, C. A. (1974) Ultraviolet spectroscopy, circular dichroism and optical rotatory dispersion, in *Basic Principles in Nucleic Acid Chemistry* (Ts'O, P. O. P., ed.), vol. 2, Academic, New York, pp. 91–169.

Gratzer, W. B. (1971) Optical rotatory dispersion and circular dichroism of nucleic acids, in *Procedures in Nucleic Acid Research* (Cantoni, G. L. and Davies, D. R., eds.), vol. 2, Harper and Row, New York, pp. 3–30.

References

1. Johnson, W. C., Jr. (1990) Protein secondary structure and circular dichroism: a practical guide. *PROTEINS: Structure, Function and Genetics* **7,** 205–214.
2. Johnson, B. B., Dakl, K. S., Tinoco, I., Jr., Ivanov, V. I., and Zhurkin, V. B. (1981) Correlations between deoxyribonucleic acid structural parameters and calculated circular dichroism spectra. *Biochemistry* **20,** 73–78.
3. Carpenter, M. L. and Kneale, G. G. (1991) Circular dichroism and fluorescence analysis of the interaction of Pf1 gene 5 protein with poly (dT). *J. Mol. Biol.* **27,** 681–689.
4. Kansy, J. W., Cluck, B. A., and Gray, D. M. (1986) The binding of fd gene 5 protein to polydeoxyribonucleotides: evidence from CD measurements for two binding modes. *J. Biomol. Struc. Dynam.* **3,** 1079–1110.
5. Culard, F. and Maurizot, J. C. (1981) Lac repressor-lac operator interaction. Circular dichroism study. *Nucleic Acids Res.* **9,** 5175–5184.
6. Wartell, R. M. and Adhya, S. (1988) DNA conformational change in Gal repressor-operator complex: involvement of central G-C base pair(s) of dyad symmetry. *Nucleic Acids Res.* **16,** 11,531–11,541.
7. Torigoe, C., Kidokoro, S., Takimoto, M., Kyoyoku, Y., and Wada, A. (1991) Spectroscopic studies on lambda cro protein-DNA interactions. *J. Mol. Biol.* **219,** 733–746.

Chapter 27

Electron Microscopy of Protein–Nucleic Acid Complexes

Uniform Spreading and Determination of Helix Handedness

Carla W. Gray

1. Introduction

There are a number of proteins involved in DNA replication, recombination, or repair that bind stoichiometrically to single DNA strands of any nucleotide sequence, and which in some cases can also bind to single-stranded RNA. The best known examples are the *ssb* protein of *E. coli*, the *gene* 32 protein of phage T4, and the *gene* 5 protein of the M13/fd/f1 filamentous viruses *(1–3)*. Complexes formed by these proteins contain protein bound to the nucleic acid at defined ratios of the number of nucleotides per molecule of bound protein; the ratios are determined by the interactive properties of the protein. These ratios, and the structures of the complexes that are formed, may vary with factors such as changes in solution conditions that alter the binding properties of the proteins.

Stoichiometric, multiprotein complexes of proteins with nucleic acids will form structural repeats, generally consisting of discrete clusters of bound proteins or the turns of a continuous nucleoprotein helix. Although the individual proteins may not be resolved, the structural repeats tend to be of a size that can be visualized by electron microscopy. "Negative" staining, in which protein masses are delineated by their exclusion of an electron-opaque stain, is a method of choice because negative staining provides a well-contrasted image at higher resolution than is attained with some other techniques, such as shadowing with refractory metals. Negatively-stained nucleoprotein complexes can also be prepared relatively quickly and examined immediately after preparation. This makes it advantageous to use studies of negatively stained preparations prior to or in conjunction with cryo electron microscopy, which offers better preservation for detailed structural studies but is a more difficult and time-consuming method.

Nucleoprotein complexes often tend to be highly flexible, however, and the complexes are easily distorted, tangled, or partially dissociated during preparation for negative staining. This author has developed procedures to overcome these difficulties *(3,4)*, so that preparations made in this laboratory consistently contain complexes having well-extended configurations that are free of gross distortions (Fig. 1). Complexes prepared in this manner are uniformly spread on a two-dimensional support film and can be used for quantitative analysis of such parameters as the number of protein clusters or helical turns in a complex, the axial length of the complex, and the extent of the local variations in interturn distances in a complex that forms a flexible helix.

These preparations can also be used for analyses of three-dimensional structures using tilted specimens *(3)*. The most likely application is a determination of the handedness of helical nucleoprotein complexes, using the approach of Finch *(5)*, which is illustrated for a general case in Fig. 2. Projections of left- or right-handed helices on a two-dimensional plane are identical, as shown in the lower part of Fig. 2. However, a left-handed helix tilted with its top toward the viewer will show deeper indentations between the helical turns along its *right* side, whereas an identically tilted right-handed helix has the deeper indentations on its *left* side. Tilting the helices in the opposite direction, with the tops of the helices away from the viewer, produces an opposite set of left- and right-side indentations. In the description below, we describe a practical means of determining the absolute orientations of helices in images of tilted specimens.

2. Materials

1. Purified water, chemically softened and then predistilled in a bulk still, is double-distilled in our laboratory using a series of two 24-in. borosilicate glass Vigreux columns. (We do not find it necessary to use a quartz still.) Alternatively, the bulk distilled water may be deionized to

Electron Microscopy of Nucleoprotein Complexes

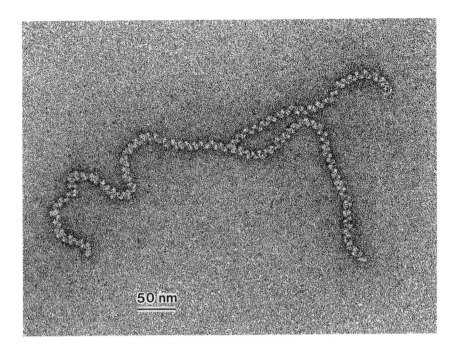

Fig. 1. A transmission electron micrograph of a helical complex of the fd *gene 5* protein with circular, single-stranded fd viral DNA. The complex was formed in vitro and was spread and stained with uranyl acetate by the methods described in this chapter. The complex is not tilted; the plane of the support film is in the plane of the page.

a resistivity of 18 MΩ-cm in a Millipore (Bedford, MA) "Milli-Q" system consisting of one Milligard cellulose ester prefilter cartridge, two ion exchange cartridges, and a 0.22-μ filter, in series. No activated charcoal filter is included in our system, because of a tendency of the charcoal filter to release minute charcoal particles that are found in our preparations for electron microscopy (*see* Note 1).

2. Buffers in which the protein–DNA complexes are to be visualized. Concentrated buffer stocks are passed through a 0.22-μm fiberless polycarbonate filter (Nuclepore/Costar, Cambridge, MA) to remove particulates and are then stored in borosilicate glass or polystyrene containers. The buffers should generally not contain high concentrations of salts or other nonvolatile components, because these components will tend to be retained on the carbon support film and can interfere with visualization of the protein–DNA complexes.

3. The protein–DNA complexes that are to be examined. These preparations should not contain significant excess quantities of noncomplexed

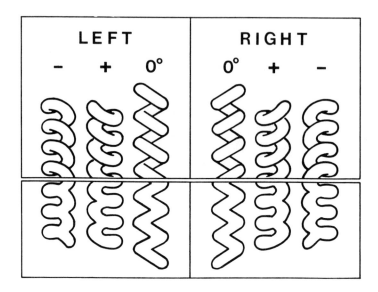

Fig. 2. Effects of tilting on the appearance of left- and right-handed helices. Three-dimensional images are drawn in the upper part and two-dimensional parallel projections in the lower part. A negatively stained complex will be seen as a two-dimensional projection. A helix at 0° tilt is parallel to the plane of the page; a "plus" (+) tilt brings the top of the helix nearer to the observer, whereas a "minus" (–) tilt brings the bottom of the helix closer to the observer. Reprinted with permission from C. W. Gray *(3)*.

or contaminating proteins, and they must generally be free of contaminating nonaqueous solvents as well as lipids, oils, salts, detergents, and other nonvolatile materials. We frequently repurify proteins and nucleic acids obtained commercially or from other laboratories, using ethanol precipitation (of nucleic acids), filtration chromatography, or dialysis. About 0.2–2 µmoles of DNA (measured as the concentration of nucleotides), together with protein added at an appropriate ratio, will be needed for a 50 µL incubation mixture from which the protein–DNA complexes are adsorbed to a single specimen grid.

4. Glutaraldehyde, purified for electron microscopy (distilled *in vacuo* and stored in sealed ampules under inert gas). The contents of one ampule are diluted to 8% (v/v) in purified water and are stored at –20°C in a tightly capped borosilicate glass tube with a Teflon-lined screw cap; this solution can be used for as long as 6 mo.

5. A suitable electron-opaque ("negative") stain, preferably analytical reagent grade. We generally use uranyl acetate, which provides good surface detail and yields satisfactory images with many proteins; however,

Electron Microscopy of Nucleoprotein Complexes   351

one should always consider the possibility that other stains may be preferable for a particular application *(6)*. A 2% (w/v) solution of a small amount of uranyl acetate in purified water is dissolved by stirring for 30 min in a borosilicate glass beaker. The beaker is sealed with a wax film and stored in the dark. The solution is used within a few days or weeks, but only if precipitates have not begun to form. Uranyl salts are weakly radioactive, so discarded solutions should be collected and properly disposed of as radioactive waste.

6. Carbon support films, 8–10 nm thick, on 500 mesh copper grids. We make our films in an Edwards E306A evaporator equipped with a liquid nitrogen trap and a quartz crystal film thickness monitor (Edwards FTM5). The chamber is evacuated just to 1×10^{-4} mbar, contaminants are burned off from the carbon rods (shutter closed, carbon rods brought to a red glow), and then evaporation is carried out over a period of several seconds. We use high purity carbon rods (Bio-Rad/Polaron [Richmond, CA], < 20 ppm impurities), that have been milled to form 1-mm tips. The carbon is evaporated onto a freshly cleaved mica film (Ladd, tested by flame spectroscopy). The carbon films are then floated from the mica onto purified water in a Teflon dish, picked up on copper grids, and dried for 30 min under a heat lamp.

3. Methods
3.1. Specimen Mounting and Negative Staining

1. Prepare stock solutions of DNA and protein, or of protein–DNA complexes, in high-quality, contamination-free, nonwetting polypropylene tubes. Care must be taken to avoid touching pipet tips or tube rims, to avoid contamination with oils and nucleases from the skin (*see* Note 2).
2. Prepare diluted DNA-protein mixtures for microscopy in a volume of 50 µL for each specimen grid. The final concentration of DNA will be about 4–40 µmoles of DNA nucleotides per mL; the optimal concentration must be determined experimentally, since it will vary with the adsorptive properties of the DNA and protein in a buffer of given composition, pH, and ionic strength.
3. Clean a Teflon surface with reagent grade ethanol and rinse it with purified water. For each specimen grid, place a row of droplets on the Teflon surface: Leave an empty space for one droplet, then deposit two 50-µL droplets of purified water, then one 50-µL droplet of 2% uranyl acetate.
4. Initiate glutaraldehyde fixation. This will be required in most (but not all) cases to maintain noncovalent protein–DNA associations during adsorption to the charged (glow-discharge-activated) carbon film. (To confirm that fixation is effective and yields an unperturbed structure,

stained preparations can be made by the method of Valentine et al. *(7),* which does not require fixation even for some relatively labile complexes.) Add 0.5 µL of 8% glutaraldehyde to the bottom of a 0.5 mL conical polyethylene tube; immediately add 50 µL of the protein-DNA mixture and gently mix by pipeting up and down once. Incubate the reaction at 20–25°C for 20 min.

5. Meanwhile, place carbon-coated grids, carbon side up, on a clean, inverted glass Petri dish and subject the grids to glow discharge. We use two parallel, L-shaped aluminum rods (6.5-mm in diameter) fitted to the high tension electrodes of the Edwards E306A evaporator. Glow discharge is carried out at 0.1–0.2 mbar (only the rotary pump is in operation), with the grids placed on the Petri dish at a distance of about 4 cm below the horizontal segments of the rods. The discharge is continued for about 50 s at 40% of maximum voltage (i.e., at approx 2 kV), and complexes are adsorbed to the grids within 10 min.

6. Place a 50-µL droplet of each protein–DNA mixture on the Teflon surface, at the beginning of a row of droplets prepared as described in step 3 above. Touch the grid, carbon side down, to the top of the droplet containing the protein–DNA mixture for 20–60 s, then wick off excess solution from the grid onto a filter paper, holding the grid perpendicular to the filter paper. Next, touch the grid to each of the two water droplets for 1 s each, and finally to the uranyl acetate droplet for 20 s, wicking off excess liquid after each step.

7. Dry the grid for 10 s by holding it within 2–3 cm of a lamp bulb (we use an illuminator having a 30-W bulb and a polished metal reflector), and then dry it for 10 min, carbon side up, on a filter paper that is under the lamp and about 12 cm from the bulb.

3.2. Specimen Tilting

Although specimen grids are readily tilted in an electron microscope that is fitted with a goniometer, when dealing with flexible helical nucleoprotein complexes one is confronted with two sets of questions. First, how does one select a specimen that will show the helical asymmetries when it is tilted, and second, how can one determine which end of a tilted helix is closer to the viewer as the helix is viewed on the fluorescent screen, in negatives, and in prints? These questions will be dealt with in the following sections.

1. Flexible nucleoprotein helices will generally not have straight helical axes and uniform turns as in Fig. 2, but a much less regular structure. It is essential to select a segment of a helical complex in which the turns

Electron Microscopy of Nucleoprotein Complexes 353

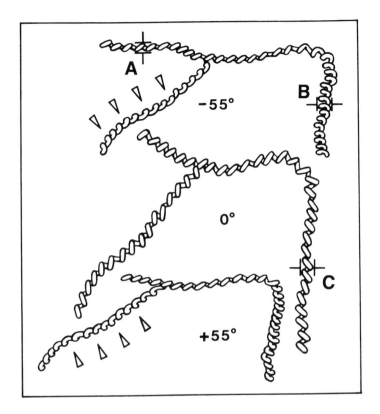

Fig. 3. Drawings made from electron micrographs of a left-handed helical complex of the IKe *gene 5* protein with single-stranded fd viral DNA. The center image is of the nontilted complex that lies in the plane of the page. The upper image is of the same complex tilted −55° around a horizontal axis, so that the top of the complex is below the plane of the page and the bottom of the complex is closer to the observer. The lower image is of the same complex tilted +55°, so that the bottom of the complex is below the plane of the page and the top of the complex is nearer to the observer. The drawings show the helical coils as three-dimensional structures to show how they can account for the projection images seen in the original electron micrographs, which are given in ref. *3*. Reprinted with permission from C. W. Gray *(3)*.

are as *regular* as can be found, with a roughly *linear* (or only gently curved) helix axis extending for 5–10 helical turns. Figure 3 contains a set of drawings taken from an actual tilt experiment *(3)*. The helix axis of a relatively linear and regular helical segment has been oriented approx perpendicular to the tilt axis (which is horizontal in Fig. 3) to maximize the changes observed on tilting. The characteristic left- and right-side indentations are seen when the complex in Fig. 3 is tilted in opposite directions (+55°, −55°), even though there is significant flexing in the helix axis. The change of indentations from one side to the other is faintly visible even in a diagonally-oriented helical segment (arrows), but the effect is more convincing in a segment that is perpendicular to the tilt axis.

2. The complexes must also have been prepared under conditions that yield a minimum of *flattening* of the helical structure; the use of a relatively deep layer of negative stain will tend to help support the three-dimensional structure. Flattening of the helix makes the left- and right-side indentations in tilted helices more difficult to observe, but they still can be seen when the helix is distorted. Note that in Fig. 3 the width at a appears to be about half of the width at b in the helix that is tilted −55°, indicating that the helix is flattened so that the height of a helical coil above the support film is roughly half of the width of the coil in a direction perpendicular to the helix axis and parallel to the support film.

3. In order to determine the absolute orientations of helical specimens as they are seen on the fluorescent screen of an electron microscope, focusing effects are utilized. An asymmetric marker (such as a macroscopic letter "R" punched on a specimen grid) can be used to correlate what is seen on the screen with the known position of the tilted grid (oriented, for example, so that the top edge of the "R" is uppermost in the column at a tilt angle of +55°). We used such a device in the top-entry stage of our Zeiss EM10C to demonstrate that when the objective lens current is adjusted to focus on the central portion of a steeply tilted grid, then the edge of the grid that is uppermost in the microscope and furthest from the upper polepiece of the objective (imaging) lens will be *overfocused* (lens current too strong to focus on it), whereas the lower edge of the grid will be *underfocused* (lens current too weak). These focusing effects can be used to determine which end of any helical complex is uppermost in the column; that same upper end of the helix will be, in effect, the end that is nearest to the viewer who is looking at the helix on the fluorescent screen.

4. The differing Fresnel patterns resulting from under- and overfocusing of a tilted specimen support film are visible in a negative bearing an image of the specimen. Hence, proof of the three-dimensional orientation of the specimens is contained in each negative. To correctly interpret the hand of a helix from an image, the negatives must be viewed *from the emulsion side,* which in the microscope faces in the same direction (toward the electron beam and toward the viewer) as does the viewing surface of the fluorescent screen. It is of course essential that the operator correctly recognize the patterns corresponding to overfocusing

as opposed to underfocusing; if in doubt, the patterns can be demonstrated by making exposures of deliberately under- and overfocused specimens at zero degrees of tilting.
5. The Fresnel patterns will also be visible in prints made at a suitable magnification. To present the helical asymmetries correctly, the prints must be made with the emulsion side of the negative facing the enlarger light source, so that the print will represent the same view of the object as that which is seen on the fluorescent screen.
6. Finally, we emphasize that it is essential to demonstrate that the indentations between helical turns *switch from one side to the other of one and the same helical nucleoprotein segment as the helix is tilted in opposite orientations.* We find that the flexing and partial compression of nucleoprotein helices can produce structures that show indentations on one side in one tilt orientation, but that do not show indentations on the other side when the helix is tilted in the opposite orientation. A satisfactory proof of the helical symmetry (hand of the helix) requires that *both* symmetrically related effects be demonstrated in the same helical segment.

4. Notes

1. The purity of the water used as solvent can be critical to the success of preparations for electron microscopy. The chemical content of the water supplied to a laboratory varies greatly with the locale, and it sometimes happens that procedures that worked in one location will fail in another, when the only reagent not carried to the new location is the water. If difficulties are encountered with experimental procedures, alternative sources of water and alternative water purification protocols should be explored.
2. The exercise of care to preclude the contamination of solutions is essential to the success of these procedures. Negative staining of specimens mounted on glow-discharge-activated grids is a widely used technique, but long and flexible nucleoprotein helices are particularly susceptible to the effects of contaminants that can interfere with the uniform adsorption of the complexes to an activated carbon film. The adsorption may fail to give satisfactory results if appropriate precautions are not taken.

Acknowledgments

The author gratefully acknowledges support of this work by National Institutes of Health Research Grant 5-RO1-GM34293-03 (to C. W. G.), by NIH Biomedical Research Support Grant 2S07-RR07133-21, by NIH Small Instrumentation Grant 1-S15-NS25421-01, and by National Science Foundation Instrumentation Grant PCM-8116109.

References

1. Bujalowski, W. and Lohman, T. M. (1991) Monomer-tetramer equilibrium of the *Escherichia coli ssb-1* single strand binding protein. *J. Biol. Chem.* **266,** 1616–1626.
2. Kodadek, T. (1990) The role of the bacteriophage T4 *gene 32* protein in homologous pairing. *J. Biol. Chem.* **265,** 20,966–20,969.
3. Gray, C. W. (1989) Three-dimensional structure of complexes of single-stranded DNA binding proteins with DNA: IKe and fd *gene 5* proteins form left-handed helices with single-stranded DNA. *J. Mol. Biol.* **208,** 57–64.
4. Gray, C. W., Brown, R. S., and Marvin, D. A. (1981) Adsorption complex of filamentous fd virus. *J. Mol. Biol.* **146,** 621–627.
5. Finch, J. T. (1972) The hand of the helix of tobacco mosaic virus. *J. Mol. Biol.* **66,** 291–294.
6. Haschemeyer, R. H. and Myers, R. J. (1972) Negative staining, in *Principles and Techniques of Electron Microscopy* (Hayat, M. A., ed.), vol. 2, Van Nostrand Reinhold, New York, pp. 101–147.
7. Valentine, R. C., Shapiro, B. M., and Stadtman, E. R. (1968) Regulation of glutamine synthetase, XII. Electron microscopy of the enzyme from *Escherichia coli. Biochemistry* **7,** 2143–2152.

CHAPTER 28

Reconstitution of Protein–DNA Complexes for Crystallization

Rachel M. Conlin and Raymond S. Brown

1. Introduction

An increasing number of structural studies are aimed at identifying the principles that govern protein–DNA recognition in gene regulation *(1)*. This work depends on the successful reconstitution of protein–DNA complexes from their purified components. X-ray crystallography and two-dimensional NMR techniques require large amounts of pure protein and DNA. These can be supplied through expression in bacteria of the cDNA coding for intact proteins or their smaller DNA-binding domains and the automated chemical synthesis of DNA in the laboratory.

Expression of the protein of interest is usually achieved at high levels in bacteria. An increasingly popular system is the combination of *E. coli* strain BL21(DE3) transformed with a pET expression vector containing a strong phage promoter adjacent to the cloning site *(2)*. This particular bacterial strain has an integrated T7 RNA polymerase gene that can be induced with IPTG *(3)*.

Target DNA sequences are easily identified by DNase I footprinting of a radioactively labeled DNA restriction fragment to which the protein is bound. Little technical difficulty is experienced in the chemical synthesis of these DNA sequences, up to about 50 bp in length, in amounts necessary to perform structural studies. Considerable progress has been made in solving three-dimensional structures of protein–DNA complexes *(1)*, largely because proteins and their isolated DNA-binding domains are able to recognize and form stable complexes with short duplexes containing the binding sequence. Indeed, protein–DNA complexes have been crystallized that contain duplexes as short as 10 bp in length *(4)*.

In this chapter we describe the preparation of a protein–DNA complex composed of a seven-zinc finger fragment (27 kDa) of *Xenopus laevis* transcription factor TFIIIA and a synthetic DNA duplex. The experimental strategy, reconstitution conditions, and technical problems discussed below are probably quite similar to those encountered with most other protein–DNA complexes. In this example the solution properties of the protein are known and the limits of the target DNA sequence are precisely defined. It has generally been assumed that association of the components takes place spontaneously to produce protein–DNA complexes of the desired molar composition. We have chosen to perform biochemical analysis in order to optimize authentic binding and ensure the efficient scaling-up of reconstitution conditions.

2. Materials

1. $(NH_4)_2SO_4$: Ultrapure enzyme grade (BRL, Frederick, MD).
2. Buffer A: 500 mM NaCl, 2 mM benzamidine-HCl, 1 mM DTT, 50% glycerol, 50 mM Tris-HCl, pH 7.5.
3. Standard protein solution: 1 mg/mL bovine serum albumin (BSA).
4. Protein assay concentrate (Bio-Rad, Richmond, CA). Dilute 1:5 with water and filter through Whatman 2^v paper.
5. 0.01% w/v Coomassie brilliant blue R250 in 20% v/v ethanol, 10% v/v acetic acid.
6. 1M Tris-HCl, pH 8.0.
7. DEAE-Sephacel (Pharmacia, Piscataway, NJ).
8. 2M Potassium acetate.
9. Repelcote solution (Hopkin and Williams, Chadwell Heath, Essex, UK).
10. Millex HA and Millex HV4 (0.45 µm) filter units (Millipore, Bedford, MA).
11. Buffer B: 100 mM NaCl, 10 mM Tris-HCl, pH 8.0.
12. Amberlite MB-1 resin (Serva, Heidelberg, Germany).
13. 150-mL (0.2-µm) NYL/50 filter unit (Sybron Corp., Rochester, NY).
14. 5X Binding buffer: 350 mM NH_4Cl, 35 mM $MgCl_2$, 5 mM DTT, 50 µM $ZnCl_2$, 50% v/v glycerol, 100 mM Tris-HCl, pH 7.5 *(5)*.
15. 1M HEPES-NaOH, pH 7.5.
16. 3.5M NH_4Cl.
17. Buffer C: 100 mM NaCl, 1 mM DTT, 1 mM NaN_3, 20 mM Tris-HCl, pH 7.5.

18. Collodion bags and 300-mL glass vacuum dialysis flask (Sartorius AG, Goettingen, Germany).
19. Magnetic Micro flea 5 × 2 mm spinbar (Bel-Art [Pequannock, NJ] products).
20. Linbro tissue culture multiwell plate with cover. 24 Flat bottom wells (1.7 × 1.6 cm) (Flow Laboratories, McLean, VA).
21. AquaSil water soluble siliconizing fluid (Pierce, Rockford, IL).
22. DiSPo plastic cover slips M6100 (American Scientific Products, McGaw Park, IL).
23. High vacuum grease (Dow Corning, Midland, MI).
24. 10-cc Plastic B-D syringe (Becton-Dickinson, Rutherford, NJ).

3. Methods

3.1. Purification of a Recombinant DNA-Binding Protein

Standard laboratory techniques for protein purification (6) will not be described in detail in this chapter. Advantage is usually taken of the rather basic nature of these proteins to isolate them from the bacterial cell extract. Few of the bacterial proteins bind so strongly to chromatography matrices such as CM-Sepharose, Bio-Rex 70, S-Sepharose, phosphocellulose, and hydroxyapatite-Ultrogel. If necessary, column chromatography can be performed in the presence of $7M$ urea to ensure solubility and prevent aggregation. The protein of interest can usually be eluted with a NaCl gradient (see Notes 1–3). Affinity chromatography with heparin-Sepharose, an immobilized dye-Sepharose, DNA-agarose, or phenyl-Sepharose often provides an adequate final purification step.

3.1.1. Preparation of Protein

The cDNA for the 27-kDa fragment of TFIIIA (amino acids 5-218) was cloned into the vector pET-11a and expressed in strain BL21(DE3). Purification of the protein is carried out in $7M$ urea on columns of Bio-Rex 70 and heparin-Sepharose (see Note 4). Workup of the protein for use in DNA binding is described in detail below.

1. Pool those fractions that contain protein after column chromatography.
2. Precipitate the protein by slow addition of solid $(NH_4)_2SO_4$ to 20% saturated in a 70-mL polycarbonate centrifuge tube at 4°C.
3. Stir slowly for 30 min and then centrifuge (15 min at 10,000g, 4°C). Discard any precipitate.
4. Increase the concentration of $(NH_4)_2SO_4$ in the supernatant to 55% saturated. After stirring for 30 min, centrifuge as in step 3.

5. Remove all of the supernatant by aspiration and allow the pellet to redissolve in ice-cold Buffer A to give a protein concentration of 5 mg/mL.
6. Store the protein at −20°C in a 1.5-mL Eppendorf tube (*see* Note 5).

3.1.2. Measurement of Protein Concentration

1. Construct a calibration curve of protein concentration versus absorbance from samples (in triplicate) containing 15, 25, 35, and 45 µL of the standard protein solution and water added to 100 µL in glass tubes (*see* Note 6).
2. Add 5 mL of protein assay reagent to each sample and read the absorbance at 595 nm with a spectrophotometer after the blue color has developed for 10 min.
3. Calculate the relative concentration from comparison of the values for the protein sample with the calibration curve (*see* Notes 7–8).
4. Examine the purity of the protein sample by standard SDS/polyacrylamide gel electrophoresis. A 5% stacking/12.5% resolving slab gel (30:0.8 acrylamide:*bis*) is suitable for this purpose (*see* Note 9). At least 0.1 µg of protein is detected in a band with Coomassie blue stain. If necessary adjust the protein concentration by the visual estimate of its gel purity.

3.2. Preparation of Synthetic Oligomers

Typically a 1 µmol scale synthesis with a commercially available machine provides 1–2 mg of a purified 35-mer. The standard laboratory methods for isolation of the 5'-dimethoxytrityl full length oligomer and subsequent chemical removal of base protecting groups will not be described here (*see* Note 10). Oligomers pure enough for this work can be obtained with a fast protein liquid chromatography system (FPLC), and suitable columns *(7)*.

3.2.1. Recovery and Concentration of Oligomers

1. Adjust the pH to 8.0 and apply the pooled column fractions containing the oligomer onto a 0.5-mL DEAE-Sephacel column equilibrated with 50 m*M* Tris-HCl, pH 8.0 at room temperature.
2. Elute the bound oligomer with 5 × 1 mL 2*M* potassium acetate into a silanized 30-mL Corex tube.
3. Add 20 mL of 95% ethanol and precipitate the oligomer overnight at −20°C.
4. Centrifuge (45 min at 7500*g*, 4°C) to recover the oligomer.
5. Wash the pellet with 25 mL of 95% ethanol to remove excess potassium acetate and centrifuge as in step 4. Pour off the supernatant and dry the pellet under vacuum.
6. Redissolve the oligomer in 1 mL of water (*see* Note 11), and then filter through a Millex HV4 unit (0.45 µm).

7. Measure the absorbance at 260 nm with a spectrophotometer (dilute the oligomer as necessary with 100 mM NaCl). The concentration is calculated with a conversion factor of 25 A_{260} U = 1 mg. Store the oligomer at 1 mg/mL at –20°C.

3.2.2. Annealing the Duplex

1. Mix equal amounts of complementary DNA strands in buffer B at 1 mg/mL in a 1.5-mL Eppendorf tube.
2. Heat at 85°C for 5 min and then slowly cool to 4°C in 4–6 h (*see* Note 12).
3. Examine duplex formation by electrophoresis at 50 V in a nondenaturing 7.5% polyacrylamide slab gel at room temperature. 0.1 µg of DNA in a gel band is easily visible after staining in ethidium bromide (1 mg/L), and illumination with UV light.

3.3. Testing the Reconstitution Conditions

The duplex is titrated with increasing amounts of protein in order to measure DNA-binding activity (Fig. 1). The resulting complexes are monitored by mobility shift on a nondenaturing 7.5% polyacrylamide gel (*see* Notes 13–15). This method is used to systematically optimize 1:1 molar complex formation with respect to incubation time, temperature, concentration of monovalent and divalent ions, pH, presence of glycerol, and nonionic detergents, such as Nonidet P-40.

1. Mix 1 µg of duplex, 2 µL of 5X binding buffer, and water (to make a final vol of 10 µL after step 2) in a 1.5-mL Eppendorf tube.
2. Add 0.5, 1, 1.5, and 2X molar excess of protein and stir in gently with the Pipetman tip (*see* Note 16).
3. Dilute with 5 µL of 1X binding buffer and incubate at room temperature for 15 min.
4. Apply the 15 µL samples, without tracking dyes, to a nondenaturing 7.5% polyacrylamide gel (*see* Note 17).
5. Load dyes (bromophenol blue and xylene cyanol FF) in adjacent tracks to indicate progress of the electrophoresis. The protein–DNA complex migrates between the dyes.
6. Stain the gel first with ethidium bromide (1 mg/L), and then with 0.01% Coomassie blue. The complex band will stain positive for DNA and for protein.

3.3.1. Scaling-Up Reconstitution of the Complex

In low salt conditions the protein shows a strong tendency to aggregate. This is detected as material trapped at the top of the nondenaturing polyacrylamide gel. An excess of protein results in formation

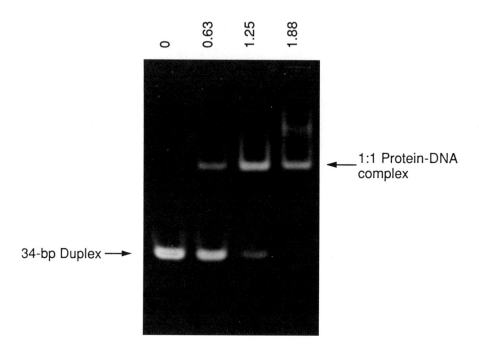

Fig. 1. Mobility shift gel for analysis of protein–DNA binding. One microgram of duplex was titrated with 0, 0.63, 1.25, and 1.88 µg of the TFIIIA 27-kDa fragment. The synthetic duplex is 34 bp in length, which spans sequence positions +61 to +94 of the *Xenopus laevis* oocyte 5S RNA gene *(8)*. An additional deoxyguanosine is present at the 5' end of the noncoding strand and a deoxycytidine on the coding strand. Following electrophoresis in a nondenaturing 7.5% polyacrylamide gel those bands that contain DNA are visualized by ethidium bromide stain and UV illumination.

of some protein–DNA complexes with higher molar stoichiometries. We have discovered that these unwanted effects can be substantially eliminated by dilution of NH_4Cl in the scaled-up reconstitution mix in steps; from 0.65–0.325M and then to 0.22M NH_4Cl *(see* Note 18).

1. Mix together 500 µg of duplex, 200 µL of 5X Binding buffer, 134 µL of 3.5M NH_4Cl, and water in a 5-mL polypropylene Nunc tube (the reconstitution mix, including the protein, is formulated to be 1 mL of 0.65M NH_4Cl).
2. Add 625 µg of protein and mix together by gentle inversion.
3. Dilute with 1 mL of 1X binding buffer, mix, and incubate for 5 min at room temperature.
4. Dilute with 1 mL of 1X binding buffer, mix, and incubate for 15 min at room temperature.

5. Dialyze for 2 d at 4°C against two changes of 1 L of buffer C (see Note 19).
6. Remove any precipitate by centrifugation (15 min at 10,000g, 4°C).
7. Concentrate the complex to 10 mg/mL by vacuum dialysis at 4°C in a collodion bag (see Note 20).
8. Store the protein–DNA complex on ice until used in crystallization trials.

3.3.2. Crystallization Trials

Over the last decade it has become apparent that the standard methods developed for protein crystallization (10) can also be used successfully to crystallize protein–DNA complexes (11). A summary of crystallization conditions, mainly employing the hanging drop/vapor diffusion method, are shown in Table 1. Conditions are screened at 4°C and at room temperature with small droplets (1–2 µL) of the protein–DNA complex in which the concentrations of salts, buffers and precipitants are systematically varied by small amounts. Crystallization often occurs at or close to the point of precipitation of the protein–DNA complex.

1. Pass all solutions of salts, buffers, and precipitants through 0.45-µm filters (Millex HA unit) before use.
2. Remove any solid material from the protein–DNA complex solution by centrifugation (15 min at 10,000g, 4°C).
3. Blow dust from plastic pipet tips and then place 1 µL aliquots of the protein–DNA complex onto silanized plastic coverslips (see Note 21).
4. Mix with 0.5 µL of appropriate salts, buffer, and precipitant or the well solution (see Note 22).
5. Apply vacuum grease to the upper rim of the wells of the tissue culture plate (see Note 23).
6. Into each well put 1 mL of suitable precipitant.
7. Carefully invert a coverslip without disturbing the droplet and place onto the greased rim to seal each well.
8. Store the tissue culture plates in closed polystyrene boxes to avoid excessive vibrations and changes in temperature.
9. After 1 wk, inspect the droplets for signs of crystallization with a stereo microscope at 50X magnification.

4. Notes

1. The selection of a particular chromatographic matrix depends on the solubility of protein in the low salt concentrations required for binding and consideration of the degree of separation achieved.
2. Losses of protein may be decreased in the presence of 10% v/v glycerol.

Table 1
Crystallization Conditions for Protein–DNA Complexes

Protein–DNA Complex	Method	Well Solution
BamHI: blunt 12-mer Aggarwal (11)	Add DNA in 1 mM EDTA,[a] 10 mM Tris-HCl to protein in 0.5M KCl, 1 mM EDTA, 1 mM DTT,[a] 10% glycerol, 20 mM potassium phosphate, pH 6.9. Mix with 1 vol of the well solution (without KCl and glycerol) at room temperature	14% PEG[a] 8000, 20 mM potassium phosphate, pH 6.9. Add 0.25M KCl and 5% glycerol later
EcoRI: 13-mer Grable et al. (12)	Protein in 0.5M NH$_4$OAc,[a] 15% dioxane, 40 mM Bis-tris propane, pH 7.4. Add DNA and 8% PEG 400 and centrifuge after 2 d at 4°C. Seed with microcrystals	16% PEG 400, 1 mM EDTA, 40 mM Bis-tris propane
EcoRV: blunt 10-mer Winkler et al. (4)	Mix protein in 0.2–0.25M NaCl, 1 mM EDTA, 10–20 mM Na/K phosphate, pH 7.0–7.5 with DNA in 20 mM MES,[a] pH 6.0. Make 2.5–3.5% PEG 4000, 90 mM NaCl at 19–22°C	80 mM NaCl, 10 mM sodium phosphate, pH 7.0
HhaII: blunt 7-mer Chandrasegaran et al. (13)	Dissolve lyophilized DNA in the protein solution in 0.2M NaCl, 1 mM EDTA, 0.1 mM DTT, 10 mM potassium phosphate, pH 7.2. Centrifuge to remove precipitate and mix with 1 vol of the well solution at room temperature	10% PEG 4000, 0.2M NaCl
DNase I: blunt "14-mer"[b] Lahm and Suck (14)	Mix DNA with protein in 0.1–0.15M NaCl, 15 mM EDTA, 0.1M imidazole, pH 6.0 and make 8% PEG 6000 at 4°C	In a closed vial
434 Cro: blunt 14-mer Wolberger et al. (15)	Mix protein and DNA in 0.7M NaCl, 1 mM EDTA, 0.1 mM DTT, 10 mM Tris-HCl, pH 8.0, with 0.5 vol of the well solution at room temperature. Crystals nucleate by deforming surface of the droplet	36% saturated (NH$_4$)$_2$SO$_4$, 10 mM hexammine cobalt(III) chloride, 1 mM MgCl$_2$

λ Cro: blunt 17-mer Brennan et al. (16)	Protein and DNA in 0.1M NaCl, 20 mM sodium cacodylate, pH 6.9	No well solution
434 repressor (residues 1–69): 20-mer Aggarwal et al. (17)	Mix DNA and protein in 0.1M NaCl, 0.12M MgCl$_2$, 2 mM spermine. Make 12–14% PEG 3000 at 4°C	
lac repressor: blunt 16-mer Pace et al. (18)	Protein and DNA in 1.4M NaOAc, 0.1M ADA,[a] pH 6.5	
λ repressor (residues 1–92): 20-mer Jordan et al. (19)	Mix DNA and protein in 1 mM NaN$_3$, 15 mM Bis-tris propane-HCl, pH 7.0 with 1 vol of the well solution at room temperature. Seed with crystals	20% PEG 400
met repressor: 19-mer Somers (20)	Protein and DNA in 15–30 mM CaCl$_2$, 6 mM NaCl, 1 mg/mL SAM,[a] 10 mM sodium cacodylate, pH 7.0. Mix with 1 vol of the well solution	30–35% MPD, 1 mM NaN$_3$, 10 mM sodium cacodylate, pH 7.0
trp repressor: 19-mer Joachimiak et al. (21)	Protein and DNA in 11 mM CaCl$_2$, 2 mM L-tryptophan, 10 mM cacodylate, pH 7.2, and 20% MPD[a] at room temperature	40% MPD, 11 mM CaCl$_2$, 2 mM L-tryptophan, 10 mM cacodylate, pH 7.2
CAP: "31-mer"[b] Schultz (22)	Protein and DNA in 0.2M NaCl, 0.1M CaCl$_2$, 2 mM cAMP, 2 mM spermine, 0.02% NaN$_3$, 0.3% n-octyl-β-D-glucopyranoside, 2 mM DTT, 50 mM MES, pH 5.0–6.0. Mix with 1 vol of the well solution at room temperature	5–10% PEG 3350, 0.2M NaCl
Engrailed homeodomain: 21-mer Liu et al. (23)	Protein and DNA in 30 mM Bis-tris-HCl, pH 6.7. Raise to pH 8.0–9.0 with NH$_4$OH. Crystals grow when pH returns to 6.7	
MATα2 homeodomain: (residues 128–210): 21-mer Wolberger et al. (24)	Mix lyophilized DNA with protein in 25 mM NaCl, 1 mM EDTA, 0.01% NaN$_3$, 15 mM Tris-HCl, pH 8.0. Add 1 vol of the well solution at room temperature. pH drops to 5.0	20% PEG 400, 0.15M CaCl$_2$

(continued)

Table 1 (continued)

Protein–DNA Complex	Method	Well Solution
Glucocorticoid receptor (residues 440–525): 19-mer Luisi et al. (25)	Protein and DNA in 25 mM NaCl, 1.8 mM MgSO$_4$, 1.8 mM spermine and 2 μM ZnCl$_2$, 18 mM sodium cacodylate, pH 6.0 at 8°C	8% MPD, 0.2M NaCl, 24 mM sodium cacodylate, pH 6.0
TFIIIA (residues 5–218): 30-mer	Protein-DNA complex in 0.1M NaCl, 1 mM DTT, 1 mM NaN$_3$, 20 mM Tris-HCl, pH 7.5. Mix with 0.5 vol of 12% PEG 6000, 0.1M NaCl, 1 mM DTT, 1 mM NaN$_3$, 20 mM Tris-HCl, pH 7.5 at 4°C	20% PEG 6000, 0.2M NaCl, 1 mM NaN$_3$
Zif268 (residues 349–421):11-mer Pavletich and Pabo (26)	Mix protein and DNA in 0.45–0.75M NaCl, 125 mM Bis-tris propane-HCl, pH 8.0 with 1 vol of the well solution	0–10% PEG 400, 0.35–0.65M NaCl, 25 mM Bis-tris propane-HCl, pH 8.0
Nucleosome core particles: (146 bp) Rhodes et al. (27)	Core particles in 60 mM KCl, 0.5 mM EDTA, 10 mM potassium cacodylate, pH 6.0. Make 40 mM MnCl$_2$ at 28°C	In a stoppered tube

[a]Abbreviations: EDTA—disodium ethylenediaminetetraacetic acid; DTT—dithiothreitol; PEG—polyethylene glycol; OAc—acetate; MES—2-[N-morpholino]ethanesulfonic acid; ADA—N-[2-Acetamido]-2-iminodiacetic acid; SAM—S-adenosyl-L-methionine; MPD—2-methyl-2,4-pentanediol.

[b]Each strand of synthetic DNA lacks 3′ and 5′ terminal phosphates. "14-mer" and "31-mer" duplexes are formed by the end-to-end stacking and overlapped base pairing of two smaller duplexes.

3. Customized gradients can be supplied to standard laboratory chromatography columns using a programmed FPLC control unit and pumps. Flow rates of 1–50 mL/h are possible without significant back pressure.
4. The protein binds to heparin-Sepharose in $0.25M$ NaCl, $7M$ urea. A reverse gradient is applied to remove urea. The protein is eluted in a $0.25M$ NaCl to $2M$ NaCl gradient.
5. The protein may be stored in a sealed tube filled with nitrogen gas to inhibit oxidation of cysteines present in the protein. No loss of DNA-binding activity or ability to crystallize occurs after several months at –20°C.
6. This assay can be used to measure protein concentrations between 50 µg/mL and 50 mg/mL according to the manufacturer's directions.
7. Data may be conveniently analyzed by linear regression using a commercial graphing program, such as KaleidaGraph or CricketGraph, and a personal computer.
8. The protein concentration obtained by the Bradford colorimetric assay is relative. An absolute value requires multiplication by a conversion factor. This factor can be derived from simultaneous amino acid analysis or MicroKjeldahl nitrogen determination *(9)*. In the case of the TFIIIA 27-kDa fragment the deduced concentration of BSA is multiplied by 0.39(±0.06).
9. TFIIIA and its fragments have anomalous electrophoretic mobilities. Reduction and carboxymethylation with 50 mM iodoacetamide restores the predicted gel mobility to the intact protein. Fragments containing zinc fingers generally run slower than expected after this treatment.
10. Chemical detritylation can be efficiently performed by passing 0.5% trifluoroacetic acid for 3 min at room temperature through a reversed phase ProRPC HR column (Pharmacia). Following deprotection of the bases (30% w/v NH_4OH, 16 h at 65°C) the oligomer is purified with a Mono Q HR column (Pharmacia), and eluted in a NaCl gradient according to the manufacturer's instructions.
11. After addition of water, the Corex tube is sealed with parafilm to avoid spillage. The water is rolled around the walls of the silanized tube as well as on the pellet.
12. This is done by switching off the heater and allowing it to cool to room temperature. The metal block is then transferred to the coldroom at 4°C to complete the annealing process.
13. A stock acrylamide solution (30:0.8 acrylamide:*bis*) is deionized with Amberlite MB-1 resin (Serva) (10 g/100 mL) for 1 h, filtered through a Nalgene filter unit (0.2-µm), and stored at 4°C. This treatment minimizes the sequestering of the protein from protein–DNA complexes during gel electrophoresis.

14. Glass plates, combs, spacers and the gel apparatus must be cleaned and washed completely free of detergent to avoid disruption of the protein–DNA complex.
15. The gel is 1.5 mm thick and contains 40 mM HEPES-NaOH, pH 8.3, and 5% v/v glycerol. The running buffer consists of 20 mM HEPES-NaOH, pH 8.3.
16. A duplex with an unrelated sequence of the same length may be used to monitor the level of nonspecific protein binding. The affinity of the protein for each of the single strands of the duplex can also be tested.
17. Samples can be applied smoothly to the gel by slowly winding the Pipetman volume control back to zero.
18. At high salt concentration, above 0.65M NH$_4$Cl, the protein–DNA complex is dissociated and both components are soluble. The salt concentration is lowered stepwise to 0.22M NH$_4$Cl to reconstitute the complex. At the same time, the protein is diluted and becomes less likely to aggregate or bind nonspecifically to the duplex.
19. It is not known initially whether any of the binding buffer components would inhibit crystallization. Thus a solution is chosen that contains the minimum number of components and lowest concentrations of salt(s), and buffer that ensure stability of the protein–DNA complex.
20. The collodion bag is preequilibrated in buffer and contains a small magnetic stirrer (5 × 2 mm) to aid dialysis and recovery of the protein–DNA complex.
21. Wash dust off the coverslips with deionized water and dry in a suitable rack at 50°C in an oven. Immerse cover slips for 15 min in dilute AquaSil (1:40), and leave to dry overnight. Wash in deionized water and dry at 50°C. Keep in a closed container to avoid contact with dust.
22. It is customary to equilibrate against a well solution that contains two or three times the concentration of the droplet components. One-half or 1 vol of the well solution containing appropriate salts, buffers, and precipitant can be added to the droplet of the protein–DNA complex.
23. The grease can be applied accurately to the rim with a 10-cc plastic syringe filled with high vacuum grease.

References

1. Steitz, T. A. (1990) Structural studies of protein-nucleic acid interaction: the sources of sequence-specific binding. *Quart. Rev. Biophys.* **23,** 205–280.
2. Dubendorff, J. W. and Studier, F. W. (1991) Controlling basal expression in an inducible T7 expression system by blocking the target T7 promoter with lac repressor. *J. Mol. Biol.* **219,** 45–59.
3. Studier, F. W., Rosenberg, A. G., Dunn, J. J., and Dubendorff, J. W. (1990) Use of T7 RNA polymerase to direct the expression of cloned genes. *Meth. Enzymol.* **85,** 60–89.

4. Winkler, F. K., D'Arcy, A., Bloecker, H., Frank, R., and van Boom, J. H. (1991) Crystallization of complexes of EcoRV endonuclease with cognate and noncognate DNA fragments. *J. Mol. Biol.* **217,** 235–238.
5. Wolffe, A. P. (1988) Transcription fraction TFIIIC can regulate differential *Xenopus* 5S RNA gene transcription in vitro. *EMBO J.* **7,** 1071–1079.
6. Deutscher, M. P. (ed.) (1990) Guide to protein purification. *Meth. Enzymol.* **182,** 309–392.
7. Oliver, R. W. A. (ed.) (1989) *HPLC of Macromolecules: A Practical Approach.* IRL, Oxford, England, pp. 183–208.
8. Zwieb, C. and Brown, R. S. (1990) Absence of substantial bending in *Xenopus laevis* transcription factor IIIA-DNA complexes. *Nucleic Acids Res.* **18,** 583–587.
9. Jaenicke, L. (1974) A rapid micromethod for the determination of nitrogen and phosphate in biological material. *Anal. Biochem.* **61,** 623–627.
10. McPherson, A. (1982) Crystallization, in *The Preparation and Analysis of Protein Crystals.* Wiley, New York, pp. 82–159.
11. Aggarwal, A. K. (1990) Crystallization of DNA binding proteins with oligodeoxynucleotides. *Methods*: a companion to *Meth. Enzymol.* **1,** 83–90.
12. Grable, J., Frederick, C. A., Samudzi, C., Jen-Jacobson, L., Lesser, D., Greene, P., Boyer, H. W., Itakura, K., and Rosenberg, J. M. (1984) Two-fold symmetry of crystalline DNA-*Eco*RI endonuclease recognition complexes. *J. Biomolec. Struct. Dyn.* **1,** 1149–1160.
13. Chandrasegaran, S., Smith, H. O., Amzel, M. L., and Ysern, X. (1986) Preliminary X-ray diffraction analysis of *Hha*II endonuclease-DNA cocrystals. *Proteins: Structure, Function and Genetics* **1,** 263–266.
14. Lahm, A. and Suck, D. (1991) DNase I-induced DNA conformation: 2Å structure of a DNase I-octamer complex. *J. Mol. Biol.* **222,** 645–667.
15. Wolberger, C. and Harrison, S. C. (1987) Crystallization and X-ray diffraction studies of a 434 Cro-DNA complex. *J. Mol. Biol.* **196,** 951–954.
16. Brennan, R. G., Takeda, Y., Kim, J., Anderson, W. F., and Matthews, B. W. (1986) Crystallization of a complex of Cro repressor with a 17 base-pair operator. *J. Mol. Biol.* **188,** 115–118.
17. Aggarwal, A. K., Rodgers, D. W., Drottar, M., Ptashne, M., and Harrison, S. C. (1988) Recognition of a DNA operator by the repressor of phage 434: a view at high resolution. *Science* **242,** 899–907.
18. Pace, H. C., Lu, P., and Lewis, M. (1990) Lac repressor: crystallization of intact tetramer and its complexes with inducer and operator DNA. *Proc. Natl. Acad. Sci. USA* **87,** 1870–1873.
19. Jordan, S. R., Whitcombe, T. V., Berg, J. M., and Pabo, C. O. (1985) Systematic variation in DNA length yields highly ordered repressor-operator cocrystals. *Science* **230,** 1383–1385.
20. Somers, W. S. (1990) Crystal structure of methionine repressor of *E. coli* and its complex with operator. Ph.D. thesis, University of Leeds, UK.
21. Joachimiak, A., Marmorstein, R. Q., Schevitz, R. W., Mandecki, W., Fox, J. L., and Sigler, P. B. (1987) Crystals of the trp repressor-operator complex suitable for X-ray diffraction analysis. *J. Biol. Chem.* **262,** 4917–4921.

22. Schultz, S. C., Shields, G. C., and Steitz, T. A. (1990) Crystallization of *Escherichia coli* catabolite gene activator protein with its DNA binding site. The use of modular DNA. *J. Mol. Biol.* **213,** 159–166.
23. Liu, B., Kissinger, C. R., Pabo, C. O., Martin-Blanco, E., and Kornberg, T. B. (1990) Crystallization and preliminary X-ray diffraction studies of the Engrailed homeodomain and of an Engrailed homeodomain/DNA complex. *Biochem. Biophys. Res. Comm.* **171,** 257–259.
24. Wolberger, C., Pabo, C. O., Vershon, A. K., and Johnson, A. D. (1991) Crystallization and preliminary X-ray diffraction studies of a MATα2–DNA complex. *J. Mol. Biol.* **217,** 11–13.
25. Luisi, B. F., Xu, W. X., Otwinowski, Z., Freedman, L. P., Yamamoto, K. R., and Sigler, P. B. (1991) Crystallographic analysis of the interaction of the glucocorticoid receptor with DNA. *Nature* **352,** 497–505.
26. Pavletich, N. P. and Pabo, C. O. (1991) Zinc finger-DNA recognition: crystal structure of a Zif268–DNA complex at 2.1Å. *Science* **252,** 809–817.
27. Rhodes, D., Brown, R. S., and Klug, A. (1989) Crystallization of nucleosome core particles. *Meth. Enzymol.* **170,** 420–428.

CHAPTER 29

Assay of Restriction Endonucleases Using Oligonucleotides

Bernard A. Connolly

1. Introduction

Type II restriction endonucleases cleave double-stranded DNA at sequence-specific sites typically 4–6 bp in length. Although large DNA molecules (viral DNA, plasmid DNA, and chromosomal DNA) are the physiological substrates for these enzymes, activity is often shown with small synthetic oligodeoxynucleotides providing that the recognition sequence is present. The use of oligodeoxynucleotide substrates often allows information to be obtained concerning the mechanism by which the particular endonuclease recognizes its cognate site so specifically and discriminates accurately against all other sequences. Excellent examples include a very thorough study with the *Eco*R1 endonuclease that revealed the energetic bases of its specificity *(1,2)*, and experiments with the *Eco*RV endonuclease using oligodeoxynucleotides containing modified bases that demonstrated several direct contacts between enzyme and substrate *(3,4)*. Central to all these experiments are methods for the assay of the activity a particular restriction endonuclease shows toward an oligonucleotide substrate. This chapter outlines three alternative assay protocols. All are based on experience in our laboratory with the *Eco*RV endonuclease. This enzyme recognizes d(GATATC) sequences cutting between the central T and dA bases to give blunt-ended products. The self-complementary dodecamer d(GACGATATCGTC) is a substrate for the endonuclease and on catalysis two hexamers d(GACGAT), and d(pATCGTC) are produced. All the assays described below are based on this 12'-mer as substrate. Longer oligodeoxynucleotides may be used providing they contain the GATATC sequence and can form a double helix. However, shorter oligomers are either very poor or not substrates. The first assay is based on high performance liquid chromatography (HPLC). The substrate 12'-mer and two hexameric products are separated on reverse-phase HPLC columns and detected by their absorbance at 254 nm. Injection of aliquots with time and determination of the amounts of substrate and products present (by integration of the areas under their corresponding peaks) gives the rate. The second method uses 5'-[^{32}P]-d(pGACGATATCGTC). The labeled substrate and labeled hexameric product (5'-[^{32}P]-d(pGACGAT) are separated on denaturing polyacrylamide gels. Determination of the counts present in the substrate and product bands as a function of time gives the rate. The final assay is a direct continuous spectrophotometric assay based on the conversion of the double-stranded substrate d(pGACGATATCGTC) into single-stranded hexameric products. This is associated with an increase in UV absorbance at 254 nm because of the hyperchromic effect. This effect comes about because bases that are poorly stacked (in a single-stranded DNA for example) have a higher extinction coefficient and so absorb light more strongly than equivalently tightly stacked bases in double-stranded DNA *(5)*. The HPLC assay allows collection of the products and after determination of their base composition verifies that hydrolysis has taken place at the correct site. It is particularly useful when base analogs *(3,4)* are introduced into the *Eco*RV recognition sequence. These are often very poor substrates and demonstration of 100% cleavage eliminates artifacts that may be caused by traces of the parent dodecamer. However, this assay is discontinuous and time consuming. The 5'-[^{32}P] assay is extremely sensitive and can be used to measure low rates and at low substrate concentrations. It is useful for K_m and k_{cat} determination. Again it is discontinuous, time consuming, and suffers from the hazards associated with ^{32}P. The continuous spectrophotometric assay is quick and very simple. Its main disadvantage is that only a limited range of substrate concentrations can be used and so K_m determination is only possible if it lies within this range.

2. Materials
2.1. HPLC Assay

1. High performance liquid chromatograph capable of running gradients and fitted with a 254-nm UV detector, chart recorder, and peak integrator.
2. Analytical size (25 × 0.45 cm) reverse-phase octadecylsilyl (C-18, 5μ particle size) HPLC column.
3. HPLC buffer A: $0.1M$ triethylammonium acetate, pH 6.5, containing 5% CH_3CN; HPLC buffer B, $0.1M$ triethylammonium acetate, pH 6.5, containing 65% CH_3CN. Both buffers should be prepared using HPLC grade triethylamine and CH_3CN, Analar acetic acid, and double-distilled water. The buffers should be degassed with a water pump before use.
4. UV spectrophotometer.
5. Quartz cuvets suitable for absorption spectrophotometry at 254 nm.
6. $0.5M$ NaOH solution.
7. 2X EcoRV endonuclease assay buffer: 100 mM HEPES-NaOH, pH 7.5, 200 mM NaCl, and 20 mM $MgCl_2$.
8. Hyperchromicity determination buffer: 10 mM potassium phosphate buffer, pH 7, and 10 mM $MgCl_2$.
9. Alkaline phosphatase (special grade for molecular biology).
10. Snake venom phosphodiesterase.
11. EcoRV endonuclease.
12. d(GACGATATCGTC); d(GACGAT); d(ATCGTC).

2.2. ^{32}P/Gel Assay

1. d(GACGATATCGTC).
2. 2X Phosphorylation buffer: 100 mM Tris-HCl, pH 8.5, 20 mM $MgCl_2$, 10 mM DTT, and 10% (v/v) glycerol.
3. T4 polynucleotide kinase.
4. ATP.
5. γ-[^{32}P]-ATP (3000 Ci/mmol, 1 μCi/μL); use within 1 wk of receipt.
6. Pharmacia (Piscataway, NJ), prepacked NAP-25 gel filtration columns (DNA grade) or equivalent.
7. 20 mM triethylammonium bicarbonate, pH 8.0.
8. Methanol (analar grade).
9. 2X EcoRV endonuclease assay buffer (see step 7, Section 2.1.).
10. EcoRV endonuclease.
11. Formamide stop mix consisting of 1 mg of bromophenol blue in 10 mL of formamide: H_2O, 9:1 (v/v).
12. Acrylamide stock solution consisting of 29% (w/v) acrylamide and 1% (w/v) bis-acrylamide in H_2O. This solution should be stirred with Amber-

lite MB1 mixed bed ion exchange resin (1 g/100 mL solution) for 1 h and then stored in the dark at 4°C. *Caution:* This solution is extremely toxic.
13. Gel buffer: 10.8% (w/v) Tris base, 5.5% (w/v) boric acid, 0.02M EDTA, pH 8.0.
14. Urea.
15. 10% Ammonium persulfate in H_2O. Make up fresh and use immediately.
16. *N,N,N',N'*-tetramethylethylenediamine (TEMED).
17. Gel electrophoresis apparatus consisting of 20 × 20-cm siliconized glass plates with 1-mm spacer, 20 tooth comb with 25 µL well volume, and power pack capable of running gels at 30 mA constant current.
18. Autoradiograph cassette, Fuji RX-X-ray film (or equivalent), Kodak X-ray film developer, and fixer (or equivalent).
19. Labscint scintillation fluid (or equivalent), scintillation vials, and scintillation counter with ^{32}P channel.

2.3. Continuous Spectrophotometric Assay

1. 2X *Eco*RV endonuclease assay buffer (*see* step 7, Section 2.1.).
2. d(pGACGATATCGTC).
3. *Eco*RV endonuclease.
4. 1-cm quartz cuvets suitable for absorbtion spectrophotometry at 254 nm.
5. UV spectrophotometer capable of 254-nm readings with temperature controlled compartments and chart recorder.

3. Methods
3.1. HPLC Assay

1. Prepare a solution (final volume after all additions including enzyme of 200 µL in an Eppendorf tube that contains 50 mM HEPES-NaOH, pH 7.5, 100 mM NaCl, 10 mM $MgCl_2$, 20 µM d(GACGATATCGTC) (*see* Note 1), and 10 U of alkaline phosphatase (*see* Note 2).
2. The oligonucleotide concentration must be accurately determined both as a prelude to studying the mechanism of the *Eco*RV endonuclease and in order to calibrate the integrator response (*see* step 8, below). This is best done spectrophotometrically but cannot be achieved simply by summing the known extinction coefficients of the constituent bases owing to hyperchromic effects. Instead proceed as follows. Dissolve d(GACGATATCGTC) in 0.5 mL of 10 mM potassium phosphate buffer containing 10 mM $MgCl_2$ in a quartz cuvet with a 1-cm light path. Note the initial absorbance at 254 nm, which should be between 0.2 and 0.5. Add 5 µg of snake venom phosphodiesterase and 10 µg of alkaline phosphatase in a small volume, typically 1–2 µL (*see* Note 3). Stopper the cuvet and note the final absorbance after it has stopped rising (this usu-

ally requires 1–2 h). The hyperchromicity is the ratio of the final absorbance divided by the initial. Values of 1.5 are typical for self-complementary 12'-mers. The true extinction coefficient of the dodecamer is the sum of the extinction coefficients of the constituent bases (at 254 nm) divided by the hyperchromicity (*see* Note 4). The extinction coefficients of the individual bases are: dA, 14.3×10^3; dG, 13.5×10^3; dC, 6×10^3; dT, 7×10^3, all at 254 nm in $M^{-1}\text{cm}^{-1}$ (6).
3. Incubate the mixture prepared in step 1 above at 25°C for 5 min.
4. Add *Eco*RV endonuclease (*see* Note 5) to give a final concentration of 0.1 μ*M*.
5. Incubate at 25°C and withdraw 15-μL samples and quench into 5 μL of NaOH at time intervals over 40 min (every 2.5 min in the first 10 min and every 5 min thereafter; *see* Note 5).
6. Inject 15-μL aliquots of the quenched samples onto the reverse-phase HPLC column. Develop with a linear gradient composed of 100% HPLC buffer A/0% HPLC buffer B at $t = 0$ min and 70% buffer A/30% buffer B at $t = 20$ min. The column should be heated to 50°C using either a column heater or by placing the column in a waterbath (*see* Note 6).
7. Detect the product and substrates at 254 nm. The HPLC elution profile is shown in Fig. 1.
8. Integrate the areas under each peak. These areas are proportional to the amounts of substrate and products present.
9. Calibrate the integrator response as follows. Inject known amounts of d(GACGATATCGTC), d(GACGAT), and d(ATCGTC) (*see* Note 1) onto the HPLC and develop with the above buffer system using 254 nm detection. Note the integrator response and thus calibrate the number of integrator units corresponding to a nmole of substrate or product. This process is best repeated at several known amounts of substrates and products for accuracy.
10. Using the calibrated integrator response evaluate the actual number of nmoles of substrate and product present at each time interval (*see* Note 7).
11. Plot a graph of nmoles of substrate consumed or nmoles of product against time. This is shown in Fig. 2. Determine rates from the initial slope (*see* Note 8).

3.2. ^{32}P/Gel Assay

1. Prepare d(pGACGATATCGTC) by incubating 40 nmol of d(GACGATATCGTC) in a volume of 100 μL containing 50 m*M* Tris-HCl, pH 8.5, 10 m*M* MgCl$_2$, 5 m*M* DTT, 5% (v/v) glycerol with 20 m*M* (2 μmol total) ATP, and 20 U of T4 polynucleotide kinase for 48 h at 30°C. Add a further 20 U of kinase and leave for another 24 h.

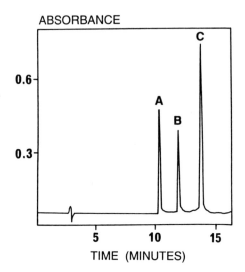

Fig. 1. The HPLC elution profile (A_{254}) of the hydrolysis of d(GACGATATCGTC) catalyzed by the *Eco*RV restriction endonuclease. This sample is midway through the digestion thus showing both the substrate and the two products. (**A**) d(GACGAT); (**B**) d(ATCGTC); (**C**) d(GACGATATCGTC).

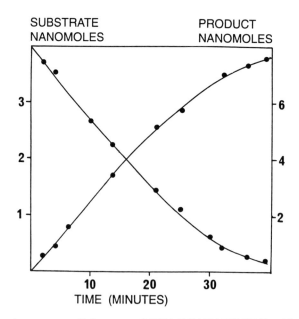

Fig. 2. The time course of cleavage of d(GACGATATCGTC) with *Eco*RV endonuclease. The rate of formation of either single-stranded product is twice the rate of hydrolysis of the double-stranded substrate.

2. Prepare [^{32}P]-d(pGACGATATCGTC) by incubating 20 nmol of d(GACGATATCGTC) in a volume of 100 µL containing 50 mM Tris-HCl, pH 8.5, 10 mM MgCl$_2$, 5 mM DTT, 5% (v/v) glycerol with 0.05 mM (5 nmol total) cold ATP, 100 µCi γ-[^{32}P]-ATP (approx 0.03 nmol total), and 10 U of T4 polynucleotide kinase for 48 h at 30°C. Add 1 µmol of ATP (nonradioactive), and a further 10 U of kinase and leave for a further 24 h (see Note 9).
3. Purify the above two oligonucleotides (separately) using Pharmacia prepacked NAP-25 gel filtration columns (DNA grade or equivalent). Elute the columns with 20 mM triethylammonium bicarbonate, pH 7.5. Pool fractions that contain the products and evaporate to dryness. Coevaporate (by addition and evaporation of 10 mL methanol) from methanol (Analar) three times to remove triethylammonium bicarbonate and dissolve the products in a small volume of water. Determine the concentration of the oligonucleotide stocks (see Note 4), and store frozen at –20°C. Radioactive oligonucleotides should be used within 3 wk of preparation.
4. Prepare a solution (final volume after all additions including enzyme of 100 µL) that contains 50 mM HEPES-NaOH, pH 7.5, 100 mM NaCl, 10 mM MgCl$_2$, and 20 µM [^{32}P]-d(pGACGATATCGTC). The oligomer should have a specific activity of about 1 µCi/nmol and be prepared by mixing [^{32}P]-d(pGACGATATCGTC), and d(GACGATATCGTC) as appropriate. Incubate at 25°C for 5 min (see Note 10).
5. Add *Eco*RV endonuclease to give a final concentration of 0.1 µM (see Note 5). Incubate at 25°C.
6. Withdraw 5-µL samples at intervals up to 10 min and quench by adding to 10 µL of formamide stop mixture at 100°C. Heat at 100°C for 2 min.
7. Prepare a polyacrylamide denaturing gel in a volume of 60 mL containing 1.08% (w/v) Tris-base, 0.55% (w/v) boric acid, 2 mM EDTA, pH 8.0 (10X dilution of gel buffer), 6M urea, and 17% (w/v) acrylamide. Add 300 µL of 10% (w/v) ammonium persulfate and 40 µL of *N,N,N',N'*-TEMED, and immediately pour into the 20 × 20 cm glass plates. Insert a 20-tooth comb of well volumes about 25 µL and leave to polymerize for 1–2 h.
8. Load 10 µL vol of the quenched samples onto the above gel and run at a constant current of 30 mA, using a 10X dilution of gel buffer, until the bromophenol blue has migrated 2/3–3/4 of the gel distance (see Note 11).
9. Remove the gel and place it on a transparent backing support. Place this in the autoradiograph cassette with the gel uppermost and cover the gel with cling film.

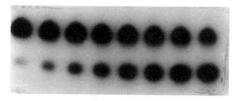

Fig. 3. Autoradiogram for the hydrolysis of [^{32}P]d(pGACGATATCGTC) to [^{32}P]d(pGACGAT) by the EcoRV restriction endonuclease. Reprinted with permission from Newman et al. (1990) *Biochemistry* **29,** 9902–9910. Copyright (1990) American Chemical Society *(3)*.

10. In the darkroom place Fuji-RX X-ray film on top of the cling film and using a needle drill holes through the film and the gel to enable the gel to be lined up with the autoradiogram.
11. Expose the gel at room temperature overnight.
12. Develop the X-ray film with Kodak X-ray developer until the bands become visible and fix with Kodak X-ray fixer (*see* Fig. 3).
13. Place the developed X-ray film under the transparent support and align the gel with the autoradiograph using the locating holes. Remove the cling film from the gel and excise the gel slices corresponding to the substrate dodecamer and product hexamer bands with a scalpel.
14. Place the gel slices in scintillation counting vials, add 3 mL of Labscint scintillation fluid, and count for 1 min on the ^{32}P channel of a liquid scintillation counter.
15. Calculate the amount of product formed using the formula:

$$\text{amount product} = \frac{I \times \text{cpm(hexamer)}}{\text{cpm(dodecamer)} + \text{cpm(hexamer)}}$$

where I = initial amount of double-stranded substrate at zero time.

16. Plot a graph of amount of product formed versus time and determine the rate from the initial slope (*see* Fig. 4).

3.3. Continuous Spectrophotometric Assay

1. Prepare a solution (final volume after all additions including enzyme of 1 mL), in a 1-cm path length quartz cuvet, containing 50 mM HEPES-NaOH, pH 7.5, 100 mM NaCl, 10 mM MgCl$_2$, and 2.7 µM d(pGACGATATCGTC) (*see* Section 3.2.1., and Notes 1 and 12).

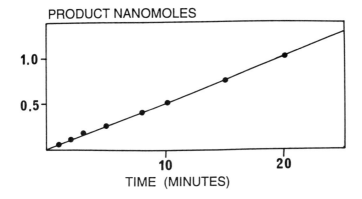

Fig. 4. Time-course for the hydrolysis of [^{32}P]d(pGACGATATCGTC) to [^{32}P] d(pGACGAT) by the EcoRV restriction endonuclease.

2. Incubate at 25°C in a spectrophotometer while recording the absorbance at 254 nm, for 10 min.
3. Add EcoRV endonuclease (see Notes 5 and 13) to give a final concentration of 0.025 µM.
4. Continue recording the absorbance at 254 nm noting the increase in absorbance at 254 nm (see Note 13). A typical trace is shown in Fig. 5.
5. Evaluate the rate of the enzyme catalyzed reaction in terms of increase in absorbance at 254 nm/U of enzyme per unit of time.
6. It is necessary to convert the increase in absorbance at 254 nm into units of substrate consumed or product produced. To do this proceed as follows. Prepare an identical mixture to that given in step 1, above, and note the initial absorbance at 254 nm. Add an excess of EcoRV endonuclease (a final concentration of 0.2 µM), and note the final absorbance at 254 nm after the increase has ceased (see Note 14). Check that hydrolysis is complete using the HPLC assay (see Section 3.1.). The difference in absorbance observed corresponds to complete hydrolysis of 2.7 nmoles of oligonucleotide and allows simple determination of the absorbance change expected per nmole of oligonucleotide hydrolyzed.
7. Convert the rate determined in step 5 from increase in 254 nm absorbance units to nmoles of substrate hydrolyzed using the factor determined in step 6 above.

4. Notes

1. Oligonucleotides can be prepared by any of the standard methods (most probably by the phosphoramidite approach using an automated DNA synthesizer). It is important that the oligomers are purified (either by HPLC

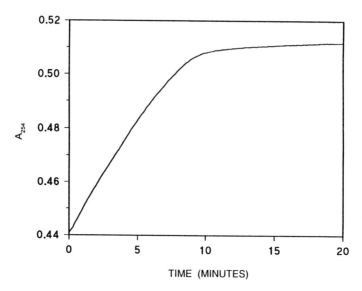

Fig. 5. Hydrolysis of d(pGACGATATCGTC) (double-stranded) to d(GACGAT), and d(pATCGTC) (both single-stranded) by the EcoRV restriction endonuclease monitored by a continuous spectrophotometric assay based on the hyperchromic effect.

or gel electrophoresis), and desalted before use *(7; see also* vol. 20 in this series).

2. The products of EcoRV hydrolysis with d(GACGATATCGTC) are d(GACGAT) and d(pATCGTC). These two hexamers are difficult to separate by HPLC. The alkaline phosphatase removes the 5'-phosphate from d(pATCGTC) giving d(ATCGTC), which is more easily separated from the other hexamer. Alkaline phosphatase is not required when product separation is straightforward (e.g., with other oligodeoxynucleotides or other restriction enzymes).

3. Snake venom phosphodiesterase and alkaline phosphatase completely digest d(GACGATATCGTC) to its constituent monodeoxynucleosides.

4. The true substrate for the EcoRV endonuclease is double-stranded d(GACGATATCGTC). Thus, in determining its extinction coefficient the sum of the extinction coefficients of the 24 bases in the double-stranded structure and not the 12 in the single-stranded should be used. Double-stranded d(GACGATATCGTC) has an extinction coefficient of $16.6 \times 10^4 M^{-1} cm^{-1}$ *(3)*.

5. The EcoRV endonuclease was purified from overproducing *E. coli* strains as described *(8,9)*. It has an extinction coefficient (at 280 nm) of $10.4 \times 10^4 M^{-1} cm^{-1}$ *(9)* (this is for a dimer of subunit mol wt 29,000,

which is the active species). Clearly, this assay can be used with any restriction endonuclease providing the oligonucleotides are designed with the correct recognition sequence. In our experience highly purified enzymes from overproducing strains are necessary to obtain good results. Commercial preparations usually contain low levels of relatively impure enzyme. Although these are suitable for molecular biology methods, they often cut smaller oligonucleotides poorly or nonspecifically. The amount of enzyme to be used and the time course of the assay should be determined by trial and error to give convenient hydrolysis rates.

6. Heating the column gives much sharper peaks, especially for self-complementary oligonucleotides, such as d(GACGATATCGTC). This effect is probably caused by the melting of this duplex at the high temperature used. Reverse-phase HPLC with the buffer combinations described can separate many oligonucleotides and their restriction endonuclease cleavage products. The best gradient conditions should be found by trial and error.

7. Because of errors in removing aliquots for quenching and injecting volumes into the HPLC columns, the sum of the number of nmoles of substrate and product present at each time point should be normalized to 100%.

8. The true substrate, the duplex of d(GACGATATCGTC), gives two equivalents of each product, d(GACGAT) and d(pATCGTC) (which are single-stranded), on *Eco*RV hydrolysis. This accounts for product being formed at twice the rate substrate is consumed.

9. This two stage phosphorylation gives the highest specific activity in the product. Clearly increasing the amount of nonradioactive ATP present in the first stage will reduce specific activity. However, reducing the levels of nonradioactive ATP here also reduces the specific activity of the final product. This is probably a result of the low levels of ATP present leading to poor incorporation. With this method almost all the ATP is incorporated in the first stage and the reaction is driven to completion by the excess of ATP in the second. The specific activity of the oligonucleotide product is about 5 µCi/nmol.

10. This assay is suitable for K_m determination because of its high sensitivity (4). Here the substrate concentration needs to be varied around the K_m value. It is most satisfactory to have the same amounts of radioactivity in each different substrate concentration assay. In practice this means making up the lowest substrate concentration with "pure" [^{32}P]-d(pGACGATATCGTC), and making up each successive higher concentration by adding more and more nonradioactive d(pGACGATATCGTC) to this amount of 5'-[^{32}P]-d(pGACGATATCGTC). This both ensures high sensitivity and that autoradiograph development takes the same

time for each substrate concentration. The approximate K_m must be determined by trial and error experimentally before accurate K_m evaluation is possible. Similarly the best levels of enzyme concentration and times of sample withdrawal must be evaluated by trial and error.

11. The gel takes about 2 h to run. In this case, the substrate 12'-mer runs just above the bromophenol blue band and the product 6-mer just below it. The exact polyacrylamide concentration in the gel can be varied depending on the lengths of product and substrate for the particular application.

12. The range of oligonucleotide concentrations that can be used is limited since the assay relies on an increase in absorption at 254 nm as the oligomers are hydrolyzed. Too low a concentration gives a very low initial reading at 254 nm, whereas too high a concentration gives a high initial reading. In both cases this makes increases in absorbance difficult to measure. We have found that initial absorbances between 0.05 and 1.3 give good results. This corresponds to d(pGACGATATCGTC) concentrations of between 0.5 and 7.5 μM. The 2.7 μM concentration used here has an initial absorbance of about 0.44. It should be possible to expand this range by using cuvets of different pathlengths (longer pathlengths for more dilute solutions and shorter pathlengths for more concentrated). Other oligodeoxynucleotides will have different concentration ranges since longer oligomers have a greater absorbance per mole at 254 nm.

13. The *Eco*RV endonuclease concentration and time of assay are selected by trial and error to find suitable values. With 0.025 μM enzyme a time of 10 min gives a suitable response. However, it is possible to greatly reduce enzyme concentrations and run the assays for times as long as overnight.

14. The actual absorbance change in this case is from about 0.44–0.51. In all cases complete hydrolysis of d(pGACGATATCGTC) leads to an increase in 254 nm absorbance of about 20%. A change in optical density (254 nm) of 0.0342 is observed for the hydrolysis of 1 nmole of double-stranded d(pGACGATATCGTC) in a volume of 1 mL with a 1-cm path length.

Acknowledgments

I would like to thank P. C. Newman and T. R. Waters, members of my group involved in developing and optimizing these assays for restriction endonucleases. The financial support of the UK SERC is gratefully acknowledged. B. A. Connolly is a Research Fellow of the Lister Institute of Preventive Medicine.

References

1. Lesser, D., Kurpiewski, M., and Jen-Jacobson, L. (1990) The energetic basis of specificity in the interaction of EcoR1 endonuclease with DNA. *Science* **250,** 776–786.
2. Jen-Jacobson, L., Lesser, D., and Kurpiewski, M. (1991) DNA sequence discrimination by EcoR1 endonuclease, in *Nucleic Acids and Molecular Biology, vol. 5* (Eckstein, F. and Lilley, D. M. J., eds.), Springer-Verlag, Berlin and New York, pp. 141–170.
3. Newman, P., Nwosu, V., Seela, F., Williams, D., Cosstick, R., and Connolly, B. A. (1990) Incorporation of a complete set of deoxyadenosine and thymidine analogues suitable for the study of protein nucleic acid interactions into oligonucleotides. *Biochemistry* **29,** 9891–9901.
4. Newman, P., Williams, D., Cosstick, R., Seela, F., and Connolly, B. A. (1990) Interaction of the EcoRV restriction endonuclease with the dA and T bases in its recognition hexamer GATATC. *Biochemistry* **29,** 9902–9910.
5. Saenger, W. (1984) *Principles of Nucleic Acid Structure.* Springer-Verlag, Berlin, pp. 49,143.
6. Fasman, G. D. (1975) Nucleic acids, in *Handbook of Biochemistry and Molecular Biology, vol. 1,* CRC, Cleveland, OH, pp. 141–147,420–447.
7. Brown, T. and Brown, D. J. S. (1991) Modern machine-aided methods of oligodeoxyribonucleotide synthesis, in *Oligonucleotides and Their Analogues, A Practical Approach* (Eckstein, F., ed.) Oxford University Press, Oxford, UK, pp. 1–24.
8. Bougueleret, L., Tenchini, M. L., Botterman, J., and Zabeau, M. (1985) Overproduction of the EcoRV endonuclease and methylase. *Nucleic Acids Res.* **13,** 3823–3839.
9. D'Arcy, A., Brown, R. S., Zabeau, M., Wijnaendts van Resandt, R., and Winkler, F. K. (1985) Purification and crystallization of the EcoRV restriction endonuclease. *J. Biol. Chem.* **260,** 1987–1990.

CHAPTER 30

Assays for Restriction Endonucleases Using Plasmid Substrates

Stephen E. Halford, John D. Taylor, Christian L. M. Vermote, and I. Barry Vipond

1. Introduction

A type II restriction enzyme purchased from a commercial supplier comes with a specified number of units of enzyme activity. The units of restriction enzyme activity are defined by the minimal amount of enzyme needed to complete the digestion of 1 µg of bacteriophage λ DNA in 1 h. These units are usually measured by making serial dilutions of the stock solution of the enzyme, adding 1 µL from each dilution to 1 µg of phage λ DNA in a suitable buffer, incubating the reactions for 1 h at 37°C, and then analyzing the DNA by electrophoresis through agarose. This is, at best, a semiquantitative assay. It cannot yield quantitative data about the rate of the reaction of a restriction enzyme on a DNA substrate.

For true reaction rates, it is essential to use a DNA that has only one copy of the recognition sequence, rather than phage λ DNA, which contains several recognition sites for nearly all restriction enzymes. DNA substrates with one recognition site can be made by oligonucleotide synthesis (*see* Chapter 29). However, the reaction kinetics of restriction enzymes on short duplexes of 8–12 bp often differ from those on longer DNA of 1 kb or more *(1)*. The difference can be a result of the protein binding to a long DNA molecule initially at any position on the chain, and then sliding along the chain until it locates its recognition site *(2)*. Short DNA molecules cannot facilitate the

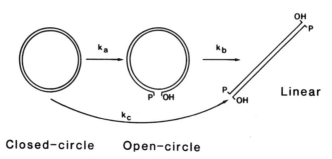

Fig. 1. Three forms of DNA may be observed during the reaction of a restriction endonuclease on a covalently closed circle of DNA that has one copy of the recognition sequence for the restriction enzyme.

location of the recognition site in the same manner. Hence, a full understanding of the mechanism of action of a restriction endonuclease demands the use of "real" DNA as a substrate.

The substrate should be a covalently closed circle of DNA rather than a linear duplex (Fig. 1). If the restriction enzyme cleaves its recognition site by cutting first one strand of the DNA and then the second strand (k_a and k_b in Fig. 1), the covalently closed circle will be converted initially to its open-circle form and only later to the linear form. On the other hand, if both strands are cut together (k_c in Fig. 1), the covalently closed DNA yields the linear form without generating the open-circle intermediate. Covalently closed DNA is naturally supercoiled and the supercoiled, open-circle, and linear forms of a DNA molecule can be separated from each other by electrophoresis through agarose *(3)*. In contrast, the initial product from the sequential pathway with a linear substrate, nicked DNA, cannot be distinguished from the intact substrate without denaturing the DNA to single-strands. Hence, by using a supercoiled DNA that has one recognition site for the restriction nuclease, it is possible to determine not only the rate of the reaction, but also the mode of action of the enzyme *(4)*.

After electrophoresis, the DNA is visualized with ethidium bromide (EtBr), from the fluorescence of ethidium-DNA on UV illumination. For a quantitative assay, the concentrations of the supercoiled, open-circle, and linear forms of the DNA must be determined. There are several ways in which this can be done. The most direct is to put the agarose gel in a scanning fluorimeter that measures the fluorescence from each band in each lane. This requires expensive equipment and, at present, the technology is not generally available. An alternative is to photograph the EtBr-stained gel and then scan the negative in a densitometer, although a major disadvantage of this method is that photographic film has a nonlinear response to light intensity. In addition, both this and the previous method need accurate controls because EtBr binds to covalently closed DNA with a different stoichiometry from either open-circle or linear DNA (5). Under standard conditions, the fluorescence from covalently closed DNA is only 77% of that from either open-circle or linear DNA. This difference means that some restriction enzymes can be assayed in a fluorimeter, by adding the enzyme to a cuvet containing a mixture of supercoiled DNA and EtBr (6). The preferred method for determining the amounts of the three forms of the DNA in an agarose gel is to use radiolabeled DNA so that slices from the gel can be subjected to scintillation counting.

The assay operates by adding the enzyme to a radiolabeled DNA in its supercoiled form, withdrawing samples at timed intervals and immediately stopping the reaction in these samples by adding EDTA. The DNA is subsequently analyzed by electrophoresis through agarose and the amounts of all three forms of the DNA, at each time-point from the reaction, are determined by scintillation counting (*see* Fig. 2 for a typical result). This procedure was first introduced many years ago for studies on the *Eco*RI restriction endonuclease (1), but it is still used extensively, not only with *Eco*RI (7,8), but also with other type II enzymes, such as *Bam*HI (9) and *Eco*RV (10,11). Moreover, essentially the same procedure can be used to assay many other enzymes that react with DNA: nonspecific nucleases, such as DNase I (12); DNA ligase (13); site-specific recombination systems (14, 15); modification methyltransferases (16,17) and other DNA methylases (18).

2. Materials

1. An electrophoresis power supply and a flat bed tank, with gel tray and well-formers, for submarine electrophoresis of agarose gels.
2. Agarose, low electroendosmosis grade.
3. Electrophoresis buffer: 40 mM Tris base, 5 mM sodium acetate, 1 mM EDTA. Adjust to pH 8.0 with glacial acetic acid (*see* Note 1).

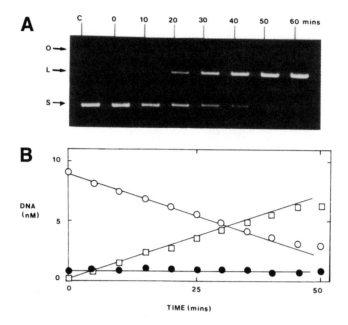

Fig. 2. A typical reaction profile for a restriction enzyme. This reaction contained the EcoRV restriction endonuclease (0.5 nM) and ^{3}H-labeled pAT153 (10 nM); initially, 90% of the DNA was in the supercoiled form. The plasmid has one copy of the EcoRV recognition sequence. Aliquots were removed from the reaction at timed intervals after the addition of the enzyme, the reaction in each aliquot stopped immediately, and the DNA subsequently analyzed by electrophoresis through agarose. The gel is shown in (**A**): The time of withdrawal of each sample is given above the gel and the electrophoretic mobilities of the supercoiled, open-circle, and linear forms of pAT153 are marked by S, O, and L on the left of the gel. The bands from all three forms of the DNA, at all of the time-points, were excised from the gel and the radioactivity in each was measured, in order to determine the concentrations of the supercoiled (○), open-circle (●), and linear (□) forms: These are shown in (**B**). This particular reaction produces no open-circle DNA, and thus it proceeds by the pathway denoted by k_c in Fig. 1. (Data from ref. *11*: reprinted with permission from *Biochemistry.* Copyright [1989] American Chemical Society.)

4. An oven set at 67°C.
5. Ethidium bromide (EtBr): Stock solution at 5 mg/mL in water (all water throughout this protocol is double-distilled). EtBr is sensitive to light: Wrap the bottle containing the stock solution in silver foil. Ethidium is also a carcinogen: *Always* wear gloves when handling materials containing EtBr.
6. Double strength reaction buffer: Make up a solution containing Tris-HCl at the required pH, NaCl, and MgCl$_2$, all at twice the concentration

needed for the reaction of the restriction enzyme that is being studied. Autoclave this solution at 121°C for 15 min to ensure that it is free of nucleases (*see* Note 2).

7. Bovine serum albumin (BSA): Dissolve BSA (standard-grade, from any supplier) in 10 mM Tris-HCl, 0.1 mM EDTA, pH 7.5, at a final concentration of 5 mg/mL, readjust the solution to pH 7.5, and place in the 67°C oven for >4 h. Store the stock solution of BSA at 4°C (*see* Note 3).
8. The DNA substrate, a covalently closed circle of DNA that has one recognition site for the restriction enzyme under study (*see* Note 4). The DNA must be radiolabeled and at least 90% of the preparation must be supercoiled (*see* Note 5). Store the stock solution of the purified DNA in 10 mM Tris-HCl, 0.1 mM EDTA, pH 7.5, at 4°C over chloroform.
9. The type II restriction endonuclease, purified at least to the extent of removing all other nucleases from the sample and, ideally, purified to homogeneity (*see* Notes 6 and 7). The enzyme stock is normally stored in a buffer containing 50% (v/v) glycerol at –20°C.
10. Stop-mix: A solution that contains all four of the following at the concentrations indicated: 100 mM EDTA, 100 mM Tris-HCl, pH 8.0, 0.5 mg/mL bromophenol blue, 40% (w/v) sucrose (*see* Note 8).
11. A UV transilluminator and a plate of either glass or plastic that is transparent at the wavelength of the transilluminator.
12. At least 40 full-size (20 mL) scintillation vials, either glass or plastic, and racks for the vials.
13. Sodium perchlorate: Make up a solution of 5M NaClO$_4$ in water. Perchlorate is potentially hazardous—it is a strong oxidizing reagent.
14. Scintillant: Some, but by no means all, commercial scintillation cocktails are suitable for this application; for example, Packard Emulsifier Scintillator 299. Alternatively, for 1 L of scintillant, mix together 333 mL of Triton X-100 and 666 mL of toluene and use this to dissolve 6 g of butyl-PDB [2-(4-biphenyl)-5-(4-*tert*-butylphenyl)-1,3,4-oxadiazole]. The homemade version has no advantages in either cost or quality over the commercial products; moreover, butyl-PDB is a powerful carcinogen.
15. A scintillation counter.

3. Method

1. Prepare a 1.2% (w/v) agarose gel by adding 100 mL of electrophoresis buffer to 1.2 g of agarose. Dissolve the agarose either by bringing to a boil while stirring on a hotplate or by placing the flask in a domestic microwave oven for 2 min at full power. Once dissolved, allow the solution to cool by leaving it in a 67°C oven for about 30 min. Then add 10 µL of EtBr and cast the gel in the gel tray. The gel will need, at

minimum, one row of 12 wells and the capacity of the wells should be at least 25 µL. The amounts given here make a gel of adequate thickness (0.5–0.7 cm) over an area of 11 × 14 cm. Obviously, the amounts are adjusted *pro rata* for larger or smaller areas.

2. After the gel has set, place it in the flat bed gel tank. Add EtBr to the electrophoresis buffer to a final concentration of 0.5 µg of EtBr/mL of buffer, and use this to fill the gel tank so that the gel is just submerged with buffer (*see* Note 9).

3. Set up the reaction by mixing together in an Eppendorf tube 100 µL of double strength reaction buffer, 10 µL of BSA solution (*see* Note 10), and the required amount of the purified substrate (typically, 5 µg of radiolabeled DNA). Add water to bring the volume of the reaction mix to 200 µL, vortex the tube, and place in a waterbath at the temperature for the reaction.

4. Remove one aliquot of 15 µL from the reaction mixture and add this to 8 µL of stop-mix: This will be the zero time-point, before the addition of the enzyme to the reaction. Remove a second aliquot of 15 µL and incubate this in a separate tube at the temperature for the reaction, for the complete duration of the reaction, before adding 8 µL of stop-mix. The latter is an essential control to check that all DNA cleavage in the reaction, following the addition of the restriction enzyme, is caused by that enzyme and is not a result of any of the other reaction components being contaminated with a nuclease.

5. To the remainder of the reaction mixture, add the restriction enzyme (*see* Note 11). Vortex thoroughly and continue the incubation at the required temperature. At timed intervals after the addition of the enzyme, withdraw aliquots (15 µL) from the reaction and, immediately after each withdrawal, add the aliquot to 8 µL of stop-mix and vortex directly. In a typical experiment, a total of 10 aliquots are taken at 2 or 5 min intervals, depending on the rate of the reaction.

6. On the completion of the reaction, all of the samples from both steps 4 and 5 are placed in a 67°C oven for 10 min and then transferred directly to an ice/water bath (*see* Note 12).

7. A portion (15 µL) of each sample is then loaded on the agarose gel prepared in step 2. Run the gel at constant voltage, at a voltage gradient of 8 V/cm of gel length, until the blue dye from the stop-mix is close to the end of the gel (*see* Note 13).

8. Remove the gel from the gel tank, place it on a transparent support plate, allow it to drain free of excess buffer, and then view it on the transilluminator. For a transilluminator operating at 366 nm, the support

Restriction Enzyme Assay

can be a glass plate, but a sheet of UV-transparent plastic is needed with a 310 nm transilluminator. If a photograph of the gel is required, take it now.

9. Use a scalpel to cut the gel into slices across the gel, so that one slice contains the supercoiled form of the DNA, a second the linear form, and a third the open-circle form. The cuts should be made as close as possible to the front of each band and about 0.5 cm behind each band. Also cut each slice down the gel into its separate lanes. At this stage, the gel consists of a large number of cubes of agarose (each about 0.5 × 0.6 × 0.7 cm), corresponding to all three forms of the DNA from all of the samples applied to the gel. Place each cube in a separate scintillation vial. Take care to record which cube is in each vial.
10. Use the scalpel to cut out three or four cubes of similar size from elsewhere in the gel, from locations where there is no DNA, and again put these into separate scintillation vials. These will be used for the background in the scintillation counting.
11. Add 0.5 mL of sodium perchlorate to each vial and leave the vials in the 67°C oven for at least 2 h (or overnight). Remove the vials from the oven, leave to cool for about 30 min, and then add to each 3 mL of water followed by 10 mL of scintillant (*see* Note 14). Immediately after adding the scintillant to a vial, shake the vial vigorously to ensure a homogeneous mixture. Wipe the outside of the vials with 70% ethanol.
12. Leave the vials in the dark for at least 1 h, to let the chemiluminesence in the samples decay to zero. Then determine the radioactivity in each vial by scintillation counting. The accuracy of scintillation counting is a function of the total number of counts accumulated during the counting period and, in this application, a count time of 10 min per vial is likely to be adequate.
13. Take the average cpm (or dpm) reading for the background vials and subtract this value from the readings for all of the other vials. It is now possible to calculate the concentrations of all three forms of the DNA (Fig. 1) at all of the time points that were selected during the reaction (as in Fig. 2).
14. Discard the contents of the vials (and the vials themselves if they are disposable) in an environmentally acceptable manner. Remember that they contain not only ^{3}H label and scintillant gel, but also sodium perchlorate.

4. Notes

1. There are many different buffers that can be used for agarose gel electrophoresis, such as Tris-borate or Tris-phosphate (*see* ref. *19* for compositions). However, agarose gels made up in Tris-acetate dissolve

readily in NaClO$_4$, as is required at a later stage in this protocol. In contrast, gels made up in Tris-borate fail to dissolve in the same way and cannot be analyzed by the methods described here.

2. The composition of the double strength reaction buffer will vary from one restriction enzyme to another. For both EcoRI *(1,6–8)* and EcoRV *(10,11)*, a suitable solution would be 100 mM Tris-HCl, pH 7.5, 200 mM NaCl, and 20 mM MgCl$_2$, but another enzyme may need a different ionic strength or a different pH *(19)*. Many nucleases function better if the NaCl in the reaction buffer is replaced by potassium glutamate *(20)*. Some, but not all, restriction enzymes are inactivated by modifications of thiol groups: This can be avoided by adding to the double strength reaction buffer, after sterilization, β-mercaptoethanol to a final concentration of 20 mM.

3. It is possible, but unnecessary, to buy nuclease-free BSA from commercial suppliers. The procedure given here inactivates the contaminating nucleases that are always present in standard-grade BSA. However, on occasion, it also precipitates the BSA. If a white precipitate is visible after heating to 67°C, start again. Alternatively, remove the precipitate by centrifugation and determine the concentration of BSA in solution by UV absorption.

4. The most convenient substrates are plasmids from *Escherichia coli*, although suitable alternatives include the replicative DNA from bacteriophage, such as φX174 or M13 and the DNA from animal viruses, such as SV40. There are now a large number of plasmids that have been fully sequenced (*see* the EMBL or GenBank databases) and the sequences can be inspected to find one that has only one copy of the recognition sequence. The archetype of plasmid substrates for restriction enzymes is pBR322 *(21)* but certain derivatives of pBR322, for example pAT153 or pUC19, have additional advantages in that they are smaller (<4 kb) and have higher copy numbers *(22)*. Multicopy plasmids in *E. coli* can undergo general recombination with themselves, with the result that one molecule of plasmid can have multiple repeats of its DNA sequence and thus multiple copies of the recognition site. For the experiments described here, it is essential to use the monomeric form of the plasmid. This can be obtained by first using the monomeric form to transform a *rec*A (recombination deficient) strain of *E. coli*, such as HB101 *(19)*, where it will be maintained as the monomer, and then purifying the plasmid from the transformant.

5. The substrate should be radiolabeled in vivo, because it is difficult to produce supercoiled DNA after labeling in vitro. The transformant of *E. coli* carrying the required plasmid is grown in M9 minimal salts

medium *(19)* to late exponential phase, 1 mCi of methyl-[^{3}H]-thymidine is added to each liter of culture along with the chloramphenicol for plasmid enrichment *(22)*, and the culture is then continued overnight. Use plastic rather than glass flasks for the culture (*see* ref. *6* for details of both the bacterial culture and the subsequent purification of the plasmid). It is important that the plasmid is purified rigorously by CsCl/EtBr density gradient centrifugations. If not, it is probable that the plasmid preparation will be contaminated by polysaccharides or other cell debris, and these contaminants can act as potent inhibitors of nuclease activity. After the purification, check that the plasmid is >90% supercoiled monomer by electrophoresis through agarose, as described here.

6. If the preparation of the restriction nuclease contains another nuclease, the extent of DNA cleavage measured by the assay described here will be the sum of both nuclease activities. A simple test to determine whether the sample of restriction enzyme is contaminated by another nuclease is to carry out the same experiment, but using a supercoiled DNA substrate that lacks the recognition sequence *(11)*.

7. If the restriction enzyme is only partially purified, such that the preparation contains several proteins in addition to the restriction enzyme, neither this nor any other assay for nucleases can yield the reaction rate in terms of moles DNA cleaved per mole of enzyme per min. At present, it is impossible to determine the concentration of a sample of restriction enzyme without purifying the protein to homogeneity: There is no "active site titration" for a restriction nuclease. The standard criterion for the purification of a protein to homogeneity is a single band on an SDS-polyacrylamide gel. However, the specific activities of pure restriction enzymes are usually greater than 1×10^6 U/mg of protein, given the units defined in Section 1. *(4)*. Consequently, the visualization of a restriction enzyme on an SDS-polyacrylamide gel requires >> 1000 U.

8. The EDTA in this solution stops DNA cleavage by the restriction enzyme, by chelating the Mg^{2+} ions in the reaction mixture. Control experiments should be carried out to ensure that the stop-mix halts DNA cleavage instantaneously. Alternative stop-mixes are phenol or SDS. The stop-mix given here also contains sucrose, to increase the density of the sample before loading it on to an agarose gel, and bromophenol blue, to act as an electrophoresis marker.

9. The concentrations of EtBr in the agarose and the electrophoresis buffer must be the same, but it is not essential to add EtBr to either. An alternative is to run the gel in the absence of EtBr, and then stain the gel afterward by immersing it for 30 min in electrophoresis buffer containing 0.5 μg/mL EtBr. Samples containing SDS must be run on a gel without EtBr.

10. The requirement for BSA stems from the fact that many restriction enzymes are inactivated by sticking to glass or plastic surfaces. This applies particularly to highly purified samples, which often contain very low concentrations of protein. BSA helps to reduce the loss. However, in many instances, BSA cannot by itself stabilize the enzyme and some extra ingredients must be added to the enzyme stock and to the reaction mixture. For example, with pure *Eco*RI nuclease, the solution of enzyme should contain a nonionic detergent, such as Triton X-100, and some gelatin is needed in the reaction mix *(1)*. With pure *Eco*RV, the enzyme should be diluted with buffer that contains 1 m*M* spermine *(10)*.
11. The amount of restriction enzyme added to the reaction depends on the nature of the kinetics that are being studied. For steady-state kinetics, the reaction should be done with at least 10 times fewer enzyme molecules than DNA molecules, but the method described here can also be used for single turnover kinetics with the enzyme in excess of the DNA *(7,9–11)*. However, in both cases, the volume of the solution of enzyme added to the reaction should be as small as possible (<10 µL), so as to minimize the effect of the enzyme buffer on the reaction mix. Restriction enzymes are usually stored in buffers that contain 50% (v/v) glycerol, and the specificity of these enzymes for their recognition sites on DNA can be perturbed by glycerol *(4)*.
12. Many restriction enzymes leave single-strand extensions on the DNA product and the purpose of the heat/quench step is to disrupt any annealing of these extensions. This step also tends to denature proteins, which sharpens the bands in the electrophoresis of DNA. However, after reactions at very high nuclease concentrations, it may be necessary to remove protein prior to electrophoresis: Add to each aliquot either 1 µL of proteinase K (stock; 20 mg/mL) or 2 µL of SDS (stock; 2%, but *see* Note 9), or extract the sample with phenol *(19)*.
13. For a given DNA, the optimal separation of the supercoiled, open-circle, and linear forms is a function of the concentration of agarose in the gel, the electrophoresis buffer, and the voltage applied across the gel *(3)*. The conditions given here, 8 V/cm across a 1.2% (w/v) gel in Tris-acetate buffer, are the optimal for separating the three forms of pAT153, a plasmid of 3.65 kb. However, a DNA substrate of another size may require either a different voltage or a different agarose concentration to optimize the separation (*see* ref. 3 for a thorough survey of electrophoresis conditions with DNA of different sizes).
14. Scintillation counting demands a homogeneous system, either a monophasic clear liquid or a translucent gel. We have not found any scintillant that gives a homogeneous system when added to these samples directly

after dissolving the agarose in sodium perchlorate. However, if the samples are diluted with water before adding the scintillant, certain scintillation cocktails give a homogeneous gel, but other cocktails produce biphasic systems or precipitates. It is important to test the scintillant, to make sure that it gives a monophasic counting system, before using it in this application. Some scintillation cocktails from commercial suppliers are much more expensive than others, but the cheaper Triton/toluene based systems are often the best.

References

1. Greene, P. J., Poonian, M. S., Nussbaum, A. L., Tobias, L., Garfin, D. E., Boyer, H. W., and Goodman, H. M. (1975) Restriction and modification of a self-complementary oligonucleotide containing the EcoRI substrate. *J. Mol. Biol.* **99,** 237–261.
2. von Hippel, P. H. and Berg, O. G. (1989) Facilitated target location in biological systems. *J. Biol. Chem.* **264,** 675–678.
3. Johnson, P. H. and Grossman, L. I. (1977) Electrophoresis of DNA in agarose gels: optimizing separations of conformational isomers of double- and single-stranded DNAs. *Biochemistry* **16,** 4217–4224.
4. Bennett, S. P. and Halford, S. E. (1989) Recognition of DNA by type II restriction enzymes. *Curr. Top. Cell. Reg.* **30,** 57–104.
5. Bauer, W. and Vinograd, J. (1968) The interaction of covalently closed DNA with intercalative dyes: I the superhelix density of SV40 DNA in the presence and absence of dye. *J. Mol. Biol.* **33,** 141–171.
6. Halford, S. E. and Johnson, N. J. (1981) The EcoRI restriction endonuclease, covalently closed DNA and ethidium bromide. *Biochem. J.* **199,** 767–777.
7. Terry, B. J., Jack, W. E., and Modrich, P. (1987) Mechanism of specific site location and DNA cleavage by EcoRI endonuclease. *Gene Amplif. Anal.* **5,** 103–118.
8. Hager, P. W., Reich, N. O., Day, J. P., Coche, T. G., Boyer, H. W., Rosenberg, J. M., and Greene, P. J. (1990) Probing the role of glutamic acid 144 in the EcoRI endonuclease using aspartate and glutamine replacements. *J. Biol. Chem.* **265,** 21,520–21,526.
9. Hensley, P., Nardone, G., Chirikjian, J. G., and Wastney, M. E. (1990) The time-resolved kinetics of superhelical DNA cleavage by BamHI restriction endonuclease. *J. Biol. Chem.* **265,** 15,300–15,307.
10. Halford, S. E. and Goodall, A. J. (1988) Modes of DNA cleavage by the EcoRV restriction endonuclease. *Biochemistry* **27,** 1771–1777.
11. Taylor, J. D. and Halford, S. E. (1989) Discrimination between DNA sequences by the EcoRV restriction endonuclease. *Biochemistry* **28,** 6198–6207.
12. Campbell, V. W. and Jackson, D. A. (1980) The effects of divalent cations on the mode of action of DNase I. *J. Biol. Chem.* **255,** 3726–3735.
13. Sugino, A., Goodman, H. M., Heynecker, H. L., Shine, J., Boyer, H. W., and Cozzarelli, N. R. (1977) Interaction of bacteriophage T4 RNA and DNA ligases in joining of duplex DNA at base-paired ends. *J. Biol. Chem.* **252,** 2987–3994.

14. Johnson, R. C., Bruist, M. B., Glaccum, M. B., and Simon, M. I. (1984) In vitro analysis of Hin-mediated site-specific recombination. *Cold Spring Harbor Symp. Quant Biol.* **49,** 751–760.
15. Castell, S. E., Jordan, S. L., and Halford, S. E. (1986) Site-specific recombination and topoisomerization by Tn*21* resolvase: role of metal ions. *Nucleic Acids Res.* **14,** 7213–7226.
16. Rubin, R. A. and Modrich, P. (1977) *Eco*RI methylase: physical and catalytic properties of the homogeneous enzyme. *J. Biol. Chem.* **252,** 7265–7272.
17. Nwosu, V. U., Connolly, B. A., Halford, S. E., and Garnett, J. (1988) The cloning, purification and characterization of the *Eco*RV modification methylase. *Nucleic Acids Res.* **16,** 3705–3720.
18. Bergerat, A., Kriebardis, A., and Guschlbauer, W. (1989) Preferential site-specific hemimethylation of GATC sites in pBR322 DNA by *dam* methyltransferase from *Escherichia coli*. *J. Biol. Chem.* **264,** 4064–4070.
19. Sambrook, J., Fritsch, E. F., and Maniatis, T. (1989) *Molecular Cloning, a Laboratory Manual* (2nd Ed.) Cold Spring Harbor Laboratory, Cold Spring Harbor, NY.
20. Leirmo, S., Harrison, C., Cayley, D. S., Burgess, R. R., and Record, M. T., Jr. (1987) Replacement of potassium chloride by potassium glutamate dramatically enhances protein-DNA interactions in vitro. *Biochemistry* **26,** 2095–2101.
21. Sutcliffe, J. G. (1979) Complete nucleotide sequence of the *Escherichia coli* plasmid pBR322. *Cold Spring Harbor Symp. Quant. Biol.* **43,** 77–90.
22. Twigg, A. J. and Sherratt, D. J. (1980) Trans-complementable copy number mutants of plasmid ColE1. *Nature (London)* **283,** 216–218.

CHAPTER 31

Assays for Transcription Factor Activity

Stephen Busby, Annie Kolb, and Stephen Minchin

1. Introduction

Most transcription activator proteins have three important features that can be probed at the molecular level: They bind to specific sequences near promoters, they can be interconverted between active and inactive forms by covalent or noncovalent modification, and, when bound at target promoters, they can stimulate the initiation of transcription by RNA polymerase *(1)*. This chapter is concerned with in vitro methods for measuring the transcription activation function of this important class of proteins. In most cases these methods will be applied to activator proteins that have been substantially purified, and in which the target promoter sequence is known and the binding site characterized. Chapters 1–11 in this volume cover methods that can be exploited to locate and investigate specific binding sites for activators. Here we are concerned with the measurement of the products of activation. Since *Escherichia coli* transcription activators have been studied more than any other, we will take *E. coli* systems as the paradigm. In principle the techniques can be applied to any organism from which in vitro systems have been developed.

The starting point of the methodology was the observation, made in the early 1970s, that purified *E. coli* RNA polymerase could initiate transcription at promoters in purified DNA *(2)*. With improvements in RNA methodology, it was found that the transcription start site in

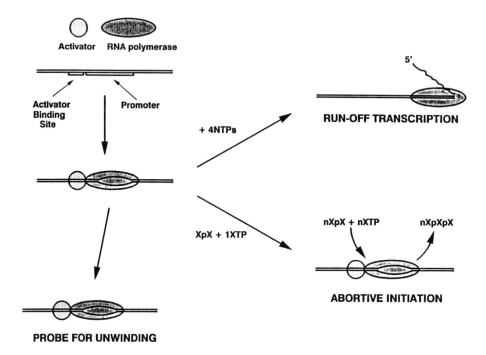

Fig. 1. Overview of techniques discussed in this chapter. *See* Section 3.1. for runoff transcription assays, Section 3.2. for abortive initiation assays, and Note 8 for probes to detect unwinding.

vitro, in many cases, was the same as *in vivo* (3). Further, at a number of promoters, interactions with specific transcription activators could be demonstrated and factor-dependent transcription in vitro occurred. The literature is now full of instances in which purified RNA polymerase plus DNA is sufficient for factor-dependent transcription initiation, thus setting the scene for studies on the mechanism of transcription activation. However, as with all in vitro techniques, it is worth noting that the conditions found in the plastic tube differ greatly from those *in vivo* and that, for some instances, the reconstituted transcription system simply may not work.

Two principal methods can be used to monitor transcription activation: runoff transcription assays and abortive initiation (Fig. 1). In runoff assays, RNA polymerase is incubated with a DNA fragment carrying the promoter of interest either with or without the transcription activator protein. Radioactively labeled nucleoside triphosphates are then added and RNA polymerase molecules that have formed "open" transcriptionally competent complexes start to make RNA. Polymerase molecules "run" to the end of the fragment making labeled RNA of a discrete length that can be monitored and quantified *(4)*. In abortive assays *(5)*, RNA polymerase forms transcriptionally competent complexes at the chosen promoter, but elongation is prevented because some of the four nucleoside triphosphates are withheld, and nucleotide precursors corresponding just to the start of the message are added. The result is that RNA polymerase is "trapped" at the promoter and can only synthesize a short oligonucleotide that is released as the polymerase cycles between a number of conformations in the abortive complex (it is termed abortive because the full-length runoff transcript cannot be made). The consequence of the cycling is that one polymerase molecule, trapped at one promoter, synthesizes the oligonucleotide continuously, and the appearance of the product can be measured directly. The rate of product formation will be dependent on the number of promoters that are occupied by polymerase in an open complex and this, in turn, will depend on the activity of the transcription factor under study. Conventionally, runoff assays are used to locate transcription starts and monitor factor activity qualitatively, whereas abortive initiation assays are exploited for quantitative and kinetic work.

2. Materials

1. *E. coli* RNA polymerase. This enzyme can be purified simply *(6,7)* or can be purchased from Boehringer (Mannheim, Germany), Pharmacia (Piscataway, NJ) or other molecular biology companies. Purity can be easily checked by denaturing polyacrylamide gel electrophoresis and activity can be verified using standard templates. Holoenzyme, which can be prepared by a number of procedures *(7–9)*, is required for the determination of kinetic constants. Preparations of RNA polymerase are usually stored at –20°C at around 1 mg/mL in buffer containing 100 m*M* NaCl and 50% glycerol.
2. DNA fragments. These can be purified from restricted plasmid DNA by polyacrylamide or agarose gels by a variety of methods *(10)*. Although any DNA can be used (*see* Note 4.1.1.), smaller fragments around 100 bp are most desirable and stock solutions of most short fragments will be adjusted to around 20 µg/mL, the concentration being checked after gel electrophoresis.

3. Heparin. Purchased from Sigma (St. Louis, MO) and made up as a 10 mg/mL stock solution in water.
4. Nucleotides. α-[^{32}P]-UTP from NEN (Boston, MA) or Amersham (Arlington Heights, IL) can be used in conjunction with the four nucleoside triphosphates from Boehringer or Pharmacia. Most workers use around 0.5 µCi α-[^{32}P]-UTP per reaction. Typically, a stock solution will contain 500 µM UTP and 2 mM of the three other NTPs, and will also include 1 mg/mL heparin to prevent reinitiation.
5. Dinucleotides. These can be bought from Sigma, and used without further purification as 10 mM stock solutions in water. The choice of the appropriate dinucleotide for priming abortive initiation assays is discussed in Section 3.2.1.
6. Transcription buffer: 20 mM Tris-HCl, pH 8, 100 mM NaCl, 5 mM MgCl$_2$, 0.1 mM EDTA, 1 mM DTT, 50 µg/mL nuclease-free BSA, 5% glycerol. This is a standard 1X buffer for many in vitro transcription assays and can be prepared as a 10X stock. Clearly there are many variations of this and the literature must be checked for any particular instance.
7. RNA gels. Standard 6% polyacrylamide sequencing gels containing urea run in TBE *(10)* can be used to separate run-off transcripts, followed by autoradiography to detect the products.
8. Loading buffer: 80% deionized formamide, 0.1% Xylene cyanol FF, 0.1% bromophenol blue in the standard gel running buffer, TBE.
9. Gel running buffer, TBE. This is usually made up in large volumes and kept as a 5X stock solution. To make up 1 L of 5X stock use 54 g Tris base, 27.5 g boric acid, and 20 mL of 0.5M EDTA adjusted to pH 8.0 with NaOH.
10. Whatman 3MM paper. Cut into strips 20-cm in length for chromatography of abortive products.
11. Chromatography buffer. 18:80:2 (v/v) water/saturated ammonium sulfate/isopropanol.
12. Transcription factors. These must be purified, at least partially, away from any nuclease activities. This can be facilitated by overproducing strains and the transcription assays can be used to monitor purification.
13. Sodium acetate. 3M solution in water with pH adjusted to 7.0 with NaOH.
14. Cold stop mix. 10 mM EDTA and 50 µg/mL crude tRNA (or any RNA) from Sigma (or homemade).
15. Ethanol. Absolute and 70% (v/v) in water.
16. Phenol. Melted and equilibrated with 10 mM Tris-HCl, pH 7.5.
17. RNA size markers. These are usually generated from runoff transcripts of well-characterized DNA fragments. Alternatively, sequence ladders can be used.
18. 0.1M EDTA.

3. Methods
3.1. Runoff Transcription Assays

1. Incubate the purified DNA fragment together with RNA polymerase plus or minus the transcription factor in a volume of, for example, 20 µL. Typically this will be made up by adding 2 µL 10X transcription buffer, 2 µL of 20 µg/mL fragment, and 1 µL of 1 mg/mL RNA polymerase giving an incubation mix containing about 100 nM RNA polymerase and 10 nM fragment. Normally experiments are run with a range of transcription factor concentrations from, for example, 0.5–10 times the promoter concentration. It is important to include any cofactor required by the transcriptional activator. For example, the cyclic AMP receptor protein (CRP) requires cAMP in the incubation mix for activity. Other transcriptional activators require covalent modification, such as phosphorylation, and therefore the active form must be used.
2. After incubation for around 20 min at 37°C, add 2 µL of a mix of the four NTPs containing 0.5 µCi of α-[^{32}P]-UTP plus 1 mg/mL heparin (*see* Section 2.4.). For each individual DNA molecule the RNA polymerase may or may not have reached an open complex on each individual DNA molecule depending on the activity (or not) of the transcription factor. At molecules where an open complex has formed, the polymerase will then "runoff" to the end of the fragment making a discrete sized RNA product. Since only one molecule of polymerase can occupy a promoter at any time, and since the inclusion of heparin prevents further initiation, the amount of any particular runoff transcript will be directly proportional to the amount of open complex formation.
3. After further incubation for about 10 min at 37°C to allow elongation to finish, add 200 µL of cold-stop mixture (10 mM EDTA plus 50 µg/mL of any RNA) and phenol extract the mixture. Typically, 500 µL of phenol is added and the sample is vortexed briefly and centrifuged in a microfuge to separate the phases. Carefully remove 200 µL of aqueous phase per tube and transfer to a fresh tube. It is not crucial that the total aqueous phase be used, but it is important to take the same proportion from each sample that is phenol-extracted.
4. Alcohol precipitate the RNA in the 200 µL aqueous phase sample by the addition of 22 µL of 3M sodium acetate, pH 7, and 660 µL cold ethanol. After centrifugation, careful removal of the supernatant, and rinsing with 70% ethanol, dry the pellet and resuspend in 10 µL of loading buffer (*see* Section 2.8.).
5. Load the samples (again quantitatively) on a sequencing gel together with size markers and perform the electrophoresis. We routinely use

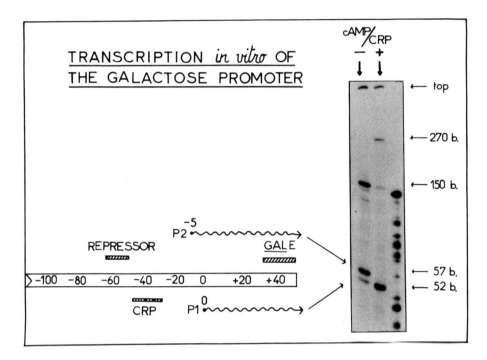

Fig. 2. Transcription in vitro of a 270 bp DNA fragment carrying the E. coli galactose operon promoter region. Run off transcripts were made as described in Section 3.1. using *Escherichia coli* RNA polymerase in the presence or absence of the cAMP-CRP complex. Transcription initiating from *gal*P1 and *gal*P2 results in products 52 and 57 bases in length, respectively. In the absence of cAMP-CRP, transcription initiates from *gal*P2: in the presence of cAMP-CRP transcription from *gal*P2 is repressed and transcription from *gal*P1 activated. The 150 base product is a result of transcription initiating from a potential promoter that is not normally used *in vivo*. Faint bands equal in length and longer than the DNA template are also seen and are discussed in Note 7.

the S2 model from BRL-Gibco (Gaithersburg, MD), running the gel for 2–3 h at 60 W constant power. After running, the gel is dried, autoradiographed, and the film is developed. From the sequence marker it will be possible to identify bands caused by transcription initiation at the promoter under study and to determine the effects of the transcription activator on the appearance of these bands. An example is given in Fig. 2.

3.2. Abortive Initiation Assays

1. Choose the nucleotides to be employed in the assay. Typically, this is done by selecting a dinucleotide appearing in the sequence anywhere

from –4 to +2 and using the next nucleotide as the labeled precursor. For example, at the *E. coli galP1* promoter *(11)*, the sequence at the transcription start is 5'-TCATA-3' with the central A as +1. The dinucleotide CpA and α-[^{32}P]-UTP can be used to give the product, [^{32}P]-labeled CpApU. It is important to ensure that no runoff product can form.
2. Before the assay is to be performed, set up a series of Whatmam 3MM paper chromatograms (typically 20-cm long). Spot the origins with 20 µL of 0.1*M* EDTA to ensure that product formation ceases the moment that the samples are loaded on the chromatogram.
3. Set up the standard assay, using concentrations of reagents as for the runoff transcription experiment. In a typical starting experiment, excess polymerase, DNA, and transcription factor will be premixed and incubated long enough to reach complete open complex formation. The experiment will be started by the addition of nucleotides. The final reaction mix will contain, for example, 0.5–3 n*M* fragment, 100 n*M* RNA polymerase, 0.5 m*M* dinucleotide, and 0.05 m*M* UTP with 2.5 µCi α-[^{32}P]-UTP in 100 µL. Run experiments both with and without the transcription factor. Perform a control with no DNA.
4. At different times after addition of the α-[^{32}P]-UTP, remove 15-µL aliquots and spot at the origin of the chromatogram. Six aliquots taken every 5 min will suffice.
5. Develop the chromatogram using chromatography buffer (*see* Section 2.11.). After the solvent front has progressed 20 cm, remove the chromatogram and dry. Cut the paper into 5-mm slices and count each slice for Cerenkov radiation to locate the bands resulting from product and unincorporated UTP. Determine the number of counts at each time-point as a result of product formation ($CPM_{product}$) and resulting from unincorporated UTP (CPM_{UTP}). From the ratio of counts in the product ($CPM_{product}$) to the total counts ($CPM_{product} + CPM_{UTP}$) the amount of product at each time-point can be deduced. A plot of this parameter vs time should be linear, and from the slope, the rate of product formation can be deduced. The rate of product formation per promoter (TON, *t*urn *o*ver *n*umber) can then be calculated from the molar amount of DNA fragment that was used in the experiment. A control run without the transcription factor will give factor-independent activity and will allow the effect of the activator to be quantified. Assuming that the rate of product formation in the presence of the activator reflects 100% occupancy at the promoter, the occupancy in the absence of the activator can be calculated (from the ratio of the TON values in the absence and the presence of the activator).
6. The analysis can then be taken a stage further (for examples *see [12,13]*). In the above experiment, RNA polymerase is preincubated with tem-

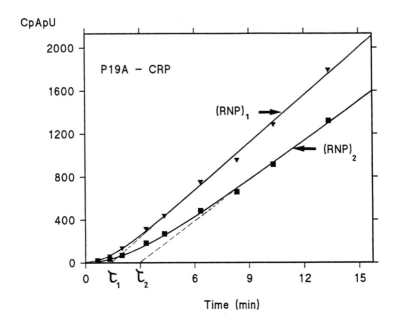

Fig. 3. Lag plots of the *gal*P1 (P19A) promoter in the presence of different concentrations of RNA polymerase (32.3 nM and 100 nM) An 80 µL sample of 4 nM DNA fragment in standard buffer solution with 200 µM cAMP was preincubated with 40 µL of a mixture containing 2 mM CpA and 200 µM UTP with 5 µCi α–[^{32}P]-UTP and the mixture incubated for 10 min at 37°C. At time 0, 40 µL of a prewarmed RNA polymerase solution was added (139 nM or 400 nM) and the reaction carefully mixed. At the indicated times, t, 15 µL portions of the reactions were removed for product quantitation. The normalized quantity of CpApU product was fitted using the Fig-P program on an IBM computer according to the equation $Y = Vt - V\tau(1 - e^{-t/\tau})$, where V is the final steady state velocity (mol of CpApU per mol of per promoter and per min). Care was taken to run the reaction until $t = 5\tau$ and to check that the final slope V was in agreement within ± 15% with the value of the TON, determined after preincubation of promoter and holoenzyme as described in Section 3.2.4.

plate prior to the addition of substrate so product formation is linear from zero time. If the reaction is started by the addition of polymerase the plot of product formation vs time shows a lag where the RNA polymerase "installs" itself at the promoter: This lag time (τ) can be easily measured and is a function of the initial binding of polymerase to the promoter and subsequent isomerisations to the open complex (*see* Fig. 3).

The interaction of holoenzyme with promoters involves at least two steps: a rapid and reversible binding to promoter DNA charac-

Assays for Transcription Activators

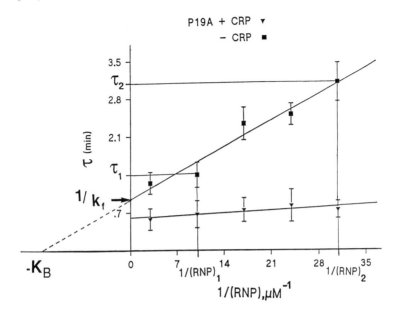

Fig. 4. τ plot for the *gal*P1 (P19A) promoter in the absence and in the presence of 100 nM CRP. The lags (τ) obtained as described in Fig. 3 (with their standard deviations) were plotted against the reciprocal of RNA polymerase concentration. The associated rate constants k_f and $K_B k_f$ are calculated respectively from the intercept and slope of the plot using the Fig-P program.

terized by an association constant K_B, which leads to the "closed" inactive complex, followed by a conformation change to the "open" complex characterized by the rate constant k_f. For most promoters the reverse of open complex formation is extremely slow. Thus, according to McClure (5), the measured lag time (τ) is related to the enzyme concentration [RNP] by the relation,

$$\tau = (-1/k_f + 1/k_f k_B [\text{RNP}])$$

To make a kinetic analysis it is necessary to perform the assays with a range of different polymerase concentrations (typically 10–200 n*M*), measuring the lag time (τ) in each case—τ is plotted as function of the reciprocal of the RNA polymerase concentration (Fig. 4). This plot can be extrapolated to infinite RNA polymerase concentrations (the intersect with the *Y*-axis) to give a value of τ, which is the reciprocal of k_f (Fig. 4). K_B can be deduced from the intercept of the τ plot with the *X*-axis. Alternatively it can be calculated from the ratio between the lag time at infinite enzyme concentration and the slope of the straight line. Data are normally fitted used a computer program, such as Enzfitter or Fig-P.

4. Notes

4.1. Runoff Transcription Assays

1. Runoff transcription assays are generally performed on short fragments in which the transcription start of interest is 50–150 bp from the end of the fragment. Often a long fragment will be chosen carrying more than one promoter: Longer transcripts can be sized by running the sequence gels further. Individual transcripts can be identified by using families of fragments that are truncated from one end. Transcription assays can also be performed using covalently closed circular supercoiled DNA: To do this use unlabeled nucleotides and extract the RNA after the elongation step. Particular transcripts can be quantified by hybridizing to synthetic oligodeoxyribonucleotides and extending using reverse transcriptase and labeled precursors (primer extension). Alternatively, the promoter to be tested can be cloned upstream of a strong transcriptional terminator (e.g., from bacteriophage λ or the *E. coli rrnB* operon) and the RNA visualized directly on a sequencing gel as for linear templates *(14)*.

2. The kinetics of open complex formation can be monitored. After addition of polymerase, take aliquots at different times and add to the NTP/heparin mix. Since heparin blocks reinitiation, the amount of runoff transcript from that sample will be proportional to the amount of open complex formed at that time (for an example, *see* ref. *15*). Typically, in the above conditions, the half-time for open complex formation ranges from 20 s to 30 min. In principle it is possible to make these measurements at different polymerase concentrations and make the τ plot analysis, as for abortive initiation; in practice this is extremely difficult and the abortive initiation assay is preferable.

3. Runoff analysis provides a simple method for monitoring the effects of transcription factors and their cofactors. However, it can also be exploited to investigate effects of conditions (e.g., temperature, salt, and so forth; *see* ref. *16*) on open complex formation. It is important to note that changes may affect elongation rather than transcription initiation. This can be checked simply by preforming open complexes and then altering the conditions. In our experience, the elongation step is usually unaltered by changes in the assay conditions and differences reflect changes at one or other step in the formation of the open complex *(16)*.

4. A number of alternative buffer systems can be used and the final choice is largely a matter of trial and error. An alternative system is 40 mM Tris, pH 8.0, 100 mM KCl, 10 mM MgCl$_2$, 1 mM DTT, and 100 µg/mL acetylated bovine serum albumin. In some cases the effects of substituting different anions or cations may be significant (17).
5. Elongation can be studied by preforming open complexes and then adding nucleotide precursors one by one. 3'O-methyl (18) or dideoxy (19) derivatives of nucleotides can be used to freeze elongation complexes at particular lengths.
6. The runoff step of the protocol is very fast and complete in minutes. Sometimes a doublet band is seen corresponding to a particular transcript. This is usually caused by hesitation by the polymerase at the end of the fragment. The relative intensities of the doublet can depend on temperature or the length of time of the elongation step (16). In some cases (20) multiple bands are caused by ambiguity in the starting base of the transcript. This can be resolved by working with γ-[^{32}P]-labeled initiating nucleotide.
7. Many, but not all, promoters are active in runoff assays and there is no way of predicting whether a particular activator will or will not work in vitro. Many workers find that runoff experiments produce more bands than "ought" to be seen. In particular, some transcripts exceed the size of the template fragment (4). This is a result of RNA polymerase molecules failing to stop when reaching the end of the fragment, turning around, and continuing to transcribe the opposite strand. This effect can be partially circumvented by lowering NTP concentrations or decreasing the temperature. Another problem may arise because any DNA sequence will contain a number of potential transcription starts that are normally not used *in vivo*, since the competition for polymerase favors the stronger promoters (21). In vitro conditions are such that there is little discrimination against weak promoters. If the appearance of bands from these weak promoters "spoils" the results, they can be reduced by using higher salt concentrations or lower amounts polymerase to increase specificity.
8. Although the appearance of a runoff transcript makes a good assay for the activity of a transcription factor, there may be situations in which the transcript cannot be found. In this case the best strategy is to attempt to monitor open complex binding directly by the opening of the strands. In most cases (22) the activity of a transcription factor will cause a measurable unwinding of the DNA duplex around the −10 sequence and transcription start. Many chemical reagents can be used to monitor unwinding, but one of the simplest is potassium permanganate that pref-

erentially attacks nonbase paired thymines. To measure unwinding, start with end-labeled DNA and make open complexes with RNA polymerase and activator proteins as in the runoff assays. Typically, then add 1 µL of fresh 200 mM potassium permanganate per 20 µL sample and incubate for 1–4 min (still at 37°C). After the addition of 50 µL stop buffer (3M ammonium acetate, 0.1 mM EDTA, 1.5M β-mercaptoethanol), phenol extract the sample and alcohol precipitate the DNA. The labeled DNA can then be cleaved at the sites of permanganate modification by using the Maxam-Gilbert piperidine protocol. The resulting fragments are run on a sequence gel to find the sites of unwinding. A typical experiment will include runs with DNA alone, DNA plus polymerase, and DNA plus polymerase plus activator. This provides a simple method for checking that the activator is functional, and provides information on the size of the region of unwinding in the open complex (23).

4.2. Abortive Initiation Assays

9. The most powerful use of abortive initiation is to determine the microscopic rate constants of individual steps during transcription initiation. Measurements of these rates in the absence or presence of a transcription factor can provide important mechanistic information about the enzymology of activation. In different situations, transcription activators can affect K_B (13), k_f (12), or TON (13). In a small number of cases, transcription factors have no effect on abortive initiation parameters. In these instances (24) the activator cannot be intervening at the level of open complex formation, but must be affecting later steps of the transcription process. Such situations can be analyzed by single or multiple rounds of run-off assays.

10. One of the advantages of the abortive initiation assay is that it can be performed on circular DNA as well as linear templates. Basically the choice of primer picks out one promoter from others. Obviously there is more chance of interference from other promoters with longer DNAs. If working with circular plasmid it is prudent to test the reaction using plasmid either with or without the insertion carrying the promoter under study. It may be possible to reduce interfering signals from the vector by altering the primers. Some primers can be used without being completely specific for the promoter tested. For instance, CpA and UTP gives the trinucleotide CpApU at galP1, but also the longer oligonucleotides CpApUpU and CpApUpUpU starting from galP2, which can be separated on the chromatogram (25).

11. Before starting any kinetics it is advisable to check chosen combinations of primer and nucleotide for specificity and for product formation: A TON value of <10/min is useless for kinetic studies. Some promoters

Assays for Transcription Activators 409

give no abortive cycling reaction *(14,26),* others may give homopolymer synthesis caused by slippage in the enzyme's active site, rendering the abortive initiation assay useless.

12. In some cases, product formation may never become linear with respect to time. Assuming that there are no contaminating nucleases, this is likely to be a result of the consumption of nucleoside triphosphates, which reduces the reaction velocity. This can be overcome by lowering the dinucleotide concentration. Ideally any time course needs to be run for at least 5 times τ.

13. The abortive initiation assay is tedious because of the chromatographic analysis of the products, which takes 2–3 h. One way to accelerate the procedure is to replace radioactive UTP with a fluorescent analog, UTP-γ-ANS (1-naphtylamine-5-sulfonic acid UTP). The assay can be measured fluorometrically by following the increase in light emission caused by the release of the pyrophosphate-ANS moiety each time a unit is incorporated *(27).* A considerable advantage of this method is that it allows the continuous monitoring of product formation. A potential disadvantage is that the fluorescent label could alter the kinetics, although, to date, this has not been reported.

References

1. Raibaud, O. and Schwartz, M. (1984) Positive control of transcription initiation in bacteria. *Ann. Rev. Genet.* **18,** 173–206.
2. Losick, R. and Chamberlin, M., eds. (1976) *RNA Polymerase.* Cold Spring Harbor Laboratory, Cold Spring Harbor, NY.
3. Irani, M., Musso, R., and Adhya, S. (1989) Cyclic-AMP-dependent switch in initiation of transcription from the two promoters of the *Escherichia coli gal* operon: identification and assay of 5'-triphosphate ends of mRNA by GTP:RNA guanyltransferase. *J. Bacteriol.* **171,** 1623–1630.
4. Zubay, G. (1980) The isolation and properties of CAP, the catabolite gene activator. *Meth. Enzymol.* **65,** 856–877.
5. McClure, W. (1980) Rate-limiting steps in RNA chain initiation. *Proc. Natl. Acad. Sci. USA* **77,** 5634–5638.
6. Burgess, R. and Jendrisak, J. (1975) A procedure for the rapid, large-scale purification of *Escherichia coli* DNA-dependent RNA polymerase involving Polymin P precipitation and DNA-cellulose chromatography. *Biochemistry* **14,** 4634–4638.
7. Hager, D. A., Jun Jin, D., and Burgess, R. R. (1990) Use of mono Q high resolution ionic exchange chromatography to obtain highly pure and active *Escherichia coli* RNA polymerase. *Biochemistry* **29,** 7890–7894.
8. Lowe, P., Hager, D., and Burgess, R. (1979) Purification and properties of the sigma subunit of *Escherichia coli* DNA-dependent RNA polymerase. *Biochemistry* **18,** 1344–1352.

9. Gonzalez, N., Wiggs, J., and Chamberlin, M. J. (1977) A simple procedure for resolution of *Escherichia coli* RNA polymerase holoenzyme from core polymerase. *Arch. Biochem. Biophys.* **182**, 404–408.
10. Maniatis, T., Fritsch, E., and Sambrook, J. (1982) *Molecular Cloning. A Laboratory Manual,* Cold Spring Harbor Laboratory, Cold Spring Harbor, NY.
11. Herbert, M., Kolb, A., and Buc, H. (1986) Overlapping promoters and their control in *Escherichia coli*: the *gal* case. *Proc. Natl. Acad. Sci. USA* **83**, 2807–2811.
12. Hawley, D. and McClure, W. (1982) Mechanism of activation of transcription initiation from the λ P_{RM} promoter. *J. Mol. Biol.* **157**, 493–525.
13. Malan, T., Kolb, A., Buc, H., and McClure, W. (1984) Mechanism of CRP-cAMP activation of *lac* operon transcription initiation: activation of the P1 promoter. *J. Mol. Biol.* **180**, 881–909.
14. Leirmo, S. and Gourse, R. (1991) Factor independent activation of *E. coli* rRNA transcription (I) Kinetic analysis of the role of the upstream activator region and supercoiling on transcription of the *rrnB*P1 promoter *in vitro*. *J. Mol. Biol.* **220**, 555–568.
15. Chan, B. and Busby, S. (1989) Recognition of nucleotide sequences at the *Esherichia coli* galactose operon *P1* promoter by RNA polymerase. *Gene* **84**, 227–236.
16. Grimes, E., Busby, S., and Minchin, S. (1991) Different thermal energy requirement for open complex formation by *Escherichia coli* RNA polymerase at two related promoters. *Nucleic Acids Res.* **19**, 6113–6118.
17. Leirmo, S., Harrison, S., Cayley, D. S., and Burgess, R. R. (1987) Replacement of potassium chloride by potassium glutamate dramatically enhances protein DNA interactions in vitro. *Biochemistry* **26**, 2095–2101.
18. Straney, D. C. and Crothers, D. M. (1985) Intermediates in transcription initiation from the *E. coli lac* UV5 promoter. *Cell* **43**, 449–459.
19. Krummel, B. and Chamberlin, M. (1992) Structural analysis of ternary complexes of *E. coli* RNA polymerase—individual complexes halted along different transcription units have distinct and unexpected biochemical properties. *J. Mol. Biol.* **225**, 221–237.
20. Spassky, A., Busby, S., and Buc, H. (1984) On the action of the cAMP-cAMP receptor protein complex at the *E. coli* lactose and galactose promoter regions. *EMBO J.* **3**, 43–50.
21. Ponnambalam, S., Spassky, A., and Busby, S. (1987) Studies with the *Esherichia coli* galactose operon regulatory region carrying a point mutation that simultaneously inactivates the two overlapping promoters. Interactions with RNA polymerase and the cyclic AMP receptor protein. *FEBS Lett.* **219**, 189–196.
22. Gaston, K., Bell, A., Kolb, A., Buc, H., and Busby, S. (1990) Stringent spacing requirements for transcription activation by CRP. *Cell* **62**, 733–743.
23. Chan, B., Minchin, S., and Busby, S. (1990) Unwinding of the duplex DNA during transcription initiation at the *Escherichia coli* galactose operon overlapping promoters. *FEBS Lett.* **267**, 46–50.
24. Menendez, M., Kolb, A., and Buc, H. (1987) A new target for CRP action at the *malT* promoter. *EMBO J.* **6**, 4227–4234.

25. Lavigne, M., Herbert, M., Kolb, A., and Buc, H. (1992) Upstream curved sequences influence the initiation of transcription at the *E. coli* galactose operon. *J. Mol. Biol.* **224,** 293–306.
26. Richet, E. and Raibaud, O. (1991) Supercoiling is essential for the formation and stability of open complexes at the divergent *malE*p and *malK*p promoters. *J. Mol. Biol.* **218,** 519–542.
27. Bertrand-Burggraf, E., Lefevre, J. F., and Daune, M. (1984) A new experimental approach for studying the association between RNA polymerase and the tet promoter of pBR322. *Nucleic Acids Res.* **12,** 1697–1706.

CHAPTER 32

An Assay for In Vitro Recombination Between Duplex DNA Molecules

Berndt Müller and Stephen C. West

1. Introduction

The RecA protein of *Escherichia coli* is essential for genetic recombination and has been extensively characterized *(1–3)*. In vitro, purified RecA protein is able to promote recombination reactions of two types: (i) strand transfer between circular single-stranded DNA (ssDNA) and homologous linear duplex DNA and (ii) strand exchange between circular duplex DNA with a defined single-stranded gap and homologous linear duplex DNA (Fig. 1A) In this chapter, we describe the preparation of substrates for reaction (ii), which occurs between essentially duplex DNA molecules. The reaction has been used extensively in studies of the mechanism of RecA-mediated strand exchange *(4,5)* and may also be used to assay for activities capable of resolving Holliday junctions in DNA *(6,7)*.

The reaction requires that the linear duplex DNA has an end homologous to the single-stranded DNA in the gap (Fig. 1A) The gap serves a dual purpose; it provides a nucleation site from which the RecA filament is assembled, and it provides the site from which strand exchange is initiated. Strand exchange then proceeds 5' to 3' with respect to the closed circular single-strand. A time-course of the strand exchange reaction described here is shown in Fig. 1B. Products of strand exchange are formed after 30–40 min of incubation (demonstrated by the appearance of [^{32}P]-labeled circular duplex DNA) At earlier times, reaction intermediates of slower gel electrophoretic

From: *Methods in Molecular Biology, Vol. 30: DNA–Protein Interactions: Principles and Protocols*
Edited by: G. G. Kneale Copyright ©1994 Humana Press Inc., Totowa, NJ

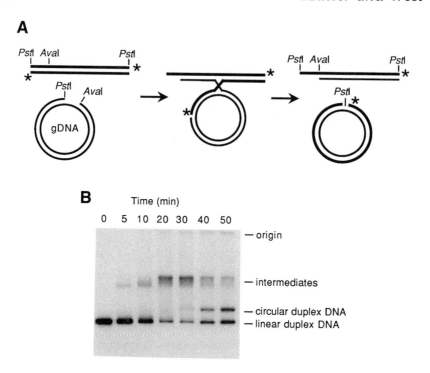

Fig. 1. (**A**) Schematic drawing showing the substrates (gDNA and *Pst*I-linearized duplex DNA of φX174) (left), intermediates (center), and the products (nicked circular duplex DNA and gapped linear duplex DNA) (right) of RecA-mediated strand exchange. The position of the 3' [^{32}P]-end-label is indicated (*). (**B**) Time-course of RecA-mediated strand exchange between gDNA and 3' [^{32}P]-end-labeled *Pst*I-linearized DNA. Aliquots of the reaction were stopped at the times indicated and reaction products were analyzed by 0.8% agarose gel electrophoresis. The agarose gel was dried and the DNA detected by exposure to a Kodak X-OMAT film.

mobility are formed. By electron microscopy, these intermediates resemble α-structures and consist of gapped circular duplex DNA molecules joined to linear duplex DNA molecules by a Holliday junction.

The substrates used for the in vitro recombination reaction described here can be prepared from the DNA of any small ssDNA bacteriophage, such as φX174 or M13 (*see* Note 1) For simplicity, we will describe in detail their preparation from φX174 DNA since it is commercially available in both single-stranded and duplex DNA forms. To form gapped circular duplex φX174 DNA (gDNA) (*see* Section 3.1. and Fig. 2A), the 5224-base pairs (bp) *Pst*I–*Ava*I fragment of duplex

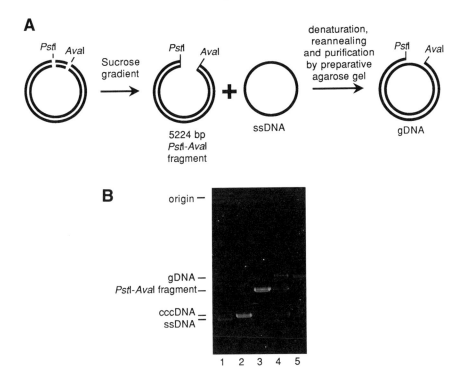

Fig. 2. (**A**) Schematic drawing showing the different steps involved in the production of gDNA from circular (+) ssDNA and cccDNA of φX174. (**B**) 0.8% agarose gel showing the substrates and products of annealing reactions during the formation of gDNA. All DNA is derived from φX174. Lane 1; 100 ng circular ssDNA. Lane 2; 100 ng cccDNA. Lane 3; 120 ng of purified 5224 bp *Pst*I–*Ava*I fragment. Lane 4; 200 ng of products of the annealing reaction between the *Pst*I–*Ava*I fragment and the circular ssDNA. Lane 5; 50 ng of purified gDNA. The DNA was visualized by staining with ethidium bromide.

φX174 DNA is produced by restriction digestion of covalently closed circular duplex DNA (cccDNA) and isolated by neutral sucrose gradients (*see* Note 2). This fragment is denatured and the complementary strand annealed to circular (+) ssDNA of φX174 by sequential dialysis against buffers containing varying amounts of formamide. Finally, the gDNA is purified from excess circular ssDNA and other annealing products by preparative agarose gel electrophoresis. The agarose gel in Fig. 2B shows the DNA substrates used for the formation of the gDNA (lanes 1 and 3), the products of annealing (lane 4), and the gDNA after purification (lane 5). The purification of the gDNA is a combination of standard molecular biology procedures. We have attempted to detail the steps that are important and for the other steps we refer to molecular biology method books.

We also describe briefly the methods used to prepare the linear reaction partner, [^{32}P]-end-labeled linear duplex DNA (*see* Section 3.2.) In addition, we describe the conditions for efficient RecA-mediated strand exchange reactions and the analysis of the reaction products (*see* Section 3.3.).

2. Materials

2.1. Stock Solutions

1. 1M Tris-HCl, pH 7.5, 1M Tris-HCl, pH 8.0; 1M Tris-HCl, pH 8.5.
2. 1M MgCl$_2$.
3. 3M sodium acetate, pH 7.0.
4. 1M NaCl, 5M NaCl.
5. 0.5M EDTA, pH 8.0.
6. 0.1M ATP adjusted to pH 7.0 with NaOH, store in aliquots at –20°C.
7. 1M dithiothreitol, store at –20°C.
8. Butan-2-ol (analytical reagent [A.R.]).
9. Diethyl ether (A.R.).
10. Absolute ethanol (A.R.).
11. Isoamyl alcohol (A.R.).
12. Chloroform (A.R.).
13. Glycerol (A.R.).
14. Glacial acetic acid (A.R.).
15. Phenol: Crystalline redistilled phenol (molecular biology grade) is melted and equilibrated once with 1 vol of 0.5M Tris-HCl, pH 8.5, and then repeatedly with 1 vol 0.1M Tris-HCl, pH 8.0, until the pH of the phenol phase is >7.0. To equilibrate the phenol, the mixture is stirred for 15 min on a magnetic stirrer at room temperature. The stirrer is then turned off to let the two phases separate. Remove the upper (aqueous) phase using a glass pipet connected to a vacuum line. After removal of the final aqueous phase add 0.1 vol of 0.1M Tris-HCl, pH 8.0, supplemented with 0.2% β-mercaptoethanol. Store at –20°C in aliquots of 50 mL.
16. Formamide: Deionize distilled formamide (molecular biology grade) by stirring on a magnetic stirrer with AG 501-X8 mixed bed resin (Bio-Rad, Richmond, CA) (1 g resin/10 mL formamide) for 1 h at room temperature. The formamide is then separated from the resin by filtering twice through Whatman No. 1 filter paper and stored at –20°C (*see* Note 3).

17. Water is deionized using a Milli-Q reagent grade water system (Millipore, Bedford, MA) or an equivalent procedure.

2.2. Preparation of gDNA

1. DNA: 250 µg circular (+) ssDNA and 200 µg cccDNA of φX174. All DNA is commercially available.
2. Restriction endonucleases *Pst*I and *Ava*I are commercially available.
3. 10X restriction buffer: 100 mM Tris-HCl, pH 7.5, 100 mM MgCl$_2$, 500 mM NaCl, 1 mg/mL nuclease-free bovine serum albumin.
4. For extractions with phenol/chloroform (1:1 [v/v]), chloroform is supplemented with isoamyl alcohol at a ratio of 24:1 (v/v).
5. Sucrose solutions 5 and 20% (w/v) for neutral sucrose gradients are prepared by dissolving sucrose (molecular biology grade) in 10 mM Tris-HCl, pH 7.5, 1M NaCl, and 10 mM EDTA, pH 8.0.
6. 95% Formamide solution: 95% (v/v) formamide, 10 mM EDTA, pH 8.0.
7. 50% Formamide solution: 50% (v/v) formamide, 200 mM Tris-HCl, 10 mM EDTA, pH 8.0.
8. TNE: 100 mM Tris-HCl, pH 7.5, 100 mM NaCl, 10 mM EDTA.
9. TE: 10 mM Tris-HCl, pH 7.5, 1 mM EDTA.
10. Agarose, type II, medium EEO.
11. Agarose gel electrophoresis buffer: 40 mM Tris-acetate, pH 8.0, 1 mM EDTA (TAE; prepared as 50X concentrated solution: 2M Tris base, 1M acetic acid, and 50 mM EDTA).
12. 5X sample loading buffer: 50% (v/v) glycerol containing traces of xylene cyanol and bromophenol blue.

2.3. Preparation of Linear Duplex DNA

1. 5' [^{32}P]-end-labeling: Alkaline phosphatase from calf intestine (molecular biology grade), T4 polynucleotide kinase, and (γ-[^{32}P]) ATP.
2. 3' [^{32}P]-end-labeling: Terminal transferase and (α-[^{32}P]) dideoxy ATP.

2.4. Strand Exchange

1. RecA protein is commercially available or can be purified as described *(8)* from *Escherichia coli* strain KM4104 containing the plasmid pDR1453 *(9)*.
2. 10X RecA reaction buffer: 200 mM Tris-HCl, pH 7.5, 150 mM MgCl$_2$, 20 mM ATP, 20 mM dithiothreitol, 1 mg/mL nuclease-free bovine serum albumin. Keep the 10X RecA reaction buffer at 4°C for 2–3 wk. For longer periods store as aliquots (100–200 µL) at –20°C.
3. 5X stop mixture: 100 mM Tris-HCl, pH 7.5, 2.5% sodium dodecyl sulfate (SDS), 200 mM EDTA, and 10 mg/mL proteinase K. Store as aliquots (100–200 µL) at –20°C and discard after use.

4. Creatine phosphokinase: 125 U/mL. Store at –20°C.
5. Phosphocreatine: 400 mM. Store at –20°C.

3. Methods

3.1. Preparation of gDNA

3.1.1. Preparation of the Linear Duplex DNA Fragment

1. Determine separately the amounts of restriction endonucleases *Ava*I and *Pst*I required to linearize 1 µg of cccDNA of φX174 within 1 h at 37°C in 1X restriction buffer in 20 µL. In our hands, approx 0.2 U *Pst*I and 2 U *Ava*I are sufficient. Digestion produces two fragments (5224 and 162 bp).
2. Digest the remaining cccDNA (180–190 µg) in 4 mL of 1X restriction buffer with the appropriate amounts of *Pst*I and *Ava*I (approx 40 U *Pst*I and 400 U *Ava*I) for 1 h at 37°C. Stop the digestion by adding 80 µL of 0.5M EDTA, pH 8.0.
3. Extract the DNA once with 1 vol of phenol/chloroform.
4. Add water to the DNA to a final volume of 10 mL and supplement with 1/10 vol of 3M sodium acetate, pH 7.0. Transfer the DNA into a polyallomer tube appropriate for an SW 28 rotor. Add 2–2.5 vol of absolute ethanol (–20°C) to precipitate the DNA.
5. Pellet the DNA by centrifugation in an ultracentrifuge (SW 28 rotor [Beckman, Fullerton, CA], 4°C, 1 h, 26,000 rpm [90,000g]) and remove the liquid. Dry the DNA pellet and resuspend the DNA in 500 µL TE.
6. Separate the 5224 and 162 bp fragments by sedimentation through a neutral 5–20% (w/v) sucrose gradient. Prepare the gradients in polyallomer tubes (SW 28 rotor) and run in an ultracentrifuge (4°C, 18–20 h, 26,000 rpm (90,000g).
7. Collect the gradient in 1.5-mL fractions. Analyze 10 µL of each fraction on a 1% agarose gel and visualize the DNA by staining with ethidium bromide.
8. Pool the fractions containing the 5224 bp fragment. Adjust with water to a volume of 10 mL and supplement with 1/10 vol of 3M sodium acetate, pH 7.0. Concentrate the DNA by ethanol precipitation as described in Section 3.1.1. steps 4–5. Finally, resuspend the DNA in 300 µL TE. The purified 5224 bp fragment is shown in lane 3 of Fig. 2B.

3.1.2. Annealing of the 5224 bp Linear Duplex DNA Fragment to Circular ssDNA

1. Mix the linear duplex DNA with excess circular φX174 ssDNA (250 µg) and adjust the mixture to 50% (v/v) formamide, 10 mM EDTA, pH 8.0 in 2–3 mL (*see* Note 4).

2. Transfer the annealing mixture into dialysis tubing using plastic Pasteur pipets and dialyze for 2 h against 200 mL 95% formamide solution in a 200 or 250-mL cylinder. Perform all dialysis steps at room temperature, with constant stirring of the buffer.
3. Dialyze for 2 h against 200 mL of 50% formamide solution.
4. Dialyze for 2 h against 200 mL of TNE.
5. Finally, dialyze for 2 h against 200 mL of TE.
6. Collect the annealing mixture from the dialysis tubing using plastic Pasteur pipets. Test for annealing by 0.8% agarose gel electrophoresis. Lane 4 in Fig. 2B shows the products of a typical annealing reaction.

3.1.3. Purification of gDNA by Preparative Agarose Gel Electrophoresis

1. Prepare three 1% (w/v) agarose gels (20 × 25 × 0.5 cm) with two long slots each (18 × 0.3 cm), one positioned near the end of the gel, one in the middle. Also make two small slots for markers.
2. Add 0.25 vol of 5X sample loading buffer to the annealing mixture and transfer the mixture into the big slots. For optimal separation apply 4–5 µg DNA per cm of slot length. Run for approx 4–5 h at 125–150 V (5–6 V/cm) (gDNA migrates slower then xylene cyanol; the bromophenol blue fronts should reach at least the middle of the gel and the end of the gel, respectively). Use linear duplex and circular ssDNA of φX174 as markers.
3. Stain the gels weakly with ethidium bromide (view as quickly as possible). Visualize the DNA in long-wave UV light to prevent UV damage. Cut out the band corresponding to the main annealing product (i.e., the slowest migrating major band shown in Fig. 2B, lane 4).
4. Separate the DNA from the agarose by electroelution. Transfer the agarose slices into dialysis tubing (1 slice per dialysis tubing) filled with TAE (use minimal amounts of buffer) and electroelute overnight (50 V). Collect the DNA from the dialysis bags using plastic Pasteur pipets (normally the DNA is in a volume of 25–50 mL).
5. Purify the DNA by two successive extractions with 1 vol of phenol followed by one extraction with 1 vol of diethyl ether (extractions are performed in 50-mL polypropylene Falcon tubes).
6. Concentrate the DNA by subsequent extractions with 1 vol of butan-2-ol (remove the upper layer consisting of butanol-2-ol and water) until the volume has been reduced to approx 10 mL. Extract once more with 1 vol of diethyl ether.
7. Supplement the DNA with 1/10 vol of 3M sodium acetate, pH 7.0, and concentrate by ethanol precipitation as described in Section 3.1.1., steps 4–5.
8. Resuspend the gDNA in 0.5–1 mL of TE and dialyze for 4–6 h against 2 L of TE with one change of buffer.

9. Determine the DNA concentration spectophotometrically by assuming $A_{260} = 1$ for a DNA solution of 50 µg/mL. Useful working concentrations are 40–80 µg/mL. Store the gDNA in aliquots (100-µL) at –20°C. Purified gDNA is shown in lane 5 of Fig. 2. Using this procedure, we normally recover between 25 and 50% of the original duplex DNA in the form of gDNA.

3.2. Preparation of Linear Duplex DNA

1. Linearize 10 µg of ϕX174 cccDNA with *Pst*I in 1X restriction buffer in a volume of 100–200 µL.
2. Extract the linearized DNA once with 1 vol of phenol/chloroform.
3. Supplement the DNA with 1/10 vol of 3*M* sodium acetate, pH 7.0, and precipitate by the addition of 2.5 vol of absolute ethanol (–20°C). Concentrate the DNA by 20 min of centrifugation in a bench-top centrifuge (16,000*g*). Remove the liquid using a drawn-out Pasteur pipet.
4. Resuspend the DNA in 250 µL 0.3*M* sodium acetate, pH 7.0, add 500 µL absolute ethanol (–20°C), and concentrate the DNA by centrifugation for 20 min in a benchtop centrifuge. Remove the liquid and dry the DNA. Resuspend the DNA in an appropriate volume of TE (normally 35–50 µL).
5. To label the DNA with ^{32}P at the 3' ends, use (α-[^{32}P]) dideoxy ATP and terminal transferase according to the instructions of the supplier. Alternatively, the DNA can be labeled with ^{32}P at the 5' ends. Dephosphorylate the DNA using alkaline phosphatase and label the 5' ends using (γ-[^{32}P]) ATP and polynucleotide kinase *(10)*.
6. Terminate the end-labeling procedure by the addition of water to a final volume of 200 µL and adjust to 10 m*M* EDTA and 0.5% SDS. Supplement with 20 µL of 3*M* sodium acetate, pH 7.0.
7. Extract the DNA with 1 vol of phenol/chloroform.
8. Add 500 µL absolute ethanol (–20°C) and concentrate the DNA by two subsequent ethanol precipitations as described in Section 3.2., steps 3 and 4. This procedure removes the unincorporated nucleotides (alternatively, separate the DNA from the unincorporated nucleotides by gel filtration using Sephadex G-50 prior to concentration by ethanol precipitation).
9. Resuspend the linear DNA in TE at a concentration of 50–100 µg/mL and dialyze for 4 h against 2 L of TE with one change of buffer.

3.3. RecA-Mediated Strand Exchange Reactions Between gDNA and Homologous Linear Duplex DNA

1. Use the 10X RecA reaction buffer, phosphocreatine and creatine phosphokinase stock solutions to prepare a reaction mixture containing gDNA (8.8 µg/mL) and RecA protein (400 µg/mL) in 20 m*M* Tris-HCl, pH

7.5, 15 m*M* MgCl$_2$, 2 m*M* ATP, 2 m*M* dithiothreitol, 100 µg/mL bovine serum albumin (BSA), 20 m*M* phosphocreatine, 12.5 U/mL creatine phosphokinase in 20–100 µL (*see* Notes 5 and 6).
2. Incubate this reaction mixture for 5 min at 37°C to form nucleoprotein filaments.
3. Initiate strand exchange by addition of *Pst*I-linearized, [^{32}P]-labeled linear duplex DNA (7 µg/mL) and continue the incubation at 37°C. To stop the strand exchange reaction, add 0.25 vol of 5X stop mixture and incubate for a further 10 min at 37°C.
4. Add 0.25 vol of 5X sample loading buffer to the stopped aliquots and analyze the samples on a 0.8% agarose gel. Run gels at 6 V/cm for 2–3 h using buffer recirculation. To detect the DNA, dry the gel onto Whatman 3MM filter paper and expose the dried gel to an X-ray film. Figure 1B shows a typical time-course of a RecA-mediated strand exchange reaction (*see* Note 7).

4. Notes

1. The method described here for the production of gapped circular duplex DNA is a modified version of that described previously *(4)*. A different method, which can be used to produce circular duplex DNA with a single-stranded gap from plasmid DNA, involves site-specific nicking of cccDNA using restriction endonucleases in the presence of ethidium bromide *(11)*. The nicks introduced into the duplex DNA are then extended into single-stranded gaps using exonuclease III. Since exonuclease III acts preferentially on free DNA ends it is important to minimize linearization of the duplex DNA during the site-specific nicking, or alternatively to purify the nicked circular duplex DNA.
2. Instead of *Pst*I and *Ava*I other restriction enzymes can be used, provided that (i) the fragments produced can be separated by a sucrose gradient and (ii) they allow for the production of a linear duplex reaction partner with an overlap (with the correct polarity) of more than 30 nucleotides into the single-stranded region of the gapped circular duplex DNA *(12,13)*.
3. The deionization of formamide is essential for the annealing procedure.
4. For the formation of gDNA, denaturation and reannealing in formamide can be replaced by heat-denaturation in 1X SSC (150 m*M* NaCl, 15 m*M* sodium citrate) at 95–100°C followed by a slow decrease in temperature. However, in our hands this reaction produces significant amounts of higher-order annealing products, probably composed of three or more DNA molecules.
5. An ATP regenerating system, here consisting of phosphocreatine and creatine phosphokinase, should be included in reactions with long incu-

bation times. This prevents the accumulation of ADP, which inhibits the action of RecA protein *(14)*.
6. The amounts of RecA protein, gDNA, and linear duplex DNA used in the strand exchange reaction described here can be reduced at least twofold, but probably even more, without loss of efficiency and without changing the time-course of the reaction.
7. Under optimal conditions, the amounts of ^{32}P label contained in the linear duplex DNA and the nicked circular duplex DNA at the end of the strand exchange reaction should be equal. If the efficiency of the strand exchange reaction is not satisfying, perform a series of pilot experiments with varied amounts of RecA protein and fixed amounts of gDNA and linear duplex DNA. Choose the best conditions and then perform a second set of pilot experiments, this time varying the amount of linear duplex DNA.

References

1. Roca, A. I. and Cox, M. M. (1990) The RecA protein: Structure and function. *Crit. Rev. Biochem. Mol. Biol.* **25,** 415–456.
2. Radding, C. M. (1991) Helical interactions in homologous pairing and strand exchange driven by RecA protein. *J. Biol. Chem.* **266,** 5355–5358.
3. West, S. C. (1992) Enzymes and molecular mechanisms of homologous recombination. *Ann. Rev. Biochem.* **61,** 603–640.
4. West, S. C., Cassuto, E., and Howard-Flanders, P. (1982) Postreplication repair in *E. coli*: Strand exchange reactions of gapped DNA by RecA protein. *Mol. Gen. Genet.* **187,** 209–217.
5. West, S. C. and Howard-Flanders, P. (1984) Duplex-duplex interactions catalyzed by RecA protein allow strand exchanges to pass double strand breaks in DNA. *Cell* **37,** 683–691.
6. Müller, B., Jones, C., Kemper, B., and West, S. C. (1990) Enzymatic formation and resolution of Holliday junctions in vitro. *Cell* **60,** 329–336.
7. Connolly, B. and West, S. C. (1990) Genetic recombination in *Escherichia coli*: Holliday junctions made by RecA protein are resolved by fractionated cell-free extracts. *Proc. Natl. Acad. Sci. USA* **87,** 8476–8480.
8. Cox, M. M., McEntee, K., and Lehman, I. R. (1981) A simple and rapid procedure for the large scale purification of the RecA protein of *Escherichia coli*. *J. Biol. Chem.* **256,** 4676–4678.
9. Sancar, A. and Rupp, W. D. (1979) Physical map of the *recA* gene. *Proc. Natl. Acad. Sci. USA* **76,** 3144–3148.
10. Sambrook, E. F., Fritsch, E. F., and Maniatis, T. (1989) *Molecular Cloning: A Laboratory Manual.* Cold Spring Harbor Laboratory, Cold Spring Harbor, NY.
11. West, S. C., Countryman, J. K., and Howard-Flanders, P. (1983) Enzymatic formation of biparental figure-8 molecules from plasmid DNA and their resolution in *Escherichia coli*. *Cell* **32,** 817–829.

12. Conley, E. C. and West, S. C. (1990) Underwinding of DNA associated with duplex-duplex pairing by RecA protein. *J. Biol. Chem.* **265,** 10,156–10,163.
13. Lindsley, J. E. and Cox, M. M. (1990) On RecA protein-mediated homologous alignment of two DNA molecules. *J. Biol. Chem.* **265,** 10,164–10,171.
14. Cox, M. M., Soltis, D. A., Lehman, I. R., Debrosse, C., and Benkovic, S. J. (1983) ADP-mediated dissociation of stable complexes of recA protein and single-stranded DNA. *J. Biol. Chem.* **258,** 2586–2592.

Index

A

1-Anilinonapthalene 8-sulfonic acid, see ANS
8-Azido adenine, 237
  synthesis of, 240–242
ANS, 327, 328
AT tracts, 104

B

Band shift assay, see Gel shift assay
Base deletion, 34
Bending vectors, 281
Bent DNA
  accessibility of, 26
  photofootprinting of, 141
Binding-site selection analysis (BSSA), 296
Bipyridine, 104

C

Chemical modification (of proteins), 151
Circular dichroism, 339–340
Competition assay, 327
Copper-phenanthroline footprinting, 43
Covalently closed circular DNA, 386, 415
Cro repressor
  CD spectroscopy of, 340
  contacts with DNA, 25
Crosslinking, see UV crosslinking
Cruciform structures, 104
Crystallization, 357

D

Degenerate oligonucleotides, 212, 295
Diethyl pyrocarbonate (DEPC), 89
Dimethyl sulfate (DMS), 79
DNA bending, 281
DNA binding domains, 161, 169
DNA deformation, 113
DNA gyrase, 162
DNA sequencing
  Maxam-Gilbert, 66
  solid phase, 295, 304
DNase I footprinting, 1

E

EcoR124 DNA methylase, 328
EcoRV endonuclease, 371
Electron microscopy, 347
Eosin, 115
Equilibrium (binding) constants, 272, 273, 313, 320, 321
Ethylnitroso urea (ENU), 125
Ethylation interference, 125, 145
Exonuclease III footprinting, 11

F

Factor Xa, 170
Fenton reaction, 21, 115
Filter binding assay, 251
Flourescence spectroscopy, 313
Flourescent light, 147
Flourescent probes, 327
Fusion proteins, 170, 176, 186
Fusion vectors, 170, 186

G

Gal repressor
  CD spectroscopy of, 340
  DNA bending by, 282
Gapped duplex DNA, 414
Gel retardation, *see* Gel shift assay
Gel shift assay, 14, 33, 45, 52, 131, 263, 270, 302
Gene 5 protein, 155, 165, 171, 347
Genomic footprinting, 79
Glutathione-S-transferase (GST), 186

H

Helix handedness, 348
Hetero-duplex DNA, 218
Holliday junction, 413
Hoogsteen base pairing, 90
Hydroxyl radical
  footprinting, 21, 145
  interference, 33

I

Immunoprecipitation, 233
Inner filter effect, 315, 331
Interference methods
  ethylation, 125
  hydroxyl radical, 33
  methylation, 79
Iron-EDTA, 22

K

Kinetics
  of DNA–protein complexes
    association, 258
    dissociation, 73, 257
  of transcription, 405

L

Lac repressor
  CD spectroscopy of, 340
  Filter binding, 252
Lambda repressor, 25, 37, 141–143
Laser irradiation, 116, 228
Limited proteolysis, 161

M

Major groove, 49, 79
Maltose binding prrotein (MBP), 170
Methionine repressor, 126, 252
Methylation
  interference, 79, 84
  protection, 79, 82
Minor groove, 25, 49, 79, 141
Mobility shift assay, *see* Gel shift assay

N

Nucleosomes
  DNA deformation, 115
  footprinting of, 25

O

Oligonucleotide purification, 202, 299
Osmium tetroxide, 97
Overexpression, 169, 185

P

PCR, 169, 177, 186, 190
Peptide mapping, 152
Phenanthroline-copper footprinting, *see* Copper phenanthroline footprinting
Photoaffinity labeling, 237

Index

Photochemical excitation, 114, 115
Photocrosslinking, 244

Polymerase chain reaction, *see* PCR
Protein–DNA titrations, 319, 330

R

RecA protein, 413
Recombination, in vitro, 413
Reconstitution (of protein–DNA complexes), 357
Restriction endonucleases, assays of
    oligonucleotide based, 371
    plasmid based, 385
RNA polymerase, 14, 25, 37
    DNA deformation by, 115
    footprinting of, 14, 25
    interference assay, 37
    photofootprinting of, 115
    transcription by, 397
RNA, hydroxyl radical cleavage of, 26

S

Singlet oxygen, 113
Site-directed mutagenesis
    cassette mutagenesis, 199
    saturation mutagenesis, 211
Stoichiometry (of DNA–protein complexes), 265, 267, 320
Storage phosphor, 275
Supercoiled DNA, 93, 98

T

Tetranitromethane, 151
Thrombin, 187
Transcription assays
    abortive initiation, 402
    run-off assay, 401
Transcription factors, 1, 36, 358
    assays of, 397
    expression of, 185
    photofootprinting of, 141
Tryptophan flourescence, 314
Tyrosine
    flourescence of, 314
    modification of, 151

U

Uranyl photofootprinting, 141
UV crosslinking, 227, 237

Z

Z-DNA, 90, 109